Cultural Planning

The late twentieth century has seen a renaissance in new and improved cultural facilities: from arts centres, theatres, museums, to multiplex cinemas and public art. Cities worldwide have sought to transform their image and economies. Industrial cities have become cultural capitals, such as 'Guggenheim Bilbao'. Barcelona and Baltimore have been emulated by New Jersey and Singapore in regenerating downtown areas through new and upgraded cultural facilities and waterfront developments. Even old world cities such as London, Paris, Berlin and Vienna have created new millennial cultural quarters.

Using an historic and contemporary analysis, *Cultural Planning* examines how and why societies have planned for the arts. From its ancient roots in the cities of classical Athenian, Roman and Byzantine empires to the European Renaissance and its global re-creation today, public culture has exhibited remarkable continuity in its location and selection of arts facilities and cultural activity, and their role in the form and function of cities. Whether as an extension of welfare provision and human rights, or the creative industries and cultural tourism, the arts are growing elements of urban, social and economic development in the post-industrial age. However, the new 'Grands Projects' and cultural resources are highly concentrated, at the cost of both local cultural amenities and a culturally diverse society. Arts audiences have been in decline as cultural venues, museum collections, orchestras and a mobile cultural milieu, have become footloose.

Cultural Planning is the first book on the planning of the arts and the relationships between State arts policy, the cultural economy and city planning. Combining cultural and economic geography with arts and urban policy, it uses case studies and examples from Europe, North America and Asia. The book calls for the adoption of a cultural approach to town planning, greater equality in distribution and integration of cultural provision and urban design, in order to prevent the reinforcement of existing geographical and cultural divides.

Graeme Evans is Director of the Centre for Leisure and Tourism Studies at the University of North London. He was formerly Director of the London Association of Arts Centres. He advises the Department for Culture, Media and Sport, the European Commission and Arts Councils on cultural policy, trends and impacts.

Cultural Planning

An urban renaissance?

Graeme Evans

London and New York

First published 2001 by Routledge
11 New Fetter Lane, London EC4P 4EE

Simultaneously published in the USA and Canada
by Routledge
29 West 35th Street, New York, NY 10001

Routledge is an imprint of the Taylor & Francis Group

Typeset in Galliard by
M Rules
Printed and bound in Great Britain by
Biddles Ltd, Guildford and King's Lynn

British Library Cataloguing in Publication Data
A catalogue record for this book is available from the British Library

Library of Congress Cataloging in Publication Data
Evans, Graeme.
    Cultural planning, an urban renaissance? / Graeme Evans.
        p. cm
    Includes bibliographical references and index.
    1. Art and state. 2. Arts and society. 3. City planning. 4. Arts facilities.
    I. Title: Cultural planning. II. Title.

    NX720 .E94 2001
    711'.57—dc21   00-054872

ISBN 0-415-20731-2 (hbk)
ISBN 0-415-20732-0 (pbk)

Contents

Plates

Figures

Tables

Preface

Like good cultural development and community planning, this book has had a long gestation. Working in an inner-city arts centre in the early 1980s gave me my first experience of how communities respond to the arts and the role of culture in education and the urban environment. From action research and model projects, which ranged from city farms, a weekend arts college, community media, and both adult and young people's touring theatres, the aspirations of and exposure to many communities, audiences and organisations naturally led to the provision of technical aid to local groups undertaking arts and cultural development and site-based facility proposals. This entailed working alongside colleagues in community architecture and planning (before it became fashionable and appropriated by mainstream design firms and politicians), in organisational development which brought together youth and social workers with local authority planners and artists, and in what was then new technology, which brought low-cost IT and media facilities into the reach of local groups and creative artists. This period coincided with a national concern and response to various forms of urban economic, social and environmental decline, which gave me the opportunity to work with communities and agencies in cities such as Liverpool (post-riots), Huddersfield and in other countries, notably the resettlement town of Ashkelon in Israel. The emergence of what became a now-established association between the arts and urban regeneration spawned in London two seminal 'think tanks': the Arts & Urban Regeneration and Planning London's Arts & Culture groups, convened by the regional arts body with voluntary members, including myself, from architecture, planning, arts policy and finance institutions. These were served by a series of case-studies developed by the British American Arts Association which provided a range of examples – good and bad – of how the arts had and could be incorporated with urban regeneration and the input of artists and local communities to this process and also by the concurrent arts and *CityPlan* being developed by Metro Toronto. The model guidelines for *Arts Culture and Entertainment* that arose from these working groups provided the basis for much thought on how the arts and town planning might better interact and the resulting guidance offered an opportunity for local boroughs to develop cultural planning within their statutory land-use development plans for the first time. When serving as Director of the London Association of Arts Centres in the late 1980s, the issues of spatial distribution, arts development and equity in cultural provision became even clearer to me, and the cumulative experience of the arts centre movement in the UK, Europe and North America has provided a foundation for much of the detailed analysis provided in this book. In particular, the notion emerged of a hierarchy of arts facilities and cultural resources through both the arts in education, com-

munity and professional practice, at small, medium to large scales and from the local community arts centre to national cultural flagship. Working with urban design action teams in
several major regeneration sites also presented insights to the fraught relationships in the
public–private development process, in local governance, and in the design and planning for
complex and often contested sites and community identities.

In the 1990s my role as director of a university research centre covering a broad spectrum of policy and planning studies in recreation and leisure, from urban and cultural
tourism and the growing concern with 'heritage', to arts plans and strategies and site-
based development schemes, has further helped me locate the various notions of culture
within a more catholic tradition of amenity and within the political economy which looks
to the arts and cultural industries as prime aspects of economic development and employment growth. Micro-level impact studies and mapping exercises have provided much
empirical data, whilst policy studies undertaken for local and central government cultural,
planning and environment departments and agencies in the UK, Europe and internationally have similarly placed culture within the public policy and ideological spheres. In
particular, research and comparative policy analysis undertaken for the Department for
Culture Media and Sport, Arts Councils, local government associations and the
European Commission has provided both access to policy formulation and implementation, and to comparative and longitudinal data.

The international perspective that I have sought to encompass from my London base
has been enabled by fieldwork and exchange with researchers, agencies and communities
in these countries – notably Canada, the USA, Brazil, the Caribbean and Mexico and
Continental Europe – as well as with colleagues in my department who have brought a
range of area studies, policy, planning and social anthropological dimensions to my
work. The basis of this book in disciplinary and conceptual terms is therefore very broad.
This in one sense reflects the approach identified with cultural planning itself, and with
the growing desire in theory if less in practice amongst the social sciences and humani-
ties to develop more interdisciplinary approaches and frameworks with which to
understand the phenomenon of cities, culture and the practice of urban planning. This
is equally valid in the field of geography and its branches of urban studies, sociology and
economic and cultural geography, and also in cultural policy studies and the wider fields
of governance, public policy and amenity resource management. My aim has therefore
been to present and interpret the range of historical and contemporary approaches to culture in its many guises, in the form and function of cities. Neither a treatise on culture
nor a thesis on town planning, I hope this book has however combined an element of
advocacy based on empirical and conceptual analysis of the relationship between the arts
and urban society. This I trust will serve as a useful and at some points thoughtful
source and tool for researchers, students and practitioners in town and urban planning,
arts policy and *cultural strategists*, and for those interested in the history and evolution
of cities from a cultural perspective. At the risk of using an opportunistic cliché, the *new
millennialism* that has seen a surge in the building of culture-houses and quarters makes
this text timely, as does a heightened political and economic concern for the *urban
renaissance* from government urban task forces, city mayors, environmentalists, to
UNESCO and the World Bank, and for local communities and creative workers who seek
to make sense of both globalised culture within their everyday lives and the continuing
aspiration for cultural amenities and opportunity for participation and pleasure.

Acknowledgements

Many colleagues, past and present, have provided inspiration, access and opportunity and the cross-fertilisation of ideas which have been drawn on and extended by myself here. Key authors on cities, urban planning and on the arts and cultural policy will be evident from the text and the extensive bibliography. I will single out several individuals who have personally (knowingly or otherwise) been both instrumental and most influential over this period: in London, Patrick Boylan, John Pick (both late of City University), Ken Worpole (Comedia *et al.*), Phyllida Shaw (Arts Research Digest), Nicky Gavron (Deputy London Mayor, former Chair of the London Planning Advisory Committee), John Montgomery (Urban Cultures Ltd), Jo Foord (UNL); Fred Coalter in Edinburgh, Tony Veal in Sydney and Hélèn Laperrière and Daniel Latouche at the INRS, Montreal.

The author and the publisher thank the following for granting permission to reproduce line figures in this work:

Blackwell Science, Oxford, for Figure 8.1 from Chang, T. C. (2000) 'Renaissance revisited: Singapore as a "global city for the arts"', *International Journal of Urban and Regional Research* 24(4): 822.

David Fulton Publishers, London, for Figure 6.1 from Burtenshaw, D., Bateman, M. and Ashworth, G. J. (1991) *The European City: A Western Perspective*, p. 165.

The Henley Centre, Consumer Consultancy, London, for the use of Figures 5.2 and 5.3 from their conference overheads.

The Orion Publishing Group Ltd for Figure 3.1 from Hobsbawm, E. J. (1977) *The Age of Capital 1848–1875*, p. 371.

Sage Publications, London, for Figures 6.4 and 6.5 from Scott, A. (2000) *The Cultural Economy of Cities*.

Every effort has been made to trace copyright holders: any omissions brought to the publisher's attention will be remedied in future editions.

1 Introduction

Those who toiled knew nothing of the dreams of those who planned.
(*Metropolis*, Fritz Lang)

The places where collective and public cultural activity occurs have an important and lasting influence – aesthetic, social, economic and symbolic – on the form and function of towns and cities. At their most integrated, the arts have played a central role in the life of different societies and in models of urban design, from various classical, renaissance, industrial and post-industrial eras the world over. Where this coincided with affluence, technological and social change, the cultural economy of cities has also supported arts and crafts production, innovation and a thriving cultural industry, which has in turn created powerful comparative advantage and helped create and reinforce a sense of identity.

Land-use and culture are fundamental natural and human phenomena, but the combined notion and practice of culture and planning conjure up a tension between not only tradition, resistance and change; heritage and contemporary cultural expression, but also the ideals of cultural rights, equity and amenity. Where public culture and 'civilisation' are celebrated and where state, ethnic or municipal pride require signification, public monuments, squares, cultural buildings and events have been used and promoted, whether motivated by ceremonial, propagandist or place-making objectives. These manifestations also symbolise, often over a long period, a place, a town, city, even a whole society or nation-state. How and why culture is planned is therefore a reflection of the place of the arts and culture in society, of the approaches to the design and planning for human settlements in the town planning tradition and therefore in the development of urban society:

> Place and culture are persistently intertwined with one another, for any given place . . . is always a locus of dense human interrelationships (out of which culture in part grows), and culture is a phenomenon that tends to have intensely local characteristics thereby helping to differentiate places from one another.
>
> (Scott 2000: 30)

Whilst the 'cities of culture' have in the past been associated with the centres of empires, city-states, trading and industrial towns and cities, the urban renaissance which incorporates culture as a consumption, production and image strategy is evident now in towns and city–regions in developed, lesser developed, emerging and reconstructing states; in historic towns and new towns; and in cities seeking to sustain their future in the so-called post-industrial age (or more accurately the *new industrial* era). The symbolic and political economies of culture have arguably never been so interlinked. This is perhaps not surprising in the context of globalisation, where late capitalism sees symbolic goods as niche markets and the arts and culture are big business – for local, domestic markets and for international and tourism trade. Planning for culture in this sense adopts industrial and economic resource planning and distribution, whilst the physical aspects of public culture – facilities, amenities, the public realm: a *cultural infrastructure* – directly contribute to urban design and the relationships between land-use, access and transport, i.e. the town planning process. Although the cultural flagships and the designated and self-styled cultural cities and industries receive most attention from both historical and contemporary perspectives, the creation, planning and support of cultural amenities for primarily local communities, and for artists themselves (e.g. education, training, small-scale production, studios), has a much wider application and tradition. This is most apparent in the twentieth century where notions of cultural equity 'rights' and growing urbanisation and cosmopolitanism looked to aspects of the arts and culture as social welfare provision. This was also evident not only in the most prescriptive socialist society models (*People's Palaces*), but also in the past where popular entertainment and common (and uncommon) culture took place in gatherings and meeting places, festivals and fairs, and pleasure gardens, as well as in buildings for arts and entertainment. It is these local *art centres, maisons de la culture, casas de cultura*, whether shared village halls, community centres, workers and association clubs, or municipal and commercial cultural facilities from the museum, theatre, civic and dance hall to the cinema and local festival, that planning for culture also encompasses. A critique of cultural planning as this book seeks to present therefore needs to consider both high-art as well as local and popular culture, in different places and in different times. An international perspective also provides a comparative basis by which culture in society and the design of urban settlements has impacted and been treated in different countries and under different regimes. How far replication, models and convergence is evident in the current and earlier examples of cosmopolitan and globalised states and empires, and how far social and planning policy has influenced this, are therefore recurrent questions considered throughout this book.

It could of course be argued that a book on planning for the arts at a time of increasing globalisation of cultural consumption and production, and the converse but not unrelated rise of individualism and *new millennialism*, is anachronistic. The technology-driven expansion of home-based entertainment and leisure activity; moves towards the twenty-four-hour city and night-time economy; the associated social atomisation of work, home and play; and fragmentation of traditionally collective forms of cultural participation might therefore render an investigation of planning for the arts somewhat redundant, or at least of historic rather than contemporary concern. Despite, and perhaps because of, the globalisation of media and cultural products, images and social expression, the late twentieth century has paradoxically seen a renaissance in the development of new and improved venues for cultural activity – from arts and media centres, theatres, museums and galleries, and centres for *edutainment*; public gatherings, raves and festivals, *Pavarotti in the Park*; to public art works, urban design and public realm schemes – as well as the promotion of cultural industries zones and workspaces to attract and support the new media and cultural economy in towns and cities world wide. This is seen in cities seeking to transform their image and appeal and thereby qualify as *cultural capitals* for the first time, such as 'Guggenheim Bilbao', to established industrial cities also undergoing re-imaging through upgraded and new cultural facilities, from Glasgow, Barcelona and Frankfurt to Baltimore, Montreal and New Jersey to name a few, with massive *fin de siècle* cultural and museum quarter developments in Berlin and Vienna and in Beijing and Singapore. As Zukin maintains: 'Rightly or wrongly, cultural strategies have become keys to cities' survival . . . how these cultural strategies are defined and how social critics, observers, and participants see them, requires explicit discussion' (1995: 271). This is not only a Western phenomenon – although its foundations may have ancient roots from the cities of the classical Athenian, Roman and Byzantine empires, to the European Renaissance – since it has been replicated and adapted in developing and emerging nation-states, from Croatia to Southern Africa. As one indication of this, the World Bank, whose mission is to provide loans to developing countries and in areas of post-conflict/reconstruction, recently initiated a Culture and Sustainable Development programme with a focus not only on conservation and heritage (e.g. sites and patrimony), but also on 'Culture and Cities' (1998). The cultural dimension to *development* – a form and function of land-use and economic planning – is therefore seen as an important component of economic and social policy, rather than an aspect of society which is peripheral or at least subsidiary to the political economy and public sphere (McGuigan 1996).

Indeed, the development and funding of cultural *Grands Projets* by national, regional and city governments, as this book will present, both emulates and parallels the urban renaissance witnessed in Europe between the fifteenth and seventeenth centuries, and subsequent public works and rational recreation policies advocated by the Georgians and later the Victorians in Britain and elsewhere. Rationales for state involvement and promotion of cultural facilities show both an historic continuity and contemporary response to economic and social change. This is not least reflected in the breaking down of traditional planning assumptions and imperatives that have in the past separated the functions of employment, leisure and housing in the dualistic industrial city, with a clear spatial divide between these social spheres (Weber 1964, Doxiadis 1968). As Charles Jencks comments on the failure of modern town planning: 'masterplans were drawn up with the city parts neatly split up into functional categories marked *working, living, recreation, circulation*', but as he goes on: 'inevitably these mechanistic models did not work; their separation of functions was too coarse and their geometry too crude to aid the fine-grained growth and decline of urban tissue. The pulsations of a living city could not be captured by the machine model' (1996: 26). Physical proximity does not however overcome social and cultural exclusion, while at the same time ambiguous transitional zones blur the edges and offer more porous boundaries that allow people to move and restructure the urban area in accordance with socio-economic change, as the post-industrial notion of the urban village and 'a complex pattern of interlinked districts takes shape' (Seregeldin 1999: 52). Cultural planning, as well as an aspect albeit an exceptional one, of amenity planning, has therefore played a role and one that is increasingly being adopted in the post-industrial era in meeting economic and physical regeneration as well as 'place-making' objectives (Ashworth and Voogd 1990, Ward 1998), and as an approach to urban design and the more integrated planning of towns and cities.

Planners, 'urban strategists' (Landry 2000) and writers on cities, urbanism and glob-alisation have of course contributed to an air of determinism and fragmentation, not quite in the manner of John Ruskin and the later Arts & Crafts movements and their planning inheritors, the Garden City and Utopian movements, but with a feeling of the failure of urbanisation and the deleterious effects of post-Fordist economic change. This is seen in the de-urbanisation and suburban sprawl evoked by Noel Garreau's *Edge City* (also Evans 1998d); Dejan Sudjic's *100 Mile City* (1993) and the *technopolis*, core and periphery divides analysed in Castells' *Information Age* (1989, 1996), as well as by mas-terplanners such as Peter Hall (1988) and others. At the same time, urban sociologists and analysts in the USA, such as Anthony King, Saskia Sassen and Sharon Zukin have linked the symbolic economy: 'the trade in signs, images and symbols . . .' (King 1990), with the post-industrial city, in terms of land-use, landscape and development, and in terms of the cultural economy itself (Scott 2000). What distinguishes the late capitalism phase and post-industrial eras from the earlier colonial and commodity trade-based glob-alisation periods is the extent to which society has become cosmopolitan, not that cultural consumption has just become homogenised and cultural facilities serially repli-cated. Some argue that the earlier period of intense globalisation that occurred in the late nineteenth/early twentieth century brought about national alliances and power struc-tures and a consequent nationalism of 'wilful nostalgia', requiring homogenised and integrated so-called common cultures and the elimination of ethnic and regional identi-ties (Robertson 1990, also Adorno and Horkheimer 1943, Adorno 1991). The heyday of the Hollywood film and movie-going was witnessed between the 1930s and 1950s, despite the resurgence of cinema attendance today, accelerated by the development of the multiplex (if not of film production and choice), whilst the culture industry which Adorno and Horkheimer (1943) railed against in Nazi Germany has exhibited important gains in cultural democracy and cultural development – the ability of people to mediate, adapt and make their own cultural forms and to access associated technology (e.g. audio-visual, desk-top publishing, photography, digital arts and multimedia) is one measure of this; the process of cultural hybridity and fusion is another. As Stuart Hall (1990) and others (Cooke 1990, King 1991) maintain, this is increasingly the norm and assumptions beneath cultural planning necessarily need to take this new reality into account. Culture, to borrow Homi Bubha's phrase (1994), has many locations: 'a dialogue in which there are many parts . . . we are forced to speak of the *cultures* of cities rather than of either a unified culture of the whole city or a diversity of exotic sub-cultures' (Zukin 1995: 290). As Willis therefore optimistically put it: 'We need to think of ourselves as only at the beginning of civilisation's historical clock. The best of what is thought, spoken, written, composed and made, must be yet to come, and come it must from our living cul-ture and not from a backwards looking, self-propagating "art"' (1991: 8–9).

Book focus and scope

The primary focus of this book is the role and relationship between cultural policy and provision and town and city planning, taking key exemplars and approaches, and presenting planning regimes and case-studies from various countries and cities – from the classical, pre-industrial periods, to the industrial and post-industrial eras. On the one hand cultural planning is considered in terms of the amenity aspects of arts and cultural facilities, or culture as an aspect of 'social welfare' and spatial approaches to such provision; and on the other, cultural planning is placed within the wider context of urban planning, regeneration and local–global relationships. The adoption of arts and urban regeneration policies and urban economic strategies from the late 1970s in Europe, the Americas and spreading to Asia presents a particular version of the urban renaissance with a hardening core–periphery and social divide in cultural activity and amenity, and an archetypal manifestation of the twin movements of globalisation and cosmopolitanism. Issues and practice of urban cultural rights, identity and the city as a shifting site for cultural production and consumption emerge from this late twentieth-century attempt to reclaim and redefine the city.

The related but distinct sub-discipline of cultural geography has also developed an approach and body of knowledge on the spatial and symbolic variations among cultural groups and the semiotics of landscape, taking Tuan's definition of culture as 'the local, customary way of doing things; geographers write about ways of life' (1976: 276). It is fair to say, however, that geographers and their urban cousins, town and city planners, have not tended to consider the arts, creative activity or cultural development – one example of this is the lack of a definition of 'amenity' in town planning legislation and practice, other than through a negative, anti-urban sentiment, and the absence of planning standards for arts facilities in contrast to the more benign areas of parks, play and recreation, and conservation and heritage, alongside housing, industry and other local amenities. Until recent times, planning has, not surprisingly therefore, avoided a deeper appreciation of the needs of arts practice and participation, or resisted engagement with 'culture' altogether, unlike other areas of social policy and urban development. This book therefore attempts to introduce and analyse some of the ways in which culture and planning have and may be integrated against these anti-planning ('Non-Plan') tendencies.

Arts/planning defined

In a book on planning for the arts and the position of arts and cultural facilities in amenity planning, the ubiquitous term 'planning' itself requires further delineation. Some core definitions of planning in these related but discrete contexts may therefore be useful at this stage. Like the term 'culture', the generic 'planning' is widely used and associated with a range of functions and disciplines, from *human geography* – the disciplinary root of modern town planning; *urban design*, as in the planning of settlements, e.g. masterplan; *planned economy* and modern political economics – 'Marshall plan', five-year plan; related social policy and public administration to business management (*corporate* and *strategic planning*) and organisation theory. Planning is the application of scientific method – dictionaries define town planning successively or cumulatively as a science and an art – however crude, to policy-making and is closely associated with 'public policy and choice' theory (Dunleavy 1991). Planning is also defined as 'a process for determining appropriate future action through a sequence of choices' (Davidoff and Reiner 1973: 11) and therefore in the case of amenity planning – as Tietz argues in his seminal work on facility location: 'public determined facilities [have a] role . . . in shaping the physical form of cities and quality of life within them' (1968: 35). The definitions below, whilst discrete, are also used in combination with each other and in practice can overlap: 'In all probability, the difficulty of achieving a closer definition of this concept is attributable to its polymorphous character: yet all would agree that in the final analysis, its purpose is to organise the city for the greater happiness of its inhabitants' (Cohen and Fortier 1988: 12). All definitions of planning therefore infer some consideration of the future and the achievement of given goals or end states, whether physical and environmental, social or economic: arguably all manifestations and impacts of culture. The terms 'strategy' and 'strategic plan' are also now widely applied, a reflection perhaps of the business and scientific management approaches exported from the USA from the 1960s and drawing on technological and military terminology – e.g. *cultural strategies* (Zukin 1995) and *urban strategists* (formerly 'planners'; Landry 2000). A specific adaptation in town planning, including the cultural sphere, is the concept of *infrastructure* – first coined by the French railways and then in military installation and public utility provision. These terms found favour and usage from the 1980s in arts administration and government policy and practice (e.g. Arts Council 1984, 1993a), as a natural terminology for both the new managerialism and rationalised public services (Pick 1988, 1991, Evans 2000b, Adorno 1991), and in local, regional and city arts plans and *strategies* – all confirming a planning approach to resource allocation and decision-making for the future.

1   **Town Planning** – in Britain, *Town & Country Planning* legislated comprehensively in town and country planning Acts in Britain from 1947 and in the USA *City Planning* and at the micro-level, *zoning*. It incorporates amenity planning – recreation, conservation, as well as economic development. Primarily a function of population, land-use and the control of development (zoning, land-use classes) and latterly heritage/area conservation. National (and supra-national, e.g. European Union) planning policy and guidance-driven, but implementation and interpretation is a local function of statutory local planning authorities, based on a local area plan (e.g. city, town, district) and regional structure or county plan (namely County of London Plan 1943, Greater London Development Plan 1969, Toronto City Plan 1991).

2   **Strategic Planning** – public sector macro-economic resource allocation, investment and long-range planning (e.g. *infrastructure*, above), and private industry corporate planning and strategic business planning. It incorporates both social welfare planning and national/regional land-use and utility development, i.e. higher level 'Structure Plans' in town planning (Point 1 above), and in the USA comprehensive *strategic* or *masterplans* (So and Getzels 1988). Hence 'strategic planning is about trying to ensure that appropriate development occurs in appropriate places and is matched and supported by the provision of required infrastructure' (Smith, in Englefield 1987:29).

3   **Arts Planning** – the allocation of resources and distribution of public subsidy and facilities for a range of designated and prescribed arts activities – 'art forms' (namely theatres, galleries, museums, concert halls, dance studios, arts and media centres, film exhibition, etc.), and the support of artists and cultural

workers, including education and training. It takes place at national (*flagship*, arts policy), regional (regional or provincial arts area) and at local community and arts amenity levels. Thus the regional or local *Arts Plan* refers to a *strategic* plan (Point 2 above) of arts resources – creative artists/workers, facilities, funding, markets/audiences and participants for a given catchment area or community. This includes the concept of arts development and access (and cultural 'rights') – often through intervention in communities and local areas to stimulate demand and participation, and in some cases to empower, e.g. notions of cultural democracy and development.

4   **Cultural Planning** – on one hand the 'art of urban planning' (Munro 1967) and also the wider integration of arts and cultural expression in urban society. It is also described as 'the strategic use of cultural resources for the integrated development of cities, regions and countries' (DMU 1995). When combined, these produce a cultural approach to Town Planning (1) which uses an infrastructure system of Arts Planning (3). Mechanisms employed include consideration of urban design, public art, transport, safety, cultural workspace and industry quarters and the linkage concept of the creative *production chain* and *scale hierarchy of facilities*. Given the role of cultural development and democracy intrinsic to a cultural planning approach, the exercise of local governance and community involvement in planning processes, facility location and urban design, also incorporates *Planning for Real*, Community Planning and delphic exercises such as Urban Design Action or Assistance Teams (UDATs) used for instance in the USA and UK for major development areas and sites.

Planning, as I have already noted, infers the planning of *resources*, present and future, and therefore cultural planning concerns activities, facilities and amenities that make up a society's cultural resources. A framework for this has been developed that goes some way to show the various spheres which a cultural planning perspective offers for policy formulation: 'a process of monitoring and acting upon the economic, cultural, social, educational, environmental, political and symbolic implications of a city's cultural resources' (Comedia 1991b: 78) (Figure 1.1).

In a recent guide for cultural planning and local development in Australia, for example, cultural planning is seen as 'simply a purposeful, strategic approach to cultural development . . . approached like any other form of planning; by a thorough assessment of the existing situation; by setting clear goals and objectives; by identifying clear issues and priorities and by formulating and implementing practical courses of action' (Guppy 1997: 8). Landry also puts this in terms of the management of cultural resources and governmentality (Bennett 1998): 'Cultural planning is the process of identifying projects, devising plans and managing implementation strategies. . . . It is not intended as the planning of culture . . . but rather as a cultural approach to any type of public policy' (Landry 2000: 173). This distinctly bureaucratic terminology perhaps overstates the 'simplicity' of such an approach and the complexities and tensions within the processes of community and cultural development and creativity itself (e.g. the role of the artist), and the selection of 'priorities' – *whose culture, whose priorities?* Later, the guidance was stated more realistically:

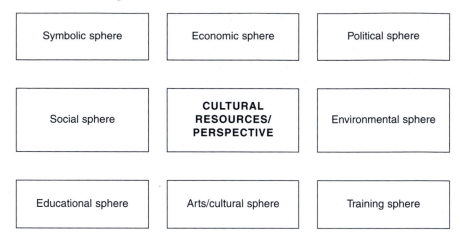

Figure 1.1 Cultural resources planning perspective
Source: Comedia (1991b: 78)

> Cultural plans and policies articulate an ongoing role for cultural appraisal and action in a competitive planning environment. They also provide a formal discourse with the statutory planning framework along with an informal and an energetic entry point for local communities eager to conserve and develop the cultural identity of their area.
>
> (Guppy 1997: 54)

Planning for what arts?

The imperatives and mechanisms of urban planning as they are applied to culture also beg at least some consideration of the arts that are and have been 'planned' in the past. Firstly, planning which infers a positive change ('development' and 'progress') also represents in practice a normative approach to public culture, the prescribed and therefore legitimated arts (Point 3 Arts Planning, above), and therefore to the place of culture within society. Environmental planning in the modern local–regional–national hierarchical sense, and the earlier planning of city-states and settlements, has influence over the nature of culture that is facilitated and promoted, whether benign or as a manifestation of ideological and/or religious foundations and their celebration and propaganda. Planning also incorporates aspects of control, censorship and therefore culture that is excluded, banned, suppressed, or even ignored. (The town planning function, it should be remembered, is also often the source of licences and permits, e.g. for public entertainment, dancing, alcohol and so on.) Distributory approaches also look to spatial equity in arts facilities and amenities and the arts that find themselves within the practice of planning and urban design therefore largely flow from the position of the state in relation to 'culture', however defined (Titmus 1974, Pick 1988). What is represented by cultural policy and municipal culture today?

Different societies have throughout European history, devised may different ways in which to find a place for the artist, ask for his work and supply him with resources and a living. The Greek City State, the Mediaeval church, the Renaissance Pope and potentate, the eighteenth century prince, the impresarios, dealers and publishers, of the nineteenth century . . . today these functions . . . are fulfilled by committees, with financial assistance from state and municipality.

(Pick 1980: 27)

As ideologies/beliefs then require policies, programmes and action, the planning of the built environment and the inclusion of cultural practice and expression within social formations, and questions about whose arts are to be represented, 'housed' and provided for, therefore arise. Concepts and definitions of culture itself are of course fraught and fluid, as Eagleton reminds us: 'Culture is said to be one of the two or three most complex words in the English language' (2000: 1). The dialectical tension between ideas of culture, between the high-arts and non-high-arts (popular culture, low-brow etc) are encapsulated in three variants: (1) its anti-capitalist critique; (2) the notion of a whole way of life (Arnold, Williams, Elliot *et al.*) and therefore culture as civilis-ation/ing; and (3) its specialisation in the forms and practices that make up the canon of the Arts (Eagleton 2000: 15). Lists and classifications of arts practice (Munro 1967), typologies of arts facilities and media, those arts eligible for state support, education and training provision and accreditation, together form the mainstream within hegemonic and intermediary power structures. However, these do not easily transfer into the planning process, which is less concerned with artistic hierarchies *per se*, but with divisions between public and private (and therefore the influence of sponsors/patrons), the polity and political, notions of the public realm and aesthetics, participation versus consumption, and the place of culture within everyday life – as defined within Point 4 Cultural Planning, above. The German sociologist Bahrdt for example saw the origin of the public sphere in the late medieval European city, with the market as the organising principle generating new forms of social exchange (1969). The anonymous social interaction that early cosmopolitan trading centres exhibited through the market-place also created locations for cultural exchange and celebration. Sennett identifies the early urban cosmopolitan with the rising bourgeoisie and the construction of public space (1986), but where the public sphere occurred not just through economic exchange but through 'much more political and social exchange . . . the debate between free citizens in the coffee-houses and the salons, the meetings in theatres and opera houses' (Burgers 1995: 151). The evolution and creation of cultural places and spaces in pre-industrial, urban renaissance, industrial and in contemporary society, is therefore discussed in the following chapters, with a recurring theme of the relationship and tension between commerce and culture; between arts-as-amenity and the cultural economy, and the dominant rationales that have effectively selected and ranked the arts that are considered worthy of planning in different times and by different societies.

Cultural equity and 'rights' versus masterplanning

The notion and practice of public planning just described may not however sit easily with cultural ideals, particularly those, on the one hand, associated with cultural rights and freedom of expression, and, on the other, the creative process that may defy if not resist prescriptive planning altogether. Critics point to the deleterious link between public planning and culture in extreme socialist and authoritarian regimes for instance, which both control cultural production and, by design, censor and limit cultural diversity or pluralist views of society. This libertarian view has therefore resisted anything more than benign state involvement in culture, promoting the structure of 'arms length' cultural agencies, as a buffer between the state and the arts, and when this is threatened, raise the spectre of communist regimes under Lenin, Castro and Mao for instance, where only state-approved art was permitted and cultural planning was both centralised, monumental and an extension of propaganda machines. The promotion of a mono-culture is also evident in nationalistic regimes, such as Ataturk's Turkey which in the 1930s sought to purify the real Turkish folk music, standardising lyrics and instrumentation in pursuit of a Westernised and sanitised version of *Anatolia*, whilst genuine folk and religious musics were all but lost (in fact kept for the enjoyment of the ruling elite, namely 'courtly culture'). Colonial influence that spawned, for instance, Ghanaan choral singing, also unwittingly ensured that tribal arts went underground and survived 'unfused', whilst in apartheid South Africa, where tribal dance was outlawed and artists imprisoned for performing in public, a post-Mandela programme of cultural development and the creation of forty community arts centres in black townships and rural areas looks to re-establish indigenous cultural practice and expression, as it did in newly independent Zimbabwe in the 1980s with a programme of village-based cultural centres.

Whilst arguments for greater spatial consideration and integration of the arts and town planning have developed, as this book will explore, the notion of equity in access and participation in the arts and cultural expression also presupposes a democratic system capable of responding to and meeting local needs – community and artistic. Unilateral declarations of cultural independence may be unrealistic (although cultural and regional independence is a late twentieth-century phenomenon), however a reassertion of ownership of cultural amenity through enabling policy and planning is both desirable and possible, and indeed a goal which arts planning standards may facilitate, as Bianchini and Shwengel assert: 'Cities will be re-imagined in democratic forms only by creating the conditions for the emergence of a genuinely public, political discourse about their future, which should go beyond the conformist platitudes of the "visions" formulated by the new breed of civic boosters and municipal marketeers' (1991: 234). Given the dualism created by the twin forces of globalisation and centralisation of power – not least in cultural production and 'free' (*sic*) trade; the reassertion of regionalism; emerging eclecticism ('global village'); cosmopolitanism and 'glocalisation' – cultural expression and the planning of urban culture in particular are central to both reconciling and locally driven responses to these potentially conflicting regimes and aspirations.

The inclusion of services within GATT following the protracted Uruguay Round (1986–93), namely the General Agreement[1] on Trade in Services (GATS), has for the first time raised the issue of cultural services and intangible 'goods' within free trade legislation – with services accounting for over 60 per cent of world production and 20 per cent of international trade (Buckley 1994: 13) and cross-border trade in services totalling over $1,350 billion in 1999. However, as Scott points out, treating culture as simply 'goods' is problematic – commenting on the US Department of Commerce's position on GATT: 'free trade in cultural products betrays a fundamental failure to grasp the full complexity of the issues at hand' (2000: 212).

In the fraught relationship between central and local government, which has been epitomised by deregulation, the imposition of internal markets (e.g. through competitive testing/tendering) and a decline in public spending during the late twentieth century, the principle and practice of 'subsidiarity', of public choice and democracy, are of fundamental importance to the continuance of the principle of public/merit goods – services that are either free or subsidised and non-excludable and accessible to all – and therefore to levels of local amenity and cultural provision. As *The Economist* therefore argued:

> One essential is to end the pretence that local taxation should pay for those services which are clearly of national importance . . . [but] to meet the cost of only those that can reasonably be allowed to vary widely in local character. . . . Within such bounds, each local authority should then be left, unfettered, to coax voters into paying for whatever it favours – whether it be a new concert hall or meditation classes.
>
> (*The Economist* 1991: 18)

The dominance of a cultural and cosmopolitan elite, described as the 'Professional-Managerial Class' by the Ehrenreichs (1979), in the consumption of high-arts and national performing and visual arts audiences, has been a perennial feature of state-legitimated culture, from Bourdieu's *cultural capitalists* and the petite bourgeoisie, to the conspicuous consumers and occupants in the post-industrial city-centre arts flagships and cultural quarters. Whilst Bilbao is celebrated as the new cultural tourism destination (Evans 1998a), the creation of a franchised Guggenheim Museum designed by American architect Frank Gehry, together with loans and exhibitions from the New York collection, provoked negative reaction among Basque artists, journalists and regional politicians alike – in this case the museum as a site of contest (MacClancy 1997). The absence of a cultural policy and planning approach here, as in many of the 1980s' versions of culture-led urban regeneration, suggests that their new-found status as cities of culture will not be sustained (or their economic development and 'trickle-down' objectives maintained) in the post-event phase. The downtown and city islands of culture celebrated by urban revitalisation advocates – public and private – have in many cases ghettoised their inhabitants and those in the often poorer adjoining districts (namely the Baltimore Waterfront, City of London, Los Angeles, and *even* Barcelona; see Chapters 7 and 8), and as Robins claims, the highly selective revitalisation of 'fragments' of cities is really about 'insulating the consumption of living spaces of the postmodern *flaneur* from the "have-nots" in the abandoned zones of the city' (1993: 323). Richard Sennett in *Flesh and Stone* also offers a commentary on the corollary of the city centre – the urban periphery, following a cinema visit to an outer New York shopping mall: 'If a theatre in a suburban mall is a meeting place for tasting violent pleasure in air-conditioned comfort, this great geographic shift of people into fragmented spaces has had a larger effect in weakening the sense of tactile reality and pacifying the body' (1994: 17). A similar socio-spatial phenomenon had also been wryly observed by

Venturi in 1966, who asserted that Americans do not need piazzas, since they should be at home watching television (and eating pizzas . . .). Notions of cultural equity therefore have to be squared with fiscal and economic development strategies as well as cultural and urban planning policies, particularly where the spatial divide and social exclusion from local amenity and cultural facilities is hardening and widening (e.g. in car ownership) and the quality of spatial relations is deteriorating.

The real and perceived 'over-concentration' of national cultural production and arts venues in cities such as New York, Los Angeles, Sao Paulo and capitals such as Paris and Madrid has not surprisingly fuelled a regional city cultural regeneration and resistance to entrenched *centrism*, and in London, for instance, to the emulation of earlier eras when:

> leisure centres frequently imported theatrical and musical performers from the metropolis . . . and their musical clubs were modelled on institutions pioneered in the capital. . . . The metropolis provided a blueprint for many other areas of provincial urban life, so much so that in 1761 it was claimed that the several great cities . . . seem to be universally inspired with the ambition of becoming the little Londons of the part of the kingdom wherein they are situated.
>
> (Borsay 1989: 286–7)

In this case, the inheritance and continued political and cultural hegemony in the location of national art institutions, policy-makers, commercial media production and headquarter operations has directly caused a cultural planning response by other cities, for instance in the UK – Sheffield, Birmingham, Glasgow and Manchester, which have all pursued cultural industries and infrastructure policies in an unusually high profile way (Fisher and Owen 1991, Bianchini *et al.* 1988, Worpole 1988). In Europe, networks of regional and 'second cities' have been established to counter the core–periphery drift, including the development of cultural policy and planning approaches and regional groupings which reflect both cultural and geographic commonalities.

As writers on the information city and technopolis have also observed (Sassen 1991, 1996, Castells 1989, 1996), the tendency for spatial concentration of the powerful trans-nationals in global city quarters, such as broadcast and print media in Times Square, New York, Burbank, Los Angeles and Soho, London, contrasts to the post-Fordist, footloose behaviour of manufacturing and other dispersed service sector activities. The suggestion is that cosmopolitan culture (and its eclectic human capital) provides a competitive advantage to these global media operators that might otherwise levitate and fragment to locations of lower labour, land and capital/entry costs. However the allocation of public cultural resources (normally in the absence of a national cultural plan) also continues to be skewed towards the capital city-state, and between larger regional cities and smaller towns and so on. In Brazil's 5,000 municipalities, over 3,000 do not have a public library, whilst the mega-cities of Sao Paulo and Rio have the disproportionate majority; similarly in Greece, where Athens dominates in professional cultural provision (Deffner 1993) and in Canada where the cities of Montreal and Toronto and the administrative capital Ottawa possess the lion's share of major cultural facilities and activity. In contrast, France and Spain where resistance to capital city and central government administration is no less strong, regional and provincial city pride in cultural investment is well established, from Barcelona and Valencia to

Grenoble, Rennes and Montpellier (Bianchini and Parkinson 1993). The resistance by the French to what is perceived, with good reason, as the American Trojan horse that goes hand in hand with *mondialisation*, therefore, also manifests itself in planning measures to restrict the growth of the multiplex and protection of francophone cultural expression and production. Before the opening of a seventeen-screen, 3,000-seat Megarama on the outskirts of Paris, the Cultural Minister announced plans to increase subsidies to small town-centre cinemas – the French government passed legislation limiting new multiplex cinemas to 2,000 seats (Evans 1998d). In contrast, the world's third largest cinema complex was planned for the site of a former powerstation near Birmingham's infamous Spaghetti Junction, with planning permission for a twenty-four-hour, thirty-screen multiplex (Star City), whilst in north London a familiar if sentimental plea from a planning officer: 'the Borough is now served by one cinema where previously there were seven' (Evans 1998d), reflects the resignation and surrender to the global market (in this case to the vertical integration and dominance of US film production, distribution and exhibition) in a liberal planning regime set against a decline in both public realm and local amenity.

Culture is in consequence inextricably linked with notions of local governance and identity, no more so than when identity and ethnicity are threatened or suppressed, as in civil wars in the Balkans, and in disempowered ethnic groups, such as the Kurds and indigenous 'fourth world minorities' (Graburn 1976), from Central America to Australasia. When in 1936 the southern Spanish town of Almùnecar thought that its republican freedom was assured, the peasants and fishermen took over the village and declared their plans for the new millennium: 'Here will be the House of Culture' along with school, health, and agriculture centre (Lee 1969: 168), but this pre-Franco cry for freedom was unfortunately short lived. This freedom is still not assured, even in Europe today where state censorship and prosecution of artists and assaults on cultural expression deemed to be at odds with the extreme right-wing political ideology, is being witnessed in Austria. Here funding is withdrawn from incumbent arts organisations only to be replaced by those more consonant with the political message (see Chapter 8). Conversely, the promotion and cultural development policies pursued in Cuba, for example, has created a celebration of national culture and identity through music and dance, as well as architecture. As Cooke observes in *Back to the Future*: 'modern perspectives undervalue . . . the consensus of minorities, local identities, non-western thinking, a capacity to deal with difference, the pluralist culture and the cosmopolitanism of modern life' (1990: 11) and this is apparent for example in the promotion of heritage sites by universalist international agencies such as the United Nations Education, Scientific and Cultural Organisation (UNESCO), the International Council on Monuments and Sites (ICOMOS), the World Tourism Organization (WTO) and the World Bank. In extreme but by no means unique cases, Shackley even warns that 'the possession of a World Heritage Site and the development of cultural tourism can create a [spurious] image of long-term stability and the basis for establishing a national identity, or may become the focus for a new nationalism' (1998: 205). The extent to which cultural heritage should be prioritised over contemporary culture and living art is a complex and ultimately political issue, as the colonial quarters of Old Havana and Spanish Town, Jamaica languish in neglect, despite intervention by international agencies and foundations (Evans 1999c). Less attention and support is afforded contemporary art, cultural expression and facility needs in these communities (Willis 1991: 8–9, as cited above) and cultural planning may offer an urban and resource planning process and framework, within which these conflicting worldviews and amenity demands may be reconciled and more equally balanced. Arguably because of the cultural and political hegemonies and global capital that drive mono-cultures and mass branding, and the benign nature of traditional planning processes which reinforce both norms and the control of development, culture-led planning might provide a fundamental response to the promotion of cultural diversity, the protection of cultural identities, and the encouragement of the local and the vernacular.

Furthermore, the planning of our towns and cities, the consideration of amenity provision within society, and the celebration and development of cultural rights – in Europe reasserted in the Maastricht Treaty (CEC 1992, cited in HMSO 1993, Fisher 1993), and the *European Urban Charter* (Council of Europe 1992) – arguably requires an element of planning: spatial, resources and 'cultural', as does meeting the changing needs of communities and creative processes. The imperatives of urban living and consumption therefore also look to a more sophisticated and integrated approach to the cultural aspect of post-industrial society, whether in developing or advanced states and the extent to which this has been achieved is therefore explored here. With this dialectic in mind, and in terms of the planning definitions presented above, this book therefore analyses the evolution of town planning in relation to public cultural amenity and arts facilities and offers a critique of arts planning approaches and the development of a *cultural planning* conceptual framework within which both urban planning and arts planning relate. Lewis Mumford's plea of 1945 is therefore just as pertinent today:

> The technical and economic studies that have engrossed city planners to the exclusion of every other element in life, must in the coming era take second place to primary studies of the needs of persons and groups. Subordinate questions – the spatial separation of industry and domestic life, or the number of houses per acre – cannot be settled intelligently until more fundamental problems are answered; What sort of personality do we seek to foster and nurture? What kind of common life? What is the order of preference in our life-needs?
>
> (quoted in Olsen 1982: 12)

More and more towns and cities, regions and countries – established and emergent – therefore look to culture to reaffirm their identity/ies; attract and retain their share of the cultural industries (and tourists); join the 'competitive city' race and contribute to the design and adaptation of the public realm and consumption in urban society. How these cultural strategies are defined and how 'we see them' (Zukin 1995, as quoted above), this book attempts to discuss explicitly, since as the *Richness of Cities* maintains over fifty years from Mumford's humanistic plea:

> Any form of urban planning is today, by definition, a form of cultural planning in its broadest sense, as it cannot but take into account people's religious and linguistic identities, their cultural institutions and lifestyles, their modes of behaviour and aspirations, and the contribution they make to the urban tapestry.
>
> (Worpole and Greenhalgh 1999: 4)

Summary of book content

Culture and its place in the planning and life of towns and cities naturally follows an incremental and evolutionary path, including the transfer and transmission of artistic products, styles and experience within and across communities and societies. In the next two chapters the book therefore presents an historical analysis and synthesis of the place and form of public culture within certain early classical societies, including Athenian, Roman and Byzantine, as well as metropolitan exemplars such as pre-Colombian Mexico. The evidence and supporting theories of the cultural influence in models of city formation and planning are considered in the context of the emerging relationship between culture, commerce and trade and the issue of population density, size and therefore cultural autonomy and the emergence of a public realm in these earlier regimes. These issues are extended in Chapter 3 in the early experience of urban culture in renaissance Europe and in the industrial age. The move from essentially elitist, private provision of the arts from court to putative state and from merchant to middle class consumer is examined through the formalisation of places and typologies of cultural facility and crafts trade. A focus on public places for drama and opera in Elizabethan London and the courts of Europe and the spread of culture-houses in the nineteenth century confirms not only the symbolic importance and continuity of location, but also an increasingly stratified audience for the arts, as class divides and state intervention in cultural activity and provision are established. Industrialisation and the move from rural to urban forms of popular culture are therefore considered in relation to state planning and programming controls and the nineteenth-century response in the rational recreation movement and its effect on new and re-created provision in the form of museums, theatres, libraries and pleasure gardens, and their inheritors, the gin palaces, music halls and precursors to the cinema. Cinema's rise and fall and resurgence is a factor in its changing building type and location, epitomised in the multiplex and like its predecessors, its forecast saturation. Cities such as London and Berlin and their emulators, and the internationalisation of cultural facilities and consumption through colonisation and trade, such as opera, theatre and libraries, serve as detailed examples, as do the Great Exhibitions which brought together culture and commerce under a global economic rationale for the first time.

As the evolving forms and locations for collective cultural activity came to be influenced by state policies towards these aspects of recreation and national identity, the beginnings of town planning and associated approaches to new town and city–regional development and decentralisation are considered in Chapter 4. This is dealt with in the context of the place of amenity in emerging town and country planning legislation and the particular place of the arts and entertainment in the post-War reconstruction and formation of welfare states and its socialist manifestation in *Peoples Palaces* and *Houses of Culture*. Concepts and case-studies of distributive policies for cultural provision are then presented, comparing French and British state arts policy and the development of the arts centre and *maison de la culture* as a gradually universal phenomenon in local and municipal cultural provision. The development of the arts centre is documented in France, the UK, USA and elsewhere as a vehicle for arts development, a network for community and social action – whether village hall or new build venue – but specifically as a *local* amenity. A theme taken up in this book is the absence of both definitions of amenity and specifically the reluctance to plan for the arts and apply planning standards and norms, as are widely used for other recreational amenities such as sports, play and open space provision. Models and techniques for developing planning standards for arts and cultural facilities are consequently outlined in Chapter 5, incorporating examples of more integrated policies for arts provision with local area development plans. The profile of cultural consumption and audiences for various arts activities confirms both their disproportionate socio-economic and spatial concentration, whilst this chapter provides evidence of the range of environmental and perceptual barriers which limit out-of-home and wider participation in cultural activities by the majority. Key concepts of the scale hierarchy and *pyramid of opportunity* are presented with case-studies from local area and city arts plans and cultural strategies. From this I argue that despite their shortcomings in implementation, arts planning standards would go some way to ensuring greater distribution and access to cultural experience and expression and counter the spatial core and periphery and institutional imbalance which the cities of culture have reinforced. From the social welfare arts-as-amenity experience, the growing attention paid to the cultural economy and the commodification of the arts as urban cultural assets is then discussed in Chapter 6. This provides a critique of the economic importance of the arts argument and the conversion of high and popular culture though cultural tourism and the cultural industries, as prime and growing elements of urban and national economies. The importance of cultural provision and other quality-of-life factors in employer re/location presents another rationale for their value and contribution to the urban environment and in post-industrial society. The tensions between traditional town/centres and the out-of-town/edge city drift considers the shopping mall and leisure–retail *pleasure periphery* which has had a radical spatial impact on cultural consumption, whilst conversely, cultural production and higher-scale facilities continue to be concentrated in the core inner urban and downtown zones. Data presented on city, national and regional cultural economies compare employment across a number of arts and cultural sectors and the clustering evident in world/cities of culture and within entertainment, touristic and cultural industry districts. Cultural activity as a universal economic development and employment strategy warrants a close look at its form and claims for its growth prospects. A definitional analysis of the cultural industries is discussed in both conceptual, economic and political terms, including the production chain as it is applied to the arts and various creative practices. Questions are raised over the politically termed creative and knowledge industries and the crude conflation of the heterogeneous cultural industries, including their creative content and employment profile and the impact of *e-commerce* and 'digital arts' on traditional forms of cultural practice and dissemination. A specific type of cultural production facility, the artists studio and workshop is then considered, with examples of this traditional and symbolic place for the arts across European and North American cities and the mixed treatment of the artist and public art in urban regeneration.

The importance of the city–region in terms of the cultural economy, identity and political aspirations towards autonomy is the subject of Chapter 7. It looks at the notion and promotion of European 'common culture' and heritage through the regional development programmes that have benefited cities and rural areas of the weaker economies of southern European, Ireland and northern industrial regions undergoing post-industrial regeneration. From an overview of European planning systems, contrast is made between the planning regimes and respective approaches to cultural amenity in different European countries. The inclusion of the arts and heritage and major culture-led redevelopment projects outside of either a national or European cultural policy or planning framework, presents a prime core–periphery and cultural capital emphasis which the examples of collaboration between the arts and urban regeneration have come to typify, from Barcelona to Birmingham. Examples such as Glasgow and Dublin, as well as cities in new EU Member States such as Helsinki and Vienna, indicate the extent that the European arts and urban renaissance has been adopted and replicated. The place of culture and the flagship arts project within major downtown, city centre and regeneration sites is the main subject of Chapter 8. Taking examples from North America and Europe, including major cultural zones in Berlin and Vienna, the arts and urban renaissance formula is reviewed here and in developing countries, notably in South East Asia and Latin America. The Westernised models of urban regeneration and architecture are evident in many of these developing countries, echoing the universalist approach to the development of heritage sites by the World Bank and others. The involvement of Western development agencies in the promotion of culture within developing and restructuring states provides another example of where cultural planning might engage with community and cultural needs, rather than the heritage tourism strategy adopted to regeneration here as in post-industrial cities in more advanced states. A comparison is also made between major regeneration areas of two European world cities – London and Paris – in the context of their contrasting city planning and governance regimes and the culture-led approaches adopted in each case and their relative outcomes. Even in the more fêted examples of regeneration and cities of culture, these strategic planning solutions, I argue, in fact reinforce the divided city at the cost of local area amenities and genuine mixed-use of buildings and sites, including more varied forms of cultural expression, production and the public realm. This is no less in the archetypal contemporary mega-event and EXPO, which are contrasted to the earlier Great Exhibitions and civic cultural monuments, with a critical review of the planning issues arising from the millennium and *Grands Projets* in Paris, London and Montreal – their sustainability and influence on the cultural maps of cities. The extent to which planning and in particular cultural and more consultative forms of planning have been evident in these cities of culture is a continuing theme which the concluding Chapter considers in terms of cultural strategies, the notion of the arts as public goods and resistance to planning that undermines many approaches to more integrated and community-based planning and resource distribution. The theme of culture-led planning and particular approaches and mechanisms offered by cultural planning in varying environments and locations runs throughout these chapters. These, the book argues, are required to counter the failure of simple economic and property-based 'solutions' to urban and cultural decline, and the explanations offered by economic and cultural geography, urban sociology and regime analysis, and to ensure the survival and growth of post-industrial society and those aspiring towards greater cultural development and diversity and greater spatial equity in cultural provision.

Notes

1   This Agreement (1993), which created the World Trade Organization (WTO) in 1995, also set up a new framework for the so-called protection of intellectual property rights (TRIPS – Trade in Intellectual Property Rights), although the import of Hollywood films to France was not resolved for fear of undermining the 1993 GATT.

2 The historical evolution of city arts and cultural planning

In the centre of Fedora, that grey stone metropolis, stands a metal building with a crystal globe in every room. Looking into each globe, you see a blue city, the model of a different Fedora. These are the forms the city could have taken if, for one reason or another, it had not become what we see today. In every age someone, looking at Fedora as it was, imagined a way of making it the ideal city, but while he constructed his miniature model, Fedora was already no longer the same as before. . . . The building with the globes is now Fedora's museum.

(*Invisible Cities*, Calvino 1979: 28)

Introduction

The extent to which conscious and deliberate planning has influenced and informed the location and provision of cultural facilities is a function of both the control of land-use and building – whether development is dictated by state or other power groups (e.g. Church, Crown or 'tribe') or a system of democratic consensus – and the place that the arts have in a particular society or community and therefore in the planning for human settlements. This book is neither a treatise on urban design and morphology through the ages nor a social historical account of 'culture' or its specific manifestations and building types, although both are touched upon in assessing the evolving relationships between urban and city planning and cultural amenity and development. Applying industrial, neo-Marxist and post-industrial (e.g. globalisation) theory and modern notions of political economy and the public realm to the pre-industrial past is dangerous and ultimately fruitless, whilst cross-cultural comparatives also suffer from both Eurocentricism and retrospective universalism, not least in the realm of 'culture' (Aitchison 1992, Schuster 1996). How past societies planned for public culture and the place of amenities within the development of cities does however offer particular insights to the inherited attitudes and paradigms of succeeding periods and societies. Arguably the 'classical' approach to urban design and the consideration and location of monumental and popular cultural forms provides a distinct example of continuity and change in the formation of cities and urban culture. This *continuity and change* in Kevin Lynch's view is needed so that the comfort of the past may anchor the future (1972).

Whilst modern and early town planning has used versions of grid or zone-based distribution of amenities and places for cultural exchange (including buildings for government, proclamations, trials, festivals, etc.), earlier settlements evolved approaches to cultural planning which reflected the degree of openness and celebration in daily life and in special events – what today we refer to as the 'public realm'. Historic settlements and town amenities have also frequently been built on the inheritance of previous societies, empires and regimes and they are therefore both incremental and more restricted in the planning for new and adapted arts facilities. This increasingly fuels the tension between the protection and conservation of heritage in urban areas, as the past 'accumulates' and conservation imperatives intensify, and the demand for new development and contemporary amenity and cultural needs. This is particularly the case in older world cities such as London and Paris and in particular design solutions (e.g. I. M. Peï's glass pyramid at the Louvre and the proposed Daniel Libeskind extension to the Victoria & Albert Museum); in contested historic areas and 'divided' cities, such as Belfast, Jerusalem, Nicosia and even ethno-linguistically in Montreal, and the cities of other post-colonised empires where heritage buildings and architecture coexist with the modern (e.g. Helsinki, Santiago de Compostela, Sao Paulo, Mexico City). Where new towns and cities have been developed, the layout and location of facilities has been less constrained, but surprisingly, higher scale cultural facilities have tended to follow traditional core, or 'hub-and-spoke' design forms, with major cultural institutions located in the central zone as part of government/institutional and public plazas, as exemplified in Le Corbusier's vision of the *Radial* and *Contemporary City* (1929), and as applied in practice, for example, in Oscar Neimeyer's federal capital of Brasilia – a sign of their reverence for the classical geometry of 'sacred' architecture. The influential German urban planner Camillo Sitte (see below), identified as a 'culturalist' due to his respect for 'community culture', elevated the central square or plaza as a major European contribution to urban design (Burtenshaw *et al.* 1991: 25) – what Diefendorf called modernism with a conservative aesthetic. Sitte toured the ancient cities of Europe seeking out vantage points from which to gather materials on the historical precedents for urban spaces: 'He spent years peering down from steeples and city walls to see how the anonymous builders of the past had fitted entrances into civic squares and positioned prominent buildings in relation to them' (Sudjic 1993: 14).

Athens – cultural capital?

The archetypal cultural planning 'model', where the integration of work, home and 'play' formed the very nature of citizenship, was of course classical Athens (750–450 BC – the 'Archaic' and 'Classical' periods; Pomeroy *et al.* 1999). The commune-ism and state provision and individual participation in art, sport and politics offers a version of cultural planning towards which municipal socialists and contemporary planners (*cf.* Mumford above) still look, as Kitto claimed: 'Religion, art, games, the discussion of things – all these were needs of life that could be fully satisfied only through the *polis* – not as with us through voluntary associations, or through *entrepreneurs* appealing to individuals (this partly explains the difference between Greek drama and the modern cinema)' (1951: 78). Whilst a linear interpretation of the evolution of cities commencing with Ancient Greece is rejected as Eurocentric, a criticism also made of Hall's selective *Cities and Civilization* (1998), Massey *et al.* (1999) also argue that the evolution of cities of the First World as an 'exemplary' (*sic*), hierarchical group, not only largely ignores civilisations from Mesoamerica to Eurasia, but also the reality that late twentieth-century urbanisation has shifted towards these regions and their mega-cities, and away from the First World (King 1990, Potter and Lloyd-Evans 1998, Seabrook 1996).

Notwithstanding diffusionist and universalist theories of human development and the evolution of cities: 'What is remarkable is how similar ancient cities everywhere were in terms of social structure, economic function, political order, and architectural monumentality' (LeGates and Stout 1996: 17). On the other hand, it must be appreciated that aboriginal societies, whether rural, urban settlements or even nomadic, recognise little or no distinction between discrete Western notions of drama, dance, music, 'art' or architecture, often with no separate concept or words for these practices or 'art forms' (and 'unlike western man, built no cities, but meticulously named every place'; Greed 1994: 74). Indeed, writing on the history of theatre, Southern observes: 'Looking at a West-End or Broadway theatre, or in Paris, Bombay or Osaka . . . it is not really believable that this all started out of a 10th-century Christian Church liturgy. *It is no more convincing to place its origin on the threshing floors of Ancient Greece*' (my italics, 1962: 37). For example, India's theatrical history predates classical Athens with the early commentaries such as the Sanskrit grammarian and scholar Panini who in the ninth century BC was already speaking about theatre as an integral part of the civic life in his *Natsuras* (Modi 1998). Bharat Muni's *Natyashastra* one of the most ancient recorded dramaturgies composed between 400 and 200 BC, is broader in scope than say Aristotle's *Poetics* and includes important discussion on performance theory, acting, aesthetics and even the construction of performance spaces. Whilst Southern questions whether primitive ritual predating the proscenium arch and Dyionisian venue can be properly called 'theatre' – where there are no words, no play, no particular place of performance, no playhouse, no scenery or even any assembled audience – the 'costumed player' even performing for a closed group in a sacred space provides the roots of theatrical performance conducted in a special place and time. In time the mask was supplemented with words and later poetry, then as the superstitious masks disappeared and human characters developed, circular acting-places came to be used for performance. This festival, conducted for instance at harvest time in Tibet, could also be seen in fourteenth- and fifteenth-century mummery and mystery plays in Britain and France, in the 'round' or central plain area – the *platea* (Latin) or *placea*. The 'place' then, meant in medieval theatre-speak, the central plain around which scaffolded tents formed the 'stages' for players.

Returning to Athens, the male elite who represented the *polis* of Ancient Greek society excluded not only women and slaves, of which there were an estimated 125,000 in Attica – over half of these in domestic service compared with about 45,000 Athenian male citizens over 18 years old – but also artisans, craftsmen and artists. The word *techne*, root of 'technique', was used for both the arts and crafts whose practitioners were manual workers, and who unlike poets and dramatists could not be considered 'gentlemen'. The architect – *arkhitetron* or 'masterbuilder' – also worked to order, with a low salary (similar to craftsmen) and without the status that a designer-architect aspires to today, although the tendency towards 'design and build' (some would argue, the 'dumbing down' of building design), has similarity to the relationship between the contemporary architect and developer. Architects also worked as sculptors, a common combination, particularly since Greek architecture was primarily a civic art (Cook

1972). The tradition of funeral urn and stone marker supported a Potters Quarter adjoining graveyards on the edge (in and outside) of the city walls (Sennett 1994: 37), an arrangement also evident in Teotihuacan, Mexico where over 150 ceramic workshops, fifteen just for the making of figurines, occupied one of the great walled compounds of the city (Davies 1982, see also below).

The housing of craftsmen within the city perimeter was a practice carried on into medieval times and the subsequent establishment of powerful craft guilds. Here the merchants could regulate quality and cost, and in turn extract rates, taxes and licence payments which eventually drove crafts (and later theatre) practitioners outside of the city walls and led to the control of the guilds in cities such as London: 'At the wall's gates the division between city residents and non-residents was sharpest. Here goods entering the city were inspected and taxed. Often non-residents were required to leave the city at dusk and seek accommodations outside the wall, as a result, suburbs – *faubourgs*, meaning "beyond the fortress" sprang up' (Jordan-Bychkov and Domosh 1999: 395). This practice of craftsmen operating in close proximity to their clients reflected the open relationship with the public, with workshops located in the heart of the city, often close to market-places. Different craft types congregated together, in 'quarters' and 'rows': 'Just as in Old London when you turned off Cheapside, you found yourself in Bucklersbury or Wood Street, Ironmonger (Row) or Leather Lane, so, in an old Greek city as you dawdled away from the Agora you could tell by the noise or the smell, by the clanging of the hammer, the grating of the saw or the pungent odour of the tannery, into whose domain you were intruding' (Zimmern 1961: 267). In touristic Athens today, Shoe Lane is a survivor of this ancient Bazaar tradition, with a cluster of shoe shops fronting workshops behind. Modern town planners and developers would have these dispersed and competitively located. In Athens these craftsmen are 'fellows' – guild members rather than competitors whilst sufficient demand exists, and all tend to benefit or suffer in times of high and low demand, whilst in thirteenth-century Paris, the *corps de metiers* consisted of around one hundred Royal-chartered crafts organisations divided into six practices: foods, jewellery and fine arts, metals, textiles and clothes, furs and building (Summerfield 1968: 58). In fourteenth-century Florence, members of established craft guilds represented the citizen-electorate itself.

Despite its over-reification (e.g. Kitto 1951), the planning of Athens, its building uses and spatial relationships, has provided a blueprint which towns and cities, not least in the neo-classical renaissance eras, have long-emulated in the location of civic and cultural spaces and buildings and which modern (master)planners such as Le Corbusier, and Alvar Aalto in Finland, took as their foundation. The Greeks themselves had not considered artistic town planning until the Hellenistic era (Zimmern 1961) – Haverfield specifically maintained that Greek town planning actually began with the 'processional way' (1913: 28) – and this therefore signified the conscious thinking about the city as a *work of art* for the first time. The 'art' of city planning was to be recognised again shortly after the enactment of full-blown town planning legislation in post-War Britain, when Munro attempted to merge a philosophical and scientific classification and listed a convenient one hundred 'visual and auditory arts' (1967). In addition to the predictable art forms, crafts and decorative work, he included city and regional planning (also Boyer 1988), whilst the German planner Sitte, whose influence reached Unwin and Geddes in Britain, Engel in Helsinki, Jansen in Madrid, and Henrici in Cologne, 'emphasise[d] town planning as a creative art' (a reaction against the 'Haussmannisation' of Vienna in 1889), so that 'Once again building and planning could be an art form again' (Burtenshaw *et al.* 1991: 27).

Public cultural facilities

The main types of Hellenic buildings were the temple, the treasury, the *tolos*, the *propylon*, the *stoa*, the fountain house, the *palaestra*, the gymnasium, the council chamber and the theatre. Many of these public amenities exist in towns and cities today, often significantly unchanged in their basic design and layout, and they therefore represent early examples of cultural facility provision. The morphology of the Hellenic city and its emulators – by 600 BC there were over 500 towns and cities on the Greek mainland and islands – also linked the functional zones represented by these differing social facilities. The two distinct use zones were the *acropolis* and the *agora*, the *acropolis* containing the temple, storehouse and seat of power; the *agora* the place for public meetings and gathering, judicial and educational exchange – the civic centre and hub of democratic life for the Athenian citizens (Jordan-Bychkov and Domosh 1999). Later, the *agora* came to incorporate commercial and retail activity, as the secular grew in advance of the sacred, and its role as place for official drama declined from the mid-fifth century, with much of the music performed in the *Odeion*, a roofed hall for musical contests (Sennett 1994: 57). Such growth was not planned in the modern sense, but ceremonial areas were designed according to prescribed sight lines, axes and with reference to the natural landscape, as had the cosmomagical cities of Egypt and Mesoamerica (e.g. Teotihuacan, see below), which were located in sacred and strategic defence positions. Colonial cities of the later Greek empire were however designed in formal plan, such as in Miletus, Turkey, using a rigid grid system imposed on otherwise irregular land areas and peninsula. The *council chamber* was a key facility in the city-state, as it still is in boroughs and city councils world wide (e.g. converted to arts centres, dance halls and theatres – Chapter 4). Other similar structures were put up for musical and religious performances. Originally a colonnaded long plan, the square plan developed to facilitate audiences that sat or stood on tiered benches or steps around three sides. However this necessitated internal columns that interfered with the audience's view – a perennial problem for theatres which only recent building technology allowing wide spanning structures has avoided. Finally, the *theatre* was a place for public meetings as well as dramatic productions, originally a convenient hollow hillside, later to be banked and excavated and with removable wooden seats. Audiences of 14,000 met in a festival atmosphere, to be partly emulated in Shakespeare's live productions in London and still in Italy today, opera performances (non-amplified) held in Verona's amphitheatre attract audiences of 16,000 each year (Evans 1999e). Only by the late fifth century BC was the first regularly planned, stone theatre constructed and then still on similar sites to their open air predecessor. Performances took place in a circular space of the orchestra (i.e. the dancing floor) and by the third century BC a platform on the 'skene' or stage building, became a stage for the actors. Other utilitarian buildings, such as the *treasury* were lavish representations of present and past glories, including those of other (city)states, much like the grander foreign embassies and occasionally their cultural centres and outposts, in capital cities today.

The incremental addition and adaptation of the artistic manifestations of former cultures and societies 'in residence' is therefore a long-standing feature of cities, ancient and modern, which war and destruction (including revolutionary, civil and 'ethnic cleansing'), rather than social change, have interrupted. In the case of India for instance (Modi 1998, see above), Sanskrit drama lost its central place after the tenth/eleventh centuries due to Greek and Mughal invaders who destroyed the centres of knowledge and culture and pushed theatre activity into the rural areas, creating a regionalisation of both place and language of performance. The Western concept of presentation was not introduced until nineteenth-century Calcutta under the Raj (now renamed by its Bengali equivalent, Kolkata), however empty halls in major cities gave way to Parsi theatre which soon lost its artistic genre and became commercialised. In the early twentieth century, Indian Classical theatre re-emerged and after independence, 'Indianness' was further established, today combining a mix of Western styles in the larger cities, indigenous production and modern adaptation of the *Natyashastra*, after centuries of rural retreat. As Massey *et al.* also note: 'the odd juxtapositions revealed by the built environment of a city also reveal its different histories. Buildings in particular have the ability to carry the traces of past interactions and how people with different cultures and memories have faced one another in the same city, if not across the same street' (1999: 75). The appropriation and reuse of buildings for worship – seen in the conversion of synagogues to Methodist halls, now to mosques in London's East End and other cosmopolitan neighbourhoods, and the conversion of music halls to cinemas then to bingo halls between eighteenth, nineteenth and twentieth centuries – are examples of this phenomenon; the occupation of redundant industrial buildings in cities of Europe and North America for use as artists' studios and galleries is another. From a more pessimistic perspective, the contested spaces and power struggles evident in Palestinian and South American cities (Selwyn 1995, Evans 1999c) over control and access to sacred sites presents a lack of cultural planning for diversity and local governance, whilst the decline in access and provision of local cultural amenities seen in the loss of music and dance halls, and then cinemas to fewer, larger (and mono-use), but more remote multiplexes, is another salutary tale. In the case of artists' 'colonies', the pressure of commercial and single-use development has not been matched by planning instruments and powers which have failed to protect cultural usage and activity in areas of gentrification and property development, as I discuss in Chapter 6.

Romanisation

City development under the Roman Empire initially followed the grid pattern of latter-day Greek cities, which later came to be broken and softened by the curved, wandering lanes of the medieval town such as in Rome itself. At the intersection of the city's two major roads – the *cardo* and *decumanus* – was the *forum* which combined the zones of both *agora* and *acropolis*, with not only temples of worship, administrative and storage buildings, but also libraries, schools and market-places, baths and theatre. The wide central avenues or *plateia* (Gr.) created covered colonnades and shops that extended to the walled fortification and terminated at the city gates. By the first century the grid was superseded by a more flexible urban 'plan' which was not planned wholly in advance of development but was shaped organically, and elaborated and extended over time. Public buildings – theatres, basilicas, amphitheatres, temples, libraries, concert halls and circuses – were 'sprinkled all across the urban fabric, so that no neighbourhood was without some public monument' (Kostof 1999: 214). As is still the practice today, theatre buildings were often built upon sites and facilities inherited from earlier times, including in larger cities the hippodrome, whilst the Greek gymnasia were largely abandoned and few amphitheatres survived. These public buildings were decorated and adorned with paintings, sculptures and fountains and were sources of civic pride, as later city fathers and burghers were to emulate seven centuries later. Theatre was still a popular activity, with programmes starting at noon and going on until evening. The notion of the *polis* and cultural integration of Hellenic Greece was not however evident in these larger, essentially functional colonial cities whose location was dictated by transportation (military, trade) and access. Whilst benefiting from the influence of Hellenistic town planning, as Hall notes: 'we know next to nothing about who produced these [city] plans, though we know there were professional *agrimensores* or land surveyors' (1998: 623).

Rome itself had avoided this grid-based city plan and had grown organically and chaotically. As London experienced 1,600 years later, a great fire in Rome whose damaging effects also resulted from its urban density, offered a rebuilding and planning opportunity and a major public works programme saw Rome emerge as a liveable city (for the elite) over the next 250 years. In AD 113 Rome had at last a coherent centre including the construction of a giant amphitheatre, the Colosseum (playing to crowds of 60,000), and by AD 356 it had twenty-eight libraries, eleven *fora*, two amphitheatres, three theatres and two circuses (Hall 1998). By this time its total population was declining from possibly over 1 million in AD 100, still by far the most populous city which even the New Rome of Constantinople could not match. By the fifth century, Constantinople's population reached over 300,000 and again a cosmopolitan existence was distinguished from the rural, which generated the demand and creation of cultural facilities, since 'the city and the city alone provided certain amenities that were considered an essential part of civilised life', and echoing Athens, 'city life was very public' (Mango 1998: 63). In the mid-fifth century, the city had been divided into fifteen regions ('districts') and contained nine princely palaces, eight public (and 153 private) baths, four *fora*, two theatres and a hippodrome, twenty public (and 120 private) bakeries and fourteen churches (ibid.: 77) – giving plenty of scope for 'bread and circuses'! The tension, moral and political, surfaced as it has done to a greater and lesser extent in societies ever since, fuelling state (Crown and Church) control and censorship of public performance and theatre in particular. In Byzantium, the power of popular entertainment, the theatre, wild beast fights[1] and the hippodrome 'were the main targets of ecclesiastical invective. . . . If only sighs our preacher, it were possible to abolish the theatre! . . . Manifestly it was the devil who had built theatres in cities' (ibid.: 63). More prosaically, the Church Fathers saw the theatre as a commercial competitor, poaching their clients and church collection funds and 'in spite of the fact that the *ecclesia Christi* drew its resources, its leaders and its rhetoric from the cities, its message was fundamentally anti-urban. It abhorred not only the theatres and the baths, the music and the dances . . . but [also] the very fact of people coming together in public' (ibid.: 229).

As urban life declined, and Rome's beauty was 'skin deep' masking extreme squalor and poverty in sharp contrast to the life of the wealthy (Hibbert 1985, Hall 1998), along with the Empire, several satellite Roman cities were redeveloped on these sites, which were later to emerge as the 'cultural capitals' of London, Vienna and Gallo-Roman Paris. Today however the Parisian fear of being 'Romanised' is associated with the threat from mass *cultural* tourism and the risk that their city is being left behind, as an 'Old World' capital which will soon exist as a magnificent relic infested with tourist buses and T-shirt shops (Connolly 1998: 49). The emerging Byzantine Empire with its Moorish influences survived this decline in the Southern Mediterranean. The first *cosmopolis* (a 'city-state comprising the world') which Alexander the Great had established and which his successors further expanded, was centred on Alexandria, a cosmopolitan emblem in the empire which the Ptolemies 'strove to make the cultural center of the Greek world' (Pomeroy *et al.* 1999: 457). This included the first 'museum', dedicated to the nine Muses, which housed a library of every Greek publication (70,000 papyrus rolls), and an effective university for scholars – all government-funded (to be re-created at the turn of the twentieth century by the Egyptian government in a new highly selective and under-stocked Islamic library). The cities of Islam – Baghdad and Cordoba – became power centres in their own right, whilst Northern cities contracted, however other civilisations such as Chinese and Mayan also maintained their great cities, perhaps the greatest being Teotihuacan in Mexico. This ancient city, home to Olmec and later Aztec civilisations, predates Athens and belies the easy claim that 'Athens was first' (Hall 1998: 24), to which earlier Egyptian and Chinese (e.g. Shang and Chow) dynasties that created walled and grid cities would also lay claim. The Olmecs that occupied much of Middle America from before 1200 BC 'invented ceremonial architecture, monumental sculpture and mural painting' (Davies 1982: 63) and in their classical phase designed ceremonial centres and buildings set in a planned complex around plazas which served religious worship, ritual ball games and a wide commercial network. The development of the seminal city, Teotihuacan, 'stood astride the Classic age of Middle America . . . and contemporary cultures will be treated in the context of their nexus with the great metropolis' (ibid.: 65). This city covered an area of 20 square kilometres, larger than Imperial Rome. Its central portion was made up of a series of rectangular plazas, separated by stairways. The southern zone contained the Great Compound, the chief market. At its zenith (between AD 350 and 650) the city had a population of 200,000, comparable with the Byzantine 'mega-cities' of Antioch and Constantinople. Murals adorned almost all wall surfaces and unlike the frescoes of Florence, mural paintings were found on the walls of rich and poor quarters and buildings alike, and in central and outer areas of the city. Whilst daily life for the populace was monotonous, mass festivals were frequently held, involving costume-wearing, dancing and singing, which celebrated crop and fertility cycles. Like Athens, Teotihuacan came to be both a cosmopolitan city and model for cities in Central America and the wider Mayan region. It was remarkable and in this sense unique in the scope of its crafts and cultural industries in which it is estimated that one-third of its residents were engaged (not including home-produced crafts such as textiles and ceramics). Its many festivals attracted thousands of visitors from other areas.

As Davies posits: 'a consensus prevails as to the concept of a home or metropolitan area, whose culture was so identical to that of the parent city' (1982: 91), and this can of course arise not only through militaristic or administrative control, but also as a result of an 'innovative' or 'creative milieu' (Hall 1998), trade (e.g. *Renaissance*; Jardine 1996), or even cultural hegemonies operating between cultural capital and provincial cities and regions (e.g. Imperial Rome, nineteenth/twentieth-century London, twentieth-century New York and Los Angeles). In reality, a *combination* of these processes maintains the dissemination and distribution of cultural practice and products to a wider region and today this of course extends globally. Cross-trading, migration and resultant cross-fertilisation is however a phenomenon long-established – for instance, the 'paradigmic shift' in philosophy and science associated with the emergence of European Baroque encouraged not only the growth of capital cities in fourteenth-century Italy and later elsewhere in the courts of Europe, such as in Madrid, Amsterdam, Paris, Copenhagen and Berlin, but also the forerunner of the Grand Tour: 'The dissemination of this cosmopolitan culture through artistic academies to which the rulers often belonged, created an international audience for the Baroque esthetic. Artists travelled among the capitals and books on architecture and cities were studied widely' (Kostof 1999: 215). As Hall asserts in the case of fourteenth-century Florence, 'the wealth-makers and the intellectual figures came from the same social groups and the same families [but the] aristocracy did not merely patronize art and learning: it was actively involved in it' (1998: 89).

Culture and commerce

The public planning of cultural facilities in the pre-industrial era as noted in Chapter 1 rested on the extent of the public realm and relative freedom of society, including of course the time and money to participate in cultural activities. As recent writers on various European 'urban renaissances' (Hall, Jardine, Borsay, Graham-Dixon, Johnson) and even the observers of antiquity and ancient cities maintain, the economic opportunities and confidence – not just surplus 'leisure' time and spending – but the expectation of economic stability and growth, were fundamental conditions under which cultural production and provision was to develop in cities. For instance '[Florentines] spent more money on luxury goods because they had more money, and because they were optimistic about their economic future' (Hall 1998: 96) and Hall also observes that the Renaissance 'marks the gulf between elite and masses. . . . Yet even men of more modest stature had a surplus to spend for "extras" including art . . . thus a market-place for decorative art came into existence' (ibid.: 96), a pattern to be replicated in the subsequent urban renaissance in France and England. As Pirenne therefore affirmed (1925), it was the economic function of the great trading towns that led to their growing power and political independence (from feudalism). Whilst the spoils of war financed the monumental outpourings of Greek and Roman empires, as colonial appropriation did for British, Iberian and other European empires, and governing tyrants established their position through the commissioning of public monuments and art works, it was a mercantile base on which cultural amenities and collective consumption were to flourish. Indeed this is Hall's main thesis in *Cities and Civilization* – that the heyday of cities of Florence, London, Vienna, Berlin and so on inextricably linked their economic ascendancy with innovation and creativity, and vice versa, a relationship echoed in Jardine's *Worldly Goods*. The capacity of a 'culture industry' – skills, innovation, scale economies, building/urban design, market demand – created both competitive and comparative advantages for these cities and nations, a cultural economy which in turn enhanced economic growth (domestic and then export) and an argument now used to promote the cultural industries in post-industrial economic development and cultural policy formulation at city/region and macro-economic levels the world over (see Chapter 6).

Hall's substantial critique and case-studies of cities as 'cultural crucibles' and the con-tribution of their *innovative milieu*, surprisingly (for a 'planner') does not seriously consider town planning and urban design or the spatial relationship and locational issues between cultural facilities and places of entertainment and consumption, and the city polity and development. Certainly there is no claim to cultural democratisation in the modern, social welfare/equity sense, or evidence of what we would identify as a cul-tural policy in these pre-/industrial cultural capitals. The elite of empires, the power of court and Church (monasteries, see below) effectively privatised professional perfor-mance, art and architecture and controlled popular 'entertainment' through licensing and designation of venues, their organisation and programming content. For instance, the rural 'fair' which was re-created in squares, bridges and *grounds* as the medieval city market-place for goods and produce, combined with religious and harvest festivals and associated entertainments, but it gradually declined, ironically as trade expanded, moving to 'stately halls for sectional or specialized trade, covered plazas and arcaded alleys' (Lopez 1971: 88), as business increasingly shifted to the workshop and private retail domain, and the power-base of the crafts guilds became established, to the ben-efit of their (pay)masters (Sennett 1996).

From late Byzantium in the fifteenth century, the public realm had also declined, to the satisfaction of the Christian Church: 'as the cities collapsed, the dream of the Church must have come true. If St. Basil had been able to come back to life and visit the *kastron* of Caesarea in the ninth/tenth century, he would have found no theatres, no mimes or buffoons' (Mango 1998: 229). For much of the medieval period follow-ing the fall of the Roman Empire, Europe was a so-called cultural backwater (LeGates and Stout 1996: 17) and urban public entertainment and civic culture had to 'wait' until the urban renaissance in fourteenth/fifteenth-century Italy and France, post-Moorish (and Jewish) Spain, spreading to Elizabethan London, and Berlin in the sixteenth and seventeenth centuries. The period in between medieval Christendom in Europe and the early modern, Renaissance epoch, has therefore been termed historically as the Middle or the Dark Ages – according to Giorgio Vasari (1550) the 'degenerate period' – and which for most people was also a period of limited leisure:

> what little leisure the common people could secure was crude like athletic contests, wrestling, ball games, cockfighting, and bull-baiting . . . the nobility themselves had ample resources and opportunity for rest, festivity, art, entertainment. They engaged in debate, oration, music, dancing, gambling. On occasion some of these leisure experiences were extended to common people but only through spectating at festivals, tournaments and events.
>
> (Searle and Brayley 1993: 12)

As Christianity spread throughout Europe, the antithetic dualism of Church and state 'hung over political, religious and cultural life as it evolved through the Middle Ages', and as Taylor and others have documented: 'the intellectual power houses of the new religion and of the new culture that accompanied it were the monasteries' (1998: 4). The contribution of the monasteries, their foundations and educational institutions undoubtedly touched literature (including as being the home of early library collections), architecture and material culture, but his was of course not accessible or open as in the pre-Christian world – much that was learnt was forgotten, preserved in unconsulted manuscripts in remote monastic libraries (Kelly 1998). However at the turn of the first millennium, medieval cities 'became true centres of commerce, culture and community' (ibid.) and the early burghers and bourgeoisie promoted and 'protected' crafts and building skills and their practitioners, as an effective buffer to the feudal power of both bishops and barons. In considering the early Renaissance city-state of Florence, Hall confirms that 'Renaissance culture was subsidized by the emerging urban bourgeoisie, who replaced the nobility and clergy as patrons' (1998: 88) and by the nineteenth century with five centuries of practice, the *creative genius* which Hobsbawm saw as a virtually bourgeois social invention, served this inexhaustible demand for material culture: 'Few have been prepared to spend money so freely on the arts and, in purely quantitative terms, no previous society bought anything like the amount of old and new books, pictures, sculptures, decorated structures of masonry and tickets to musical or theatrical performances' (1977: 327).

The relationship – constructive, censorious or benign – the first of which Hall sees as a clear 'cause and effect' between creativity: the *innovative milieu*, and the cultural economy of dominant cities, is neither linear nor one that is easy to prove, not least since there is no opportunity for a genuine 'control experiment'. Hall also notes Lopez's suggestion (1959, also 1971) that artistic development also occurs during times of economic recession, as art works are perceived to maintain their real terms value in comparison with and as a hedge against devaluing currency or other investments (borne out in the buoyant art and antique markets centred on London and New York today), to which one could add the cliché of the impoverished artist in their garret, and the quality of art produced in times of oppression, hardship (Lopez 1959) and social upheaval, compared with more settled and static periods and societies. The commodification of culture has always has its severe critics, the *cultural pessimists* (Cowen 1998) – William Blake felt that commerce in fact killed creation: 'where any view of money exists, art cannot be carried on' (quoted in Porter 1982: 260), but he was in the minority since in England 'most writers thought it natural that as in Augustan Rome or Renaissance Italy comfort should succour culture' (ibid.). As Porter goes on, citing Hogarth, Pope and Johnson: 'Lions of culture were unashamed about turning art to advantage'. There is of course a serious argument that creative talent, the emergence of artists, innovators, etc. follows no easily identifiable environmental, social or economic determinants, although the commodification and exploitation of cultural expression and its outputs (Evans 2000b) is of course so linked, i.e. it is demand led. This is also the case in terms of urban development and planning, as Wheatley observed: 'It is doubtful if a single, autonomous, causative factor will ever be identified in the nexus of social, economic and political transformations which resulted in the emergence of urban forms' (quoted in Kostof 1999: 32). As Taylor, writing on the cultural development of Berlin also maintains: 'The relationship between "culture" . . . and the historical, political, social circumstances from which it emerges is a complex one. Conditions may favour the cultivation of arts or they may hinder it. Artists may rise to the challenge of their age or ignore it. The arts have their own inner momentum and may respond to self-generated pressures that pay scant heed to the world outside' (1998: 119). This argument is often used in response to calls for greater state funding and intervention in the arts and culture, with contrast made between the artistic output and quality of centrally planned societies ('state-approved art'), and the *arms length* arts policies of liberal (i.e. capitalist) regimes. Pick therefore argues that 'The greater number of arts activities in Britain had never been dependent upon state aid, but had been sustained by a great variety of other economic means . . . at least 150 years of support from the private sector' (1991: 75), and private support – individual and collective – and patronage of the arts has of course been a feature of 'culture and commerce' with the performing and visual arts, architecture and crafts, being reliant upon individual commissions and corporate sponsorship, from the Medicis to Mobil Oil. Then, like today, the relationship between art and trade is not benign: 'the power of the purse shaped the content of art and letters. Many commissions were to glorify rank, wealth, and status' (Porter 1982: 264). In New York, the Metropolitan Opera, recipient of sponsorship from Mobil Oil amongst others, is accused of 'dumbing down' their repertoire, or 'playing safe' (namely West End 'musical theatre') to appeal to a low-brow audience, corporate hospitality guest and subscriber alike (Evans 1999e), although arguably all have played safe to please their patrons and subscribers long before the advent of corporate sponsorship. Social structures, as sociologists would maintain, are the determinant of the place and relationship of the artist with society and which generated the Romantic notion of art in the nineteenth century. Wolff notes two factors in this development: 'the rise of individualism concomitant with the development of industrial capitalism. The second the separation of the artist from any clear social group or class and from any secure patronage' (1981: 11). Therefore, culture from this well-argued position 'is an immanent construct whose form and substance are comprehensible only in terms of the wider systems of human relationships with which it is bound up' (Scott 2000: 31).

This of course does not assume that the arts are necessarily subservient to the mainstream social order. The reaction to and rejection of the traditional and the hegemony of both the market and forerunners of the 'ministries of culture' is also represented by what came to be termed the *avant-garde*, and oppositional and alternative artistic and social movements, from the Dadaists to Walter Benjamin, observing that culture was steeped in blood and war: 'there is no document of civilisation which is not at the same time a document of barbarism' (Glancey quoted in Jones 2000: 5). On the one hand, as with science which views 'progress' as logically an improvement on the past, contemporary art also looks to new forms of expression, and thereby rejects what Baudelaire coined as 'the essential character of being the present'. Whether avant-garde art is associated with revolutionary and political movements or not, as Hobsbawm maintained, their emergence 'marks the collapse of the attempt to produce an art intellectually consistent with (although often critical of) bourgeois society – an art embodying the physical realities of the capitalist world, progress and natural science conceived by positivism' (1977: 346). The notion of culture that represented an 'economic world reversed' also forms an important element of Bourdieu's concept of *cultural capital* in relation to high-, middle- and low-brow art consumed by Parisians a century later, as expounded in his *Distinction* (1984) and other works (1993:

306–9). Ironically, but not surprisingly, the bohème of nineteenth-century Paris formed clusters or specialised districts such as the Latin Quarter, Montparnasse and on the periphery of the city, Montmartre – centres which provided both the producers and consumers of what would, a century later, be called an underground, or 'counter-culture', and a source of the avant-garde arts which formed a key part of Bourdieu's stratification of culture, taste and the predeterminants of consumption in the 1960/1970s. As Hobsbawm noted, 'the growing desire of the bourgeoisie to clasp the arts to its bosom multiplied the candidates for its embrace – arts students, aspiring writers . . . in what was now the secular paradise of the western world and an art-centre with which Italy could no longer compete' (1977: 347). These alternative cultural quarters also created the foundation of artist colonies which the city of Paris has continued to protect and control through planning and zoning legislation. Hall refers to 6,000 artists in Paris *c*.1870 (1998: 232), one-quarter of these in Montparnasse, supplemented by art suppliers, dealers and academics; however, Hobsbawm (1977: 347) quotes between 10,000 and 20,000 people 'calling themselves artists' in Paris, and this is compared with another concentration of bohemian artists in Munich at this time – the 4,500 members of the *Munchner Kunstverein* (ibid.: 387). In the city fringe crafts quarter of Clerkenwell, London, the census of 1861 recorded 877 men who were clock and watchmakers, 725 goldsmiths, 720 printers, 314 bookbinders, 164 engravers, 97 musical instrument makers and twenty surgical instrument makers and 1,477 women were milliners/dressmakers, 267 bookbinders and thirty-three embroiderers (Olsen 1982). One hundred and thirty years later, this cultural industry quarter still maintained over 900 small arts and crafts-based firms, nearly 50 per cent in the print/design and jewellery/metal craft trades (Evans 1990). However, with successive property and change of use pressures since the 1970s to higher use-values, e.g. offices, private apartments – 'loft-living' – by 1993, the number of firms had reduced overall by 15 per cent with the highest reduction in these traditional crafts and studio workspaces, only partially offset by increases in media and design practice and the soft crafts of designer-making (e.g. textiles – milliners, weaving). As one of the pioneering managed workspace companies observed: 'One can chart the migration of artists across London's map over the last two hundred years, driven not by pleasant environments or fashion, but simply in pursuit of cheap space' (ACME 1990: 7), a phenomenon explored further in Chapter 6.

City formation and planning

In terms of urban planning and cultural development, therefore, and this is in contrast to Hall's thesis, Taylor's viewpoint is pertinent: 'any chosen periodisation of the historical or political continuum will not necessarily coincide with a reasonable periodisation in terms of the development of art' (1998: 120). Art history as a distinct discipline owes its practice to the ethical and philosophical concerns of the Renaissance and Enlightenment (Smith 2000), and the obsession (e.g. in museums and gallery curatorship) with a linear chronology and categorisation of styles-periods – what Edensor calls a 'classifying mentality' (1998: 184). Trevelyan however recognised the limits of conjecture, of cause and effect: 'The spirit bloweth where it listeth: the social historian cannot pretend to explain why art or literature flourished at a particular period or followed a particular course. But he can point out certain general conditions favourable to a high level of taste and production' (1967: 411). The term *medieval* itself is largely a Western, historicist construct, and the stylistic periods used by art historians such as 'Renaissance', 'Gothic', 'Romanesque', 'Baroque' do not adequately reflect urban, religious or social change movements (Graham Dixon 1999), or indeed city development and urban planning forms, although they may coincide with changes and milestones in technological (e.g. building), political and economic change. The Baroque city for instance is said to date from the Rome of Sixtus V and Carlo Fontana's commission in 1585 to design the street-plan which gave the city a grandeur and regularity it had never known before – Fontana's Rome served as a model for many neo-classical cities, not least Paris and the creation of the Royal axis from the Louvre to the Tuileries gardens (Cohen and Fortier 1988). Writers earlier in the twentieth century such as Weber and more recently Braudel (1981, 1985) have typified cities in a historicist manner, as 'open' (namely Greek and Roman); 'closed' (medieval) and towns subjugated by Crown or state, from the Renaissance era onward (Kostof 1999). Sjoeberg (1960) applied the *pre-industrial, industrial* and *socialist/centrist* city distinction (to which we must add the *post-industrial* and, arguably, Castell's *informational* city), relating these phases to population size, a binary class divide and, critically, land-*use* versus *exchange* value-systems. Whilst modern industrial cities are therefore traditionally associated with the industrial revolution, cities prefigured by capitalism and segregated land-use and the emergence of a rentier class, can be seen from the late fifteenth century onwards: 'Sjoeberg saw cities in pre-industrial Europe as the product of their societies whether they be the community of merchants at a market point, or an agricultural-based primary civilisation, or the quarters of a medieval city created by guilds and political rivalry' (Burtenshaw *et al.* 1991: 8–9). Urban planning traditions likewise conform to no chronological order or sociological parallel, although a lack of both real democratic consensus and a cultural planning approach are common to all. Five separate but increasingly conflated planning types have been identified and which can in part be matched to Sjoeberg's city formations: the *authoritarian, organic, romantic, technocratic utopian* and *utopian*, and associated *utilitarian* and *socialist* movements. Many of these planning typologies are personified (for authoritarian and utopian read 'visionary', see below), whether linked to majesties or masterplanners, plutocrats or philosophers (e.g. Plato, Thomas More), but as Burtenshaw *et al.* note, 'It is the underlying thesis of some commentators that many movements are the product of their time and that emphasising the individual who fashions a movement is placing the wrong emphasis on the individual' (1991: 13). A discussion of planning theory and urban design movements is beyond the scope here, but in addition to the seminal urban designers/planners and writers highlighted in this book, notably Kostof (1999), see Faludi (1973) and on the European City: Burtenshaw *et al.* (1991), Choay (1969) and in Britain: Bell and Bell (1972), Olsen (1982) and Cherry (1972).

The *socialist city* is of course identified with central planning regimes of the Eastern Bloc and other communist states where land-use reflected the state's priorities, rather than a hierarchy of use (location) and exchange values: 'It is the government which decides the size and look of the public spaces, the amount of housing, the size of living units, patterns of transportation and questions of zoning' (Kostof 1999: 27). Central planning applied literally in cities such as Sofia and East Berlin created a monumental administrative/governmental core with echoes, albeit atheistic, of Athens and Imperial Rome (including high levels of community participation): 'A vast public space of a ceremonial nature, in addition a park of Culture and Rest for the recreation of the working people, with promenades, tea rooms, picnic areas, and the obligatory socialist monuments. . . . The prominent presence is an architecture of public welfare goods and services' (ibid.: 29). Public provision and the time and space to participate in cultural and recreational amenities, including those provided at the work place and vacation areas (e.g. spas), are features of socialist amenity planning; the provision of a network of arts education institutions feeding state-supported performing arts venues is another. Versions of cultural planning for participation and arts education and training are however familiar in both liberal and socialist states – how far they form part of a cultural democratic and 'access' policy, or serve an effective elite (i.e. private and highly selective), distinguishes the level and location of such provision (i.e. a hierarchy of facility and opportunity) – from community arts centre to conservatoire, irrespective of the prevailing political philosophy.

The palaces of Church and monarch were for example re-christened the *Palace of Labour* in the Soviet Republic, designed for the new collective 'ruler' – the worker – these centres served as locations for large-scale congresses, rallies, meetings and theatrical productions, more seriously undertaken than their predecessors, the so-called *people's palaces* in nineteenth-century Britain. One of the first of these was subject of an architectural competition in Moscow in 1923, and the format was replicated in other cities and regions, coming to be known as 'clubs'. Existing civic buildings were converted and new facilities developed although the conglomeration of theatre, cinema, corridors and other unrelated uses created problems for the new order. These places of culture, recreation and rest came to be located at the centre of settlements, on two perpendicular axes or converging arterial roads leading to the public core, hosting sport and amusement facilities, and large and small theatre auditoriums in a radial and spiral system, with open and closed rooms and circulation areas for flexible public and small group usage (Lissitzky 1970). Whilst residential and work places formed one functional building and location type (i.e. separated), the commune served as a focus where the local community unites to perform all of its activities in one place, so that work, 'club' (cultural and community centre), restaurants, and dwellings are combined into a single complex – a model adapted in the Israeli kibbutz. The population and production (agriculture, manufacturing, public services) distribution in communist Russia rejected the concentric development of the Romanised, capitalistic city centred around the market-place with distinct land-use and socio-economic class separation, and in the words of German Planner Ernst May writing in 1931:

> locating people as close as possible to their respective place of work, the task consists in the equitable distribution of all communal functions, for everybody's equal enjoyment . . . nurseries, kindergartens, schools, stores, laundries, ambulances, hospitals, clubs, cinemas, and other facilities should be apportioned in a manner as to be within a comfortable and functionally optimum distance from the dwellings.
>
> (quoted in Lissitzky 1970: 198)

Monumentalism in public culture and urban design is also seen under totalitarian regimes and dictatorships (e.g. Saddam Hussain's 'Victory Arch' in Baghdad), often building upon former colonial times and occupation, and subsequently mediated and transformed by post-independence urbanisation and industrialisation. As Cowen maintains: 'Totalitarian regimes can teach us something about the liberating nature of capitalist art. Both Marxist and fascist governments have repeatedly placed a tight grip on cultural markets. Hitler and Goebbels devoted much of their time to planning the new artistic order of the Reich' (1999: 210). Nineteenth-century colonial occupiers also 'consciously manipulated the urban landscape to symbolize and reinforce their claims to legitimate rule' (Jordan-Bychkov and Domosh 1999: 412). Examples are evident in capital and regional cities of Latin America for example combining the characteristics of pre-industrial/pre-Colombian colonial cities, which have been adapted to modernisation such as in Bogota and Sao Paulo. Most Spanish cities in the 'New World' (*sic*) were laid out in terms of the Laws of the Indies (drafted in 1573), specifying the grid-iron street plan already implemented in cities such as Puebla in the 1530s, with central plaza and church/cathedral and walled building plots, replicated on a smaller scale throughout the city, facilitating religious worship and control. Griffin and Ford's model of such a city structure presents this hybrid in which 'traditional elements of Latin American culture have been merged with the modernisation processes altering them' (1980: 397–42). This model highlights the powerful commercial/spine sector that extends from the central business district (CBD), along which key economic, social and cultural amenities are located: 'Residential areas, facilities such as theatres, hotels, restaurants, prestigious offices, private hospitals, museums and leisure facilities are located on, or near to a "tree-lined boulevard"' (Potter and Lloyd-Evans 1998: 129). The combination and location of high-art venues, up-market residential, office and hotels in 'tree-lined boulevards' is a familiar cultural cluster in many cities, from Athens to Amsterdam. However, the relationship between 'residents' (*sic*) in cultural and historic areas is now complex, but an observation is that few (if any) residents inhabit the core historic centres, and certainly few of the workers who service the tourist and cultural facilities who are drawn typically from a peripheral and urban fringe zone (Evans 1998a: 13).

As residential suburbanisation gradually decentralises and creates the core/inner–outer urban and periphery divides and 'zones' in mature cities, levels of amenities increasingly reflect the social class and land-use-values with middle-income groups inhabiting areas of gradual improvement. These attract a certain level of leisure amenity and facilities, in contrast to areas of decline and disamenity, the most extreme seen in the shanty towns, *favelas* and squatter settlements all too familiar in developing country mega-cities. The centre/spine and concentric development of expanding (in population and geographic area) cities is no more evident than Mexico City. From a population of 345,000 in 1900, now standing at 19 million and forecast to reach 30 million by 2010, the traditional and symbolic core is centred on the Zocalo, which was built on the site of the Aztec's Tenochtitlan of the 1300s, based on a rectangular grid which survived Mexican independence in 1821. The Zocalo and plaza housed the National Palace and Cathedrals, with the Paseo De La Reforma leading to the Zona Rosa west of the Zocalo, forming the emerging CBD corridor in which the city's main commercial offices/headquarters, hotels, theatres, branded retail stores and cosmopolitan restaurants are located. This leaves the historic Zocalo as both a heritage and symbolic location for political demonstrations, fiestas, proclamations and tourists, and as a residential area of converted tenements for low-income groups. In the most classical Spanish colonial city of Puebla, the Zocalo, with surrounding arcades dating from the sixteenth century, served as the venue for market-place as well as for hangings, bullfights and theatre, but today performances are confined to street entertainers for visitors. Occupying the whole block on the central plaza, facing the cathedral-south, the former bishop's palace is shared between governmental and tourist offices and the Casa de la Cultura, and like the capital city, the up-market entertainment and nightlife area is west of the Zocalo, the *Zona Esmerelda*. This pattern is familiar in historic centres of architectural heritage, once functional obsolescence occurs, such as traditional hospital buildings, government offices, convents and churches, old libraries and even railway stations (e.g. the Musée D'Orsay, Paris; Cabana Cultural, Guadalajara). When upper-income householders departed the historic core, commercial and service activities exploiting large-scale floor space and cheap rents also moved west and to the second growth ring of the city, which gave way to the marginal usage of low-income residents and slum dwellings. Post-industrial gentrification now follows this cycle such as in Paris where the 1970 *Plan de Sauvegarde et de mise en valeur* codified the restoration techniques which were adopted in the conversion of the *hôtel particulier* to cultural and civic uses, from museums, galleries and archives, as well as prestigious office and bank premises, now the common reusage of so many historic city conversions (e.g. Venice).

   The conservation movement and heritage regeneration schemes are now gaining hold in developed and developing country historic centres alike. In Europe this 'conservation ethic' can be attributed to the growing demand – domestic and from overseas – for heritage tourism and 'a mix of psychological needs, social and intellectual fashions and consumer prosperity . . . a concern with the quality of urban life and a reappraisal of the value of the form of the city' (Burtenshaw *et al.* 1991: 145). This pressure to conserve within a universalist framework (patrimony, free trade, e.g. tourism), is exported from the European headquarters of international cultural and heritage organisations such as UNESCO and ICOMOS, to the many developing World Heritage Sites or aspirants to this designation, who together with intervention from development banks (World Bank *et al.*) and powerful private foundations, are effectively assisting, whether consciously so or not, in a repeat of the gentrification seen in the nineteenth century in the boulevards of Haussmann's Paris and in many city cultural quarter developments today. This is evident in the historic zone of Quito, Peru and Pelurinho, Bahia (Salvador) where tourism activities priced-out the residents and craftspeople that used to live in this historic centre (Rojas 1998: 7–8). Here the Cultural and Historic Heritage Institute of Bahia finances free performances by music groups and theatre companies to attract customers to the area, in the style of waterfront and festival market-places in post-industrial cities worldwide, in a self-conscious effort to recapture the *agora* and *fora* – however this phenomenon is more *real-estate* than *city-state*. This pattern is evident in the westward corridors of wealthy residential and associated uses which extend, in the case of Mexico City, to the equally symbolic Chapultepec Park district. This contains over eight museums including the National Museum of Anthropology and the famous collection of pre-Columbian art, the National History Museum, Museum of Modern Art, and the 'People's History Museum' – Museo del Carcol, in the familiar cluster or 'museum island' located in park settings (see Chapter 3). This east–west (or equivalent) divide has been an established feature in the development of European cities such as Paris and London, with poorer housing and industrial property and activity confined 'down wind' to the eastern side, and western expansion of middle-class residential and commercial property and 'public' (*sic*) cultural facilities, the *beaux quartiers*. This entrenched divide has driven, for example, the public sector-led regeneration programmes in East London's Docklands and major development projects in the Villette basin and at Bercy, north-east and south-east Paris respectively (see Chapter 8).

The public sphere

The incidence and location of public culture and the evidence of cultural planning in classical, pre-industrial and emerging industrial cities therefore offers some indication of the imperatives that drive the location and rationale for higher scale cultural facilities, whether new or in fact inherited from former urban societies. How far these are *public* facilities and amenities; the extent to which public cultural activity and consumption is communal or effectively private/ised, rests in large part on who or more importantly, how included the 'public' is in these historic examples. For example, the celebrated English theatre that was supported by the court, as it had done in Italy and France, also took place in private homes, notably during the interruption of public performance when the Puritan closure of theatres drove this activity even more behind closed doors, and which came to influence the nature of performance itself through the evolution of the 'interlude' – short, single-theme plays and comedies with a small cast. Even during the heyday of Elizabethan theatre and the emerging impresarios, actor-managers and proponents of the new 'playhouses', private performances in stately homes and aristocratic houses provided a loophole to the otherwise stringent control of public venues by the Lord Chamberlain. Moreover, whilst Athens created a holistic environment for citizenship and cultural exchange, as McGuigan states in comparison, this was 'without the ostensible universalism of the bourgeois public sphere…(which) assumed in theory, boundless equality' (1996: 23-24). Whilst in the eighteenth century of Dr Johnson's England, 'art was a part of ordinary life and trade' (Trevelyan 1962: 412), at the height of the bourgeoisie's power in the nineteenth century Hobsbawm maintains that 'The hegemony of the official culture, inevitably identified with the triumphant middle class, was asserted over the subaltern masses. In this period there was little to mitigate that subalternity' (1977: 353).

The patronage of public culture and art works and crafts, which created and maintained a pre-industrial cultural industry and a certain level of architectural building, therefore had little connection to forms of popular entertainments, pleasures and pastimes which the 'lower orders' enjoyed or in which they participated. Western history (as opposed to traditional oral history) gives the informal and non-powerful scant coverage and archaeology finds little 'hard' evidence of human cultural agency. We therefore simply know less about this aspect of urban life, with the particular exception of the genre of social history which emerged in Britain from the nineteenth century (Evans 1997), notably with Macaulay's *History of England*, and the work of Plumb and later Trevelyan (Macaulay's great-nephew), as well as Porter and Briggs. From the 1970s the *history of society* as coined by Hobsbawm (1971) was established amongst many countries, including in Europe where the genre was more closely associated with social and revolutionary movements. As Evans observes: 'Not just the proletariat, but other social classes, from the landowning aristocracy and the propertied and professional bourgeoisie to the mass of the peasantry and the criminal underclass, came into its purview. Social institutions such as the family, clubs, societies, leisure organizations and the like, entered the picture' (1997: 168). With the advent of broadcast media: radio, film/television and most recently the Internet/web, visual and oral history has extended the scope and dissemination of social history, with major events (e.g. Millennium celebrations – Evans 1998f, 1999d) warranting their own extended broadcast coverage and documentation, and no end to the historic make-over (e.g. *Renaissance* by Graham-Dixon 1999).

The extent to which the dichotomy of high-art and popular culture came together in mass cultural spectacles – versions of the *agora* and *odeon* – can be taken as one measure of cultural planning, as in the *polis* and *bread and circuses* of classical, Roman and Byzantine city-states. The growth of a decorative and material cultural economy which encompassed a wider market than the Church, aristocracy, guilds and merchant princes, may be another test, particularly as the role and status of artist and craft person became established (including home-produced crafts and entertainment) and the possession of disposable income and the propensity of cultural spending widened. In the emerging municipal town halls, however, as Hobsbawm claims, 'Secular public authorities were almost the only customers for those gigantic and monumental buildings whose purpose was to testify to the wealth and splendour of the age in general and the city in particular' (1977: 329). This is a perspective and sentiment which the late twentieth-century *Grands Projets* perpetuate (Home 1986: 184), and a phenomenon and paradox that I will also return to later in Chapter 8. In Greed's words: 'The aim was to create fine buildings and squares for the upper classes, as a stage set to urban life' (1994: 85). Seldom utilitarian in purpose, this largely civic movement in the nineteenth century (even when restoring church and cathedral monuments), was writ large in cities such as Haussmann's Paris whose tree-lined boulevards, public buildings and monuments displaced thousands of working-class residences and potential beds of insurrection (Haussmann has been described by Walter Benjamin as an 'artist of demolition' in Buck-Morss 1995: 89). Previously, under Louis XIV (1638–1715), 'His patronage of art, architecture, music, dance, literature and *town planning* was intended to create an illusion of civilised culture' (Greed 1994: 85), but in retrospect this represented the *defeat of town planning* (Sutcliffe 1970). Haussmann produced an urban landscape in which streets were defined by blocks disposed as sculptural elements and focused on landmarks and distant views. The lesson was noted enthusiastically as far away as Chicago and Buenos Aires, and 'it provided the model for half-baked megalomaniacs like Nicola Ceausescu' (Sudjic 1993: 16). The building of Rome's processionary way to St Peter's Square similarly displaced residents, and earlier, in the mid-1800s, Vienna had razed its fortifications and subsequently replaced the space in the *Grand Manner* with a massive circular boulevard and public buildings – a stock exchange, a cathedral (*Votivkirche*), three university colleges, a town hall, justice and parliament building, and *eight theatres*, *museums* and *academies*. The rationale for the nineteenth-century public cultural monuments, as Miles says, 'within a programme of education and betterment', is a 'process of persuasion in which the dominant class seems to naturally inherit history' (i.e. what Gramsci termed *hegemony*) (1997: 66).

Where the opportunity to plan completely new cities arose, the grand or monumental axis was used both as a symbolic and functional divide between activities, as in the new federal capital Brasilia. Inaugurated in 1960 and the antithesis to the favelas and chaos of the old colonial capital, Rio, Brasilia followed the radial city ideology and symmetry of earlier cities, with, on one side of the esplanade the important functions of embassies, commerce, hospitals, public services, banking and hotels, and, on the other, 'leisure'. In practice, however, as Rykwert maintains: 'All this inevitably means that most pedestrians in any zone have to walk, sometimes quite a long way, to enter a different milieu' (2000: 178).

What notion of *planning* in the cultural *amenity* sense may have been considered in the redesign of major European and colonial cities is not clear, as Kostof maintains: 'city form is neutral until it is impressed with specific cultural intent' (1999: 11). For instance, whilst Imperial Rome is fêted for the scope of its public culture and places of exchange, at its population peak of around 1 million, 'there were so many large public basilicas, temples, circuses, baths and theatres, so many acres of imperial gardens, so much land that could not be inhabited for fear of offending the gods that most people were compelled to live in tall apartment blocks, *insulae*, which towered as many as six storeys high' (Hibbert 1985: 53). In the case of Paris – one of the more controlled (if not 'planned') cities – development projects and plans seldom originated in collective decisions or those of elected assemblies (Cohen and Fortier 1988), but under the influence of 'visionary men' with both the political will and financial backing, from Philippe Augustus, Louis' XIV to XVI, Napoleon and Baron Haussmann, to Georges Pompidou and François Mitterand ('Mitteramses I'), to whom the term *megalomaniac* was applied: 'In every city, I feel like an emperor or an architect; I decide, I arbitrate' (Scalbert 1994: 20). However, who these civic museums and theatres served – and public subsidy was a long-standing factor, whether from government, patron or public subscription – was not limited to the great and the good, as with crafts and other luxury goods since: 'A handful of competing merchant-princes is enough to make the fortunes of a handful of painters and art-dealers, but even a numerically modest public is enough to maintain a substantial artistic output' (Hobsbawm 1997: 330). The success of the early Great Exhibitions in London, Glasgow, Liverpool and Paris between 1851 and 1901 and turn-of-the-century fairs and expositions in the USA was also due in part to their popular appeal and repeat visits across the socio-economic classes (Greenhalgh 1988, Rydell 1993). Public art and entertainment which grew in early theatrical performance and live venues, notably in Elizabethan London, and the urban adaptation of rural gatherings such as the produce fairs and pageants, and later the early music and dance halls and inns of London, were later to be celebrated in festival grounds, people's palaces and pleasure gardens. These provide both a counterbalance to the monumental and succeeding civic emblems of cultural celebration, and a profile of popular pastimes and places of entertainment in the city.

Public culture also infers both physical as well as economic access, and as already noted, politically the notion of 'rights'. Few of these pre-industrial cities encompassed systems of democratic planning and resource allocation, or of amenity provision outside of basic infrastructure imperatives of population growth, which generated certain zoning and early examples of structure planning. As Habermas maintained, the (net-works of) the public sphere should 'make it possible for a public of art-enjoying private persons to participate in the reproduction of culture, and for a public of citizens of the state to participate in the social integration mediated by public opinion' (1987: 319, also McGuigan 1996: 176). Public culture is therefore associated with notions of civil society and the welfare economics concept of 'public goods', and a cultural *planning* approach would seek to apply resource, facility and land-use allocation and distribu-tion – including what cultural geographer Crang identifies as 'ideas of spaces to which everyone has access in which people can meet as formal equals' (1998: 164), harking back to the Roman market-place. Other significant and symbolic meeting places such as coffee-houses have arguably fulfilled such a role – over 2,000 having been set up in London by 1700: 'In Britain, the London coffee-houses were the equivalent of the Parisian *salons* and became the gentlemanly settings for "rational critical debate"' (McGuigan 1996: 25), and as Porter notes: 'If the Puritan chapel had been the citadel of seventeenth-century freedoms, by the eighteenth the coffee-house was the seat of English liberty. . . . Unlike churches they were open to all denominations' (1982: 244). But again this café society was highly exclusive and elitist in their membership (few admitted women). The new city *cyber-cafés* also claim to be engendering the role of the coffee-houses through 'virtual debates', however given the technological divide between those with access to the Internet and those marginalised or 'excluded', this could be again viewed as an elite place and practice. As Crang maintains, true equality of 'access' is undermined by educational and other constraints and predeterminants (*cultural capital*), as well as the economic and environmental barriers to participation to which I shall return later in this book. Moreover, 80 per cent of Internet communication (Information Communications Technology – ICT) is in English, although spoken by only 10 per cent of the world's population. In the USA, for instance, 62 per cent of urban households earning over \$75,000 were found to have access to the Internet, compared with only 2.9 per cent of poorer/rural households. Downey sees this in terms of a distinct spatial divide, echoing Castells' core–periphery polarisation (1996):

> While it is likely that the greater use of ICT will have significant benefits in terms of productivity, GDP [Gross Domestic Product] growth and employment, it is also probable that these benefits will not be equally distributed. Inequalities between core and peripheral regions will grow as core regions increase their grip on the global economy; and inequalities within cities will widen.
>
> (Downey 1999: 137)

The spatial and physical inheritance of these cities in the 'pre-planning' era, seen in the nature and location of popular and public performance and exhibition, is evident in the classical and Byzantine city-states, to the medieval cities and their incremental rebuilding and life-cycle of growth, decline and regeneration. The centre-core, CBD/corridors and concentric layout and growth of these early industrial cities is mir-rored in new towns and settlements, whether driven by monumental and ceremonial objectives, or the effective separation of workplace, residential and recreational activity, and of increasing significance, public transport systems. This evolution of the singular monumental city and its public realm, through urbanisation, segmentation – class, employment, land-use – and cosmopolitanism, is usefully reflected in the modernist ('classical-humanist') architect-designer from Finland, Alvar Aalto. His thoughtful pronouncements during his own evolution of style and architectural purpose – 'form and function' – echoes this shift in emphasis, writing first in 1921 (Museum of Finnish Architecture: 1978: 107–8):

> In the old days a nation needed huge and above all beautiful buildings to meet their longing for beauty and to symbolize their spiritual aspirations. Temples, cathedrals, forums, theatres and palaces recounted history more clearly and more sensitively than old rolls of parchment. There was only one art in the world – architecture – and painting and sculpture in all their many forms were harmonious parts of it. Even music was like part of the construction of a vault in a Gothic cathedral.
>
> (*Painters and Masons*, Jousimies)

In 1930, writing at the micro-level:

> A minimum dwelling is made possible by having some of the activities of those living in it shifted outside – to public areas such as schools, sports fields, libraries, cinemas, concert and lecture halls…. The old imperialist demands for representation from civic buildings give way before another kind of function. Before, Abbot Coignard sat in solitary splendour on his heavily ornamented chair in the library of the Bishop of Sez: now it is the public library shared by all those who have no library, or even space for one, at home.
>
> (*The Minimum Dwelling*)

By the 1950s, he returns to the higher civic order, echoing Mumford (1945) and fifty years on, Hobsbawm (1995):

> However, the standing of civic buildings in society must be as important as the main organ in the human body if we do not want our society to foul up in its own traffic, and become psychologically unpleasant and physically exhausting. Society must get back to a proper sense of order (a better term would be 're-create' the order which is so vital for a socially organized community). The society which is currently taking shape in the name of shall we say, classless society is even more sensitive than the bourgeois society set up by the French revolution, for it is made up of human masses whose physical wellbeing, civic education and increasing cultural strength are closely dependent on having properly ordered institutions and areas serving the general public.
>
> (*The Decadence of Civic Buildings*, Arkkitehti 1953).

Aalto's own influence was instrumental in the design and layout of city plans and cultural facilities – both buildings and interiors/furniture – in Helsinki, Jyväskylä and other towns and cities (including in Austria), although like Le Corbusier, e.g. Chandigarh, India, and more recently Richard Rogers, e.g. South Bank/Thames, few of his complete *masterplans* and larger site plans have been fully realised, such as the Finlandia arts complex in Helsinki (Plate 2.1) (see Chapter 7).

Plate 2.1 Finlandia Hall, Helsinki, by Alvar Aalto (1999)

Population size and urban density

Population size and its concentration is also a factor that cannot be ignored in amenity provision and cultural planning, and modern town planning as an extension of human geography has been predicated on both demographic and population distribution, in assessing social need and applying this in a spatial dimension. Notwithstanding the early mega-cities of Rome, Constantinople and Mesoamerica, the public participation in a 'common culture' achieved in Athens rested on a very small (male-only) citizenry, and the classical city-states were based on what today are very small resident populations of under 10,000. In the Attica of Pericles there were an estimated 120,000 people, of whom 55,000 were slaves (20,000 of these miners) and 24,000 'outlanders', so little over 40,000 were actually 'citizens' (Zimmern 1961). Sjoeberg's *pre-industrial city* very rarely contained over 100,000 people; the average Byzantine provincial city contained between 5,000 and 20,000 and, as Hall notes, Florence, albeit smaller than other Italian cities at the time, peaked at 95,000: 'the population of a smallish English town, or a Californian rural town such as Bakersfield' (1998: 69), whilst 'many cities had genuine municipal independence with as few as somewhere between 5,000 and 10,000' (ibid.: 78). As already mentioned, less than 10 per cent of this populace was 'citizens' for whom cultural and other civic amenities were created and maintained – Plato's ideal *polis* of 5,000: 'the basis of the folksy, neighbourhood community concept applied in twentieth-century new towns' (Greed 1994: 79), extrapolates to a gross population, given its social and spatial structure (public/private and gendered space), of ten times this amount. Lenin's Five Year Plan foresaw cities as no larger than 150,000 to 200,000 (the population of an average London borough) seeking population and economic diffusion and a balance between the isolation of villages and the over-concentration of major cities. Ernst May proposed population 'quarters' of between 8,000 and 10,000 people (quoted in Lissitzky 1970).

Urban density is another factor that together with the notion and practice of an elite suggests the secret of the success of the small city-state. Here the close proximity and interaction enables a degree of cultural exchange and 'ownership' (or 'belonging'), which larger cities can never achieve, despite attempts through both socialist central planning and amenity planning standards (e.g. scale hierarchy of facility provision), as well as the recent rediscovery of the 'urban village' (Aldous 1992). The psychological and planning imperatives dictated by the 'walled city' also literally fall down as this control factor is broken by expansion into satellite, suburban and rural areas and transport mobility reaches beyond the old city itself, and beyond hegemonic and fiscal control (guilds, taxes, development). As Sacco claimed (1976), the relationship between socio-cultures and the urban landscape is symbiotic: 'the morphology of the city is not only a product of the civilisation that it houses but also a factor in the creation of that civilisation' (Burtenshaw *et al.* 1991: 8). Applying a common culture and homogeneous citizenship model to the *cosmopolitan* state which has been the norm for far longer than most writers recognise (Said 1978, 1994, Hall 1990, Bubha 1994), also leads us to conclude that the classical cultural plan is of historic (even sentimental) rather than contemporary relevance. Sennett quotes historian Maurice Lombard who describes the *bourgeois* (and German *burgher*) of the medieval town as *cosmopolitan*, thanks to commerce and trade in the city: 'a man at a crossroads, the crossroads at which different urban centers overlap, he is a man open to the outside, receptive to influences which end in his city and which come from other cities' (1996: 186).

In masterplanning his *Contemporary City* (1929), Le Corbusier took as his model a city of 3 million inhabitants but varying densities were distributed within skyscrapers housing between 10,000 and 50,000 employees. These habitations supported parks and gardens, restaurants, cafés and shops housed in buildings in which theatres and public halls were also located. His *Radial City* consisted of small plots and sections of forty acres with a population of 50,000 down to 6,000 depending on their primary business or residential nature: 'Thus the elite were to reside near the city centre in high-rise apartments surrounding the administrative, cultural and entertainment centre, while the rest of the population lived in satellite towns on the outskirts. The grid-iron city, which is characterised by green space with 85 per cent of its surface area devoted to parks, was to be a city of leisure as well as production' (Burtenshaw *et al.* 1991: 31). Such was the seductive quality of Le Corbusier's plans that the Danish writer Rasmussen in updating his third edition of *London: The Unique City*, wrote: 'Le Corbusier is a modernist in his artistic form but a conservative in his planning of a city. When he plans to rebuild Paris . . . he is merely keeping up the old tradition of the Bourbons and the Bonapartes' (1948, quoted in Sherlock 1991: 14). Finally, Ebenezer Howard's model of 1898 for the garden city of 30,000 people – developed by him shortly afterwards at Letchworth – was divided into neighbourhoods of 5,000 (*polis* above) that were to be designed around local amenities such as schools, community centre and surrounded by green belt, providing a low density and high environmental quality living space. The issue and practice of local governance and identity inevitably raises its head once natural population 'groups' and densities are established, for instance an early twentieth-century consociational democratic model was adopted in Estonia in 1925 by the 'Law of Cultural Self-Government' for national minorities. This gave any minority larger than 3,000 the right to claim cultural autonomy and to set up elected councils, which had the right to legislate in the educational and cultural fields including schools, libraries, arts and heritage, and to raise taxes (Lipjhart 1977).

Conclusion

This brief overview of how the arts and culture, and the planning of past and present cities, have interacted and been manifested through places of public performance, institutional and civic centres for cultural exchange, has highlighted a number of common features and factors. These will be further explored in the following chapters in the urban industrial period and in the immediate pre- and post-town planning eras between the eighteenth/nineteenth and early twentieth centuries, when public culture and amenity became more formalised and responsive to population growth and social change. Common and inherited features include the core/centre and periphery divide in terms of a hierarchy of cultural activity and provision; the notion of the public realm and 'goods' (and by association, local governance); 'town and country' and tensions between rural community culture carried out in the city (urban and rural, in Greek *asteios* and *agroikos* – 'witty' and 'boorish'); the significance of the cultural (industry) economy, and finally spatial effects such as the close proximity of cultural facilities and places of consumption and production. The clustering of similar or complementary production and consumption activity is now a strategy being adopted and fostered by post-industrial cities in promoting cultural and cultural industry quarters, in the first case around locations of visitor-based activity (retail/markets, arts venues, restaurants) and, secondly, through managed workspaces which house a number of small cultural producers from pre-industrial crafts and designer-making, to media and cultural industries activity as discussed in Chapter 6. The tendency for the formation of Marshallian districts (1925) where a concentration of specialised industries cluster in a particular place, is therefore long-established and now evident in services and post-industrial forms of production and cultural consumption.

   All of these features, to a greater or lesser extent, and changing in their impact over time, have resonance with twentieth-century urban culture and society, and the post-industrial city. Their inheritance is clear on the one hand in the heritage buildings and sites such as the museum and gallery quarters of London and New York – South Kensington and Central Park ('Museum Mile'; Rosenzweig and Blackmar 1992); to Madrid's Prado Museum and Mexico City's Museum of Anthropology – all of these extensions of their respective grand city parks, which still serve as places of display and exhibition (and a continuing rediscovery and re-creation) – and on the other, the critical mass of 'houses of culture' such as theatre-lands of London's West End and New York's Broadway; Rio's cinema-land; and artist districts in old and new cities from Paris and Berlin, to Helsinki and Toronto. The significance of venues such as theatres, museums and galleries warrants particular attention, since they have both symbolic and economic importance in the development of the metropolis and a long history in the form and function of city culture.

The planning consideration, and the intervention by state institutions in their provision and programming, provides a particular example of how cultural planning has bridged the political, economic and social spheres, and the pursuit of popular entertainment and rational recreation. Chapter 3 therefore considers this aspect of urban cultural development in more depth and will explore the early development of planning for the arts from this inheritance. This will include approaches to amenity planning in the formative industrial and welfare planning eras; the place and space for popular arts and entertainment on the one hand and 'educative' cultural provision on the other, including the islands of culture and pleasure gardens and palaces which respectively reflect the urban(e) and rural pursuits of an emerging high-art and popular culture dialectic.

Notes

1 Before the Elizabethan development of public theatres, London had retained two circular venues, the Bull-ring and Bear-garden on the South Bank of the Thames (both place names still exist in this expanding cultural quarter of Bankside, which includes the re-created Globe Theatre and Tate Modern gallery). Animal-baiting was to be expunged from the City of London, as was to be the temporary fate of theatre at the hands of the Puritans (Hall 1998: 129).

3 Urban culture and the early industrial city

Introduction

Before modern town and city planning, as distinct from *building* planning and regulation, became formalised in the period before the post-War reconstruction era, the nature and location of arts facilities had derived from early city design and prevailing social and economic systems. The degree to which, on the one hand, a natural diversity is evident in cultural activity and city development, or whether according to Hall (1977) and Cheshire and Hay (1989) 'convergent development theory' places urban societies at differing points in an inevitable, linear path towards industrial urbanisation; suburbanisation, post-industrialisation and finally, *globalised* states, as Burtenshaw *et al.* observe: 'this distinctiveness is a result of the variety of historical experiences which have contributed to the physical fabric in which citizens live, work and play' (1991: 1). As power over development, land-use and public culture shifts amongst Crown, Church, merchant/patron and putative 'state', dispersing horizontally (e.g. amongst the same class/groups – aristocracy,[1] guilds), vertically (e.g. devolution, subsidiarity), and finally democratically, in an increasingly urbanised and secular society the notion of *public* and *private* also gains importance in terms of both the cultural economy, 'market' and cultural democracy (i.e. 'identity', 'rights'). As Wall, writing on Restoration London for instance, observes: 'The emphasis on the urban in eighteenth-century literature charts the intersecting boundaries of public and private interests, commercial and recreational space, domestic trade and domestic life' (1998: 150). As such power is redistributed not only as a result of periodic ruptures such as political, revolutionary and nation-state movements, but also particularly in proportion to population growth and city expansion, cultural equity and planning imperatives intensify. The separation and tensions between the *polity* and the political *economy* are therefore a growing feature of cultural policy and practice, and therefore of the planning for culture in emerging industrial cities.

As Chapter 2 highlighted, commerce and civic culture have also coexisted and together played a major hand in drawing the map of cultural activity and facilities which towns and cities have inherited today. Even where opportunities for new town and urban settlements have presented themselves, they have tended to emulate rather than radically depart from the scale hierarchy and central administrative and cultural districts epitomised in classical city formation discussed in Chapter 2. The approach to and treatment of arts and cultural amenity has however differed between countries, societies and cities and it would therefore be an over-generalisation to place this aspect of urban development as universal (Burtenshaw *et al.* 1991). Notions of citizenship and constitutional rights that in turn inform land-use and planning controls are one key aspect, which distinguishes say Anglo-Saxon from Latin and Napoleonic ('code') society – from plan-led to more liberal, 'light touch' planning regimes. The celebration or castigation of urban living is another, stereotyped in the Continental versus Anglo-American response to urbanisation and city-life (Jacobs 1961, Bianchini and Parkinson 1993). The relative popularity and success of certain art forms and practices also depends on the extent of their support and provision, whilst the comparative advantage in certain cultural production and performance already highlighted rests on a critical mass of cultural activity, often historically based, e.g. art and fashion in Paris, opera in Italy and Germany, drama and literature in Britain, film and pop music in the USA, ballet in Russia, and in more 'indigenously complete' societies, integrated performance (e.g. Gamelan – Bali, African musics) and pre-industrial crafts production. In Molotoch's view therefore: 'The positive connection of product image to place yields a kind of monopoly rent that adheres to places, their insignia, and the brand names that may attach to them' (1996: 229). In terms of maintaining this advantage economically, the strength of internal markets and *cultural milieux* are also essential ingredients, although as Hall documents (1998) this 'agglomeration' does not guarantee perpetual success – witness the heydays of cultural cities that have come and gone. As Hobsbawm points out, cultural hegemony also has its limits: 'Think of Italy's dominance in music in the seventeenth and eighteenth centuries. It had no political, military or economic support, yet it was total. In the end, however, it disappeared' (2000: 47). The associated forces of late twentieth-century globalisation and post-Fordism have however impacted on cultural activity, albeit unevenly, as it has in other production spheres, although the lyric arts have been less subject to this process than has been experienced by the mass produced arts and media (Lacroix and Tremblay 1997). The 'creative crucible' (Hall 1998) has however survived to date in the case of Hollywood, Los Angeles; West End and Soho, London, and Manhattan, New York – despite tough competition – where the high priests, the cultural intermediaries still operate. However the hardware that carries their creative products has long 'gone east', e.g. Sony (du Gay 1997). The American cultural hegemony in popular culture is of course reinforced by the increasing role of the 'English' language (e.g. in standardised computer technology), which may extend its dominance beyond what might have been its cultural shelf-life. This does not mean however that the creative origins will necessarily lie in the USA (or cultural cities, e.g. New York – see Chapter 6), even if the universal expression is anglophone, as in the case of Britain's nineteenth-century hegemony in sport and male fashion: 'Today, people still play football everywhere in the world and men dress in the English manner, yet Great Britain is no longer the leader in either football or fashion' (Hobsbawm 2000: 48).

City drama

'More than any other form of art, drama depends on the city' argues Renaissance critic Anne Barton, for in the city 'drama can afford to build a house of its own' (1978). As Wall maintains, 'The city supplies the audience that fills and therefore finances the theatre houses' (1998: 150). In planning-terms, Kostof observes that 'The city as theater is not the exclusive preserve of the Grand Manner. In every age urban spaces – streets and squares – have served to stage spectacles in which the citizenry participated as players and audience. Urban life is nothing if not theatrical' (1999: 222). The growth of theatre-building was also apparent in Berlin at this time, but 'less to accommodate vibrant new plays than to satisfy a rising demand for places of entertainment' (Taylor 1998: 196), a demand-led pattern which was to be repeated in the music and dance hall, and then picture-palace building booms in successive centuries. Not only an expanding population, but also the growth of a privileged class fuelled the demand for theatres and live entertainment in Elizabethan London: 'Crowding into London, they had both time and money to patronize the theatre' (Cook 1981). As Hall points out, Keynes also observed this economic influence on cultural consumption in *A Treatise on Money*:

> We were just in a position to afford Shakespeare at the moment when he presented himself . . . by far the larger proportion of the world's great writers and artists have flourished in the atmosphere of buoyancy, exhilaration and freedom from economic cares felt by the governing classes
>
> (1930: 4, quoted in Knight 1937)

and in the classic example, 'Florence became Florence because its artists found within the city walls all the patrons and audience they needed' (Laperièrre and Latouche 1996: 1).

The history and evolution of the theatre in the pre-eminent *World City* London (Fox 1992) is helpfully well-explored in general and social history, in writings on the city and urban development (e.g. Rasmussen, Mumford, Hall) and specialised texts on theatre (Southern 1962, Pick 1988, Mulryne and Shewring 1995). Their planning, in terms of the expansion from court, private house to public venue, follows a pattern from Renaissance and Restoration periods to the growing urbanisation of the industrial revolution, an evolution closely linked to the social and economic changes already noted. Their location and rationale for development rested in part on existing sites and places of public gathering, already established from earlier times (see Chapter 2), with the medieval market-place and guild still betraying the origins of theatres even today – the Haymarket, Guildhall, Cornmarket, Exchange, and those with grander, classical aspirations – the Apollo, Palladium, Coliseum, Hippodrome and so on. For instance the Hope Theatre built in the half decade of new public theatre ventures in London between 1576 and 1623 was a 'dual-purpose' playhouse and bear-baiting arena – quoting from the contract in 1613 this required the builder 'at own cost, pull down the Bull- and Bear-baiting house and stable on Bankside, and build a Playhouse fit for players and for baiting, and also fit a Tiring House, and stage to be removable and to stand on trestles . . . he will build this playhouse or gamehouse near or upon the place where it stood before' (Southern 1962: 176). Post-Restoration theatre development looked to the reuse of existing recreational facilities, such as tennis courts, riding schools and symbolic public–private areas, for example Lincoln's Inn Fields (previously known as St Giles), although Crown property the Fields were 'on the edge of legal London', and surrounded as they are today by lawyers chambers (Thorold 1999).

When James Burbage dismantled his Theatre in Shoreditch in 1598 after a dispute over the lease, his new theatre (reusing the old timbers) was also situated at Bankside, outside of the city walls. His Globe Theatre seated up to 3,000 in a cluster of venues: 'to play before the motley and critical audience of the capital; while citizens with their wives and apprentices with their sweethearts walked over London Bridge to see the play, men of rank and fashion came over by boat' (Trevelyan 1962: 217–18). As estate agents say, location is all: 'These Bankside theatres enjoyed an even better location for the privileged audience via the river. . . . By 1614 it was calculated that there were 40,000 watermen, most catering for the theatres . . . the watermen claimed they transported some three or four thousand people to the playhouses every afternoon' (Hall 1998: 135). (By 1867 there were under 30,000 watermen but twice that number working on the railways; Best 1979.) Public venues were also determined by a form of planning control – namely licensing and permissions held by Crown and later state and city/local authority bureaucrats, again indicated by the Royal assignation – from Royal Opera House, 'Theatre Royals', King's, Queen's and Duke's theatres and playhouses. A prefix that is still sought by national companies such as the Royal Shakespeare, Ballet and National Theatre companies in Britain, and 'Royal' companies in monarchies the world over – from Cambodia and Thailand to Sweden and The Netherlands – even when royal patronage has long passed to the state. As an entrepreneurial activity, planning for theatres and places of entertainment has therefore been influenced by the responses to these licence and other controls and the opportunism of the actor–manager, player-companies and impresarios themselves. The new sites and effective public realms opened up by these theatre ventures also had a physical dimension in the city, which as Manley perceived had a 'spatial orientation because the mobilities it creates open for some the possibility of choice at any critical moment in time' (1995: 394).

The informal and sacred sites in which performance was carried out had by the fifteenth century moved from the medieval rounds and *place*, and in larger towns to religious festival sites made up of tented structures or within town squares such the Mystery and Easter plays enacted in Lucerne, Switzerland. Mobile productions also met the demand for plays outside of major towns and cities with 'Pageant Waggons' used by, for example, the English Mystery cycles which were primarily guild performances, i.e. amateur productions. The emergence of a professional troupe or players probably used this stage-on-wheels as well as the forerunner to the proscenium theatre, the 'booth stage' (a temporary structure): 'a solution devised purely by the theatre people themselves, almost as a solution is arrived at in folk art, and without any outside specialists – whether architects or painters or engineers' (Southern 1962: 160). In England, one of the first professional companies, the Earl of Leicester's men, built the first permanent theatre structure, under their leader James Burbage, who: 'beside being a player, was a business man and a joiner' and whose decision to build a specific wooden playhouse was 'to accommodate a regular, paying, ordinary public audience, and in which to present to them a policy of new, highly attractive, five-act plays, independent of any princely command, or of any widely-spread religious intention' (ibid.: 171). This secular break from court and religious approval lay the foundation of the Elizabethan playhouse tradition that spread nationally and internationally, enabling playwright and playgoer alike to develop a direct relationship for the first time.

The three forms of theatre operating by this time were therefore:

1   the court theatre
2   the private house theatre
3   the public playhouse.

Private houses, as already mentioned, had served as locations for performance both before and after the Restoration drama periods, as Southern points out: 'in many early theatre ceremonies, the background to the action is, quite normally, the facades of the ordinary houses of the community' (1962: 98). In 1574 an Act of Common Council was passed by the City of London which laid down strict limits on performance of plays in their jurisdiction. However, provided no ticket fee was charged to the 'public', private productions were not controlled, and weddings and friendly gatherings were also allowed, providing a loophole that early actor–managers such as Burbage exploited. As well as plays, music was commonly performed, for example in Berlin from the mid-1700s: 'the cultivation of music in prosperous houses and the formation of amateur music societies flourished as never before' (Taylor 1998: 74). As a private pleasure, form of patronage and avoidance of licence control and censorship, private performance was also similar to court performances that took place wherever the court sat – both town and country. Elizabeth I employed a 'Master of Revels' who organised court performances, including their location, staging, scenery, etc., and her father Henry VIII had 'maintained two troupes with eight men' (Hall 1998: 129). As Trevelyan maintains: 'a way to wealth and honour had been opened to the actor and the playwright. . . . The travelling companies had the patronage of literary noblemen, whose castles and manors they visited as welcome guests, acting in hall or gallery' (1962: 217). The picture of a court performance is described by a French production of 1581, known as the Ballet comique de la Reine: 'Here the King sits at one end of the hall, with audience down either side in two galleries and also possibly behind the king. The "houses" – a wood and a bower of clouds – are planted midway down the hall, one either side; a fountain is planted near the far end' (Southern 1962: 150). The court had also been the location for pageants, balls and masquerades, celebrating weddings and birthdays such as Mummings and Maskings, but 'such occasions as these were in no sense public entertainments. They were private and ceremonial amusements; most of them were performed by the courtiers themselves' (ibid.: 144). The distinction between court, private and public activity did however begin to blur, largely due to the aspirations of the professional players themselves, and changing tastes.

For instance, in the German court at the turn of the seventeenth century, the wedding of the Crown Prince was attended by a French company of actors, forty-strong, who 'were given a long-term engagement in the town which committed them on the basis of a repertoire of French plays and ballets, to give two performances a week for the court . . . and other summer residencies', but as Taylor goes on to say: 'On other weekdays they had permission to play in a public theatre set up in the rear of a large Renaissance house . . . and to charge for admission. Performances at court, on the other hand, cost nothing to the privileged circles entitled to attend them' (1998: 49). The subsidy of high-art performances of 'imported' productions still persists today in European and other cities, with the state superseding the Crown's role of funnelling taxpayer's money to opera, drama and dance 'houses' and companies, largely to the benefit of the well-heeled – *from aristocracy to meritocracy.* In the Renaissance era, the court also sought to impress foreigners: 'Visiting dignitaries from other courts, who needed to be suitably impressed, offered convenient stimuli . . . and poets and composers were retained for the purpose' (Taylor 1998: 48), whereas today, it is cultural tourists – business and leisure – who need to be wooed, and who account for between 25 and 33 per cent of the London theatre audience (Gardiner 1998, MORI 1998).

The latent and supply-led market for live entertainment, (relatively) free of the city wall and outside of the private court and stately houses, was epitomised in sixteenth-century London, and significantly this city almost completely dominated theatrical development, with few exceptions in other towns such as King's Lynn, Norfolk (see below). Whilst the Italian then French court had been the prime exporters of opera, music and drama 'fit for royalty' and their extended entourage, for instance in the court of Friedrich I, 'the advance of French cultural values, soon to become irresistible, finally made itself felt in the court which had persisted mainly with things German' (Taylor 1998: 49), it was the conditions that created and supported drama and the new London playhouses that in turn found a willing audience elsewhere in the country (Borsay 1989) and on the Continent. For example, back in Germany 'The Englische Comodianten, with their Pickelhering plays and their improvisatory *commedia dell'arte* manner were still in some demand and German troupes followed in their footsteps' (Taylor 1998: 48). The public playhouse therefore influenced both performance and playwriting, however the heyday of Elizabethan theatre building covered less than fifty years and less than a dozen theatres, although 'By 1629 when Paris was building its second public playhouse, London already had seventeen' (Hall 1998: 136). A chronology of this period of theatre building reveals the extent of its concentration in space and operation:

1576	*The Theatre* built by James Burbage in Shoreditch, East London
1576/7	*The Curtain* theatre built nearby
1587	*The Rose* built on Bankside (south bank of the Thames)
1595	*The Swan* built by Francis Langley on Bankside
1598–9	*The Globe* built on Bankside from timbers of the demolished *Theatre*
1600	*The Fortune* theatre built in Cripplegate (City of London) in the *Globe* design
1600	*The Red Bull* built in Clerkenwell (just outside of city walls)
1613	*The Globe* burnt
1614	*The Hope* theatre built on Bankside based on the *Swan* design
1614	*The Globe* rebuilt
1621	*The Fortune* burnt and rebuilt in 1623

Source: adapted from Southern (1962: 172)

These building ventures were neither cheap nor crudely constructed (unlike their lighting, hence the propensity for destruction by fire): 'These theatres cost daunting sums of money to build. . . . *The Theatre* was valued at £666; *The Globe* cost £1,400 to construct' (Hall 1998: 136), and as Southern revealed, 'the interiors of the Elizabethan playhouses were beautiful to look at, were painted and were marbled in a fashion skilful enough to deceive even a curious bystander' (1962: 181). The conditions under which this cultural development emerged were also significant. An urban renaissance in

Continental Europe during the sixteenth century spread to London from the early seventeenth century, although it was arguably 'small fry when compared with the great Italian Renaissance' (Borsay 1989: viii), whilst in Germany the Italian Renaissance had 'only a stunted influence, and then rather in Catholic than in Protestant states' (Taylor 1998: 37). Early Italian Renaissance artistic influence (e.g. church, cathedrals, painting) was however evident in Poland and Russia (e.g. the Kremlin) by the early 1500s (Kauffman 1995). This period had also witnessed the secularisation of the theatre with the licensing of theatres to perform Shakespeare and 'legitimate drama' – the first London theatres in Shakespeare's day had been to the east of the city and later on the Southwark (south) bank of the Thames. However, the theatrical world had been devastated by Oliver Cromwell's puritanical closure of playhouses and pleasure palaces, with public theatres only regaining their licences in 1660 following restoration of the monarchy (although surreptitious activity took place at private houses during this clampdown, including Sir William Davenant's; see below). In this year, Charles II issued two theatre patents to members of his household, Sir William Davenant and Thomas Killigrew, creating an effective duopoly. The nature of theatre presentation and building had also changed from that of Shakespeare and his contemporaries – roofed-over, artificially lit with painted scenery and curtains and, for the first time, women-players whereas before well-trained boys took women's parts: 'It was to a large extent a new theatre and a new dramatic art, with new possibilities, and new dangers' (Trevelyan 1942: 275) (Table 3.1).

The Theatre Royal Drury Lane and Covent Garden were the effective birthplace of the West End and this area 'attracted not only playgoers but also hordes of prostitutes . . . gangs with robberies galore and fighting in the streets at night. . . . The rich continued to visit Covent Garden but they came as visitors to its gaming houses or as customers to its brothels' (Thorold 1999: 102). A hostility to the theatre also persisted amongst 'many pious and decent-minded families, High Church as well as Low. . . . Till late in the nineteenth century not a few well-brought up young people were never allowed to visit the theatre' (Trevelyan 1942: 276). In 1737 the licensing of plays and theatres passed to the Lord Chamberlain's Office and from 1757 the justices also issued music and dancing licences. These comprised a different but confusing set of regulations from the those applied to the earlier Patent theatres (Southern 1962), the precursors to the 'Theatre Royals' and Edwardian and Georgian theatre ventures (Pick 1985, 1988, Fox 1992), already reflecting the divide between high-art (e.g. drama) and popular culture (e.g. music and dance halls). This urban renaissance also spread to provincial towns and cities such as Bath in the West Country and King's Lynn in East Anglia during the late sixteenth and early seventeenth centuries (Bell 1972, Borsay 1989), often emulating London's development and attractions. This relationship between (cultural) capital and provincial cities is one that resonates today, both as a 'model' for replication and as a source of resistance which rising regionalism has sought to exploit through its own cultural renaissance and reassertion of regional identity and self-reliance – from Birmingham and Barcelona to Lyons and Pittsburgh (see Chapters 7 and 8).

In King's Lynn, a medieval guildhall had been converted to house a complete Georgian playhouse in 1766. Many of the provincial eighteenth-century theatres were dual-purpose, as precursors to the multi-use community centres and halls that the

Table 3.1 Restoration theatre development and duopoly in London

Killigrew Company The King's Men	Davenport Company The Duke's Men
1660 Opened at Gibbon's tennis court, called the Theatre Royal (platform stage)	1660 Opened at Salisbury Court
	1661 Moved to Lisle's tennis court (scenic stage) called Duke's Theatre, Lincoln's Inn Fields
1663 Opened at a disused riding school, Theatre Royal (scenic stage), Bridges Street	
1665–6 Closed due to the Plague and Fire of London	
	1668 Davenport died
	1671 Opened a new playhouse designed by Christopher Wren called the Duke's Theatre, Dorset Garden
1672 Bridges Street burnt. Company move to abandoned Duke's Theatre	
1674 Opened at a new playhouse designed by Wren, called Theatre Royal Drury Lane	
1682 The two companies united at Drury Lane	1682 Duke's Theatre abandoned
	1695 Companies separate again, new theatre at Lincoln's Inn opened by lead-actor T. Betterton
	1732 Patent transferred to Covent Garden

Additional licences were awarded to the King's Theatre (John Vanbrugh's Opera House, today Her Majesty's Theatre) in 1705 and the New or Little Theatre in 1720 (which became the Theatre Royal in 1766), both in the Haymarket, and another at Goodman's Yard (1729) near the Tower of London.

Sources: adapted from Southern (1962: 238); see Wall (1998: 151–2), Weightman (1992) and Hall (1998)

amateur and touring arts share with other amenity uses in small towns in the absence of dedicated arts centres – the *village hall,* as Porter says: 'smaller barns to be stormed by strolling players' (1982: 256). Companies were established in the larger provincial towns, and strolling players were always moving around the countryside, acting in barns and town halls before rustic audiences (Trevelyan 1962). These early provincial theatres were largely seasonal, to be used only when travelling companies visited on touring circuits, and lying vacant the rest of the year. Many however were levelled over with a temporary floor, creating a flattened area of stage and pit, and on which dances and other events were held. Following King's Lynn, other regional city and market town playhouses were established, nearby in Norwich, 1769 in Nottingham and the Manchester Theatre Royal in 1775, and whilst the Theatre Royal Drury Lane seated over 3,000, these provincial theatres, many still surviving today in refurbished form, seated over 1,000 people. Popular and largely commercial entertainment, from pleasure gardens to music halls and theatres (Bailey 1986, Crowhurst 1992), therefore evolved within the private sector under the growing influence of state licensing and control, which directed and restricted their location, operation and programming (Pick 1988, Weightman 1992). The growth of a public realm on the one hand and an entrepreneurial cultural producer on the other combined to widen access and participation in arts, entertainment and material culture. As Porter maintains, 'Market forces – affluence, leisure, the book trade – led to high culture becoming available if not to the masses at least to the many' (1982: 248), and he summarises thus:

> The popularization of what had been once reserved for the cognoscenti often developed by stages. For instance in the C17th collections of antiquities and natural history had been privately owned, but their owners had displayed them to visitors. By 1759 the British Museum had been opened, at the bequest of Sir Hans Sloane, as the first publicly owned free-entrance museum in Europe. The second half of the century saw privately-run museums opening their doors as commercial ventures in London and the provinces. Similarly all private-domicile, Stately Home proprietors threw them and their grounds open to visitors, strangers included (ibid.: 248).

Continental opera houses

Whilst the eighteenth century was the great period of European baroque opera house building – Italy of course produced a wealth of opera theatres – a longer history of opera house development, albeit court and privately patronised (e.g. in Sweden), is shown in the following selection straddling three style periods of opera's development (Horowitz 1989):

1618   Aleotti's Teatro Farnese, Parma (one of the first 'proscenium arch' theatres)
1670   Celle, Germany
1731   Manoel Theatre, Valletta, Malta (still in use)
1737   San Carlo, Naples
1742   Royal Opera House, Berlin
1748   Galli-Bibiena, Bayreuth
1752   Castle Schwetzingen, Germany
1752   Residenztheater, Munich
1766   Krumau, Czechoslovakia
1766   Drottingholm, Sweden (private court theatre)
1770   Gabriel's, Versailles
1781   Gripsholm Castle, Sweden (private court theatre)

As Southern points out, however, 'these are specialist theatres, and reserved for occasional, princely performances. . . . The culmination came, perhaps, when state and people banded together to give Paris the world-famous Opera House by Charles Garnier in 1875' (1962: 249). Tension and competition between spoken drama, musical theatre, operetta and full-blown 'classical' opera has therefore been a feature of the court versus popular theatrical traditions noted above: 'Drama was in competition with opera for both royal and popular approval' (Taylor 1998: 66). But for instance, in Berlin, home of the German court, by 1700 'spoken drama and play-with-music – operas, Singspiele [opera in the vernacular] – existed side by side in one and the same theatre' (Taylor 1998: 48). The development of popular drama and theatres to house it was irresistible, as is it had been in London earlier: 'change which, as so often in the performing arts, forced itself through by a coincidence of energies pressing towards a common goal. From one side came theatrical producers and policy makers resolved to gain for their medium a new level of respect; from the other came superior materials on which to base this rise in status' (ibid.: 67).

In 1760 Berlin followed other cities and constructed its first purpose-built commercial theatre in the courtyard of a house behind the Unter den Linden seating 800 spectators, at first presenting translations of Shakespeare and other English plays but later indigenous writers such as Goethe (who himself had been an advocate of an outward-looking cultural influence, rather than a parochial and nationalistic view of the arts). Its location was also symbolic, in close proximity to, but not actually situated on, the royal mile of Unter den Linden where the King's new opera house was built in 1742 on the southern side of the avenue that linked the royal palace, through the Tiergarten, to the Palace of Charlottenburg. The site of the Royal Opera House 'had been chosen by the King and was not open to discussion' (Taylor 1998: 69) and the building was a grand mix of rococo and neo-classical styles, adorned by motifs and sculptures depicting Apollo, the Muses, Aristophanes, etc. This was thus a courtly extravagance whose capital and operating costs were met by the monarch. Tickets were free, although strict segregation within the auditorium was maintained, but admission was not aimed at the 'lower classes and common masses' (ibid.: 70). It was to be short-lived, by 1760 following the Seven Years' War, the opera house was starved of funds and suffered at the Russian and Austrian bombardment of the city. Frederick the Great's theatre ceased to operate in 1778 and reopened eight years later under the reign of Friedrich Wilhelm II, who renamed it the Nationaltheater: 'Theatre was now a matter of serious national concern, with a mission to address the state of public morality and public culture . . . [and it] became a nationalised enterprise. Its funds provided by the state' (ibid.: 102). A different but still nationalistic attitude to culture resurfaced following the collapse of the Second Reich in 1871, when the new Republic proclaimed that 'The royal opera houses, theatres and museums had become the property of the Prussian state' and, as Taylor observes, the new Prussian Ministry of Science, Arts and National Education adopted a cultural policy in which 'culture belonged to the people' (1998: 254). In less than 130 years, the high-art venue had moved from courtly plaything, nationalistic emblem, to *people's palace*. The extent of opera's expansion however reached not only European cities, but also by the third quarter of the nineteenth century, the Americas, Australia and even Cairo (Figure 3.1).

From playhouse to the music hall and pleasure garden

The distinction between spoken, 'straight' drama and other forms of performing arts – opera, music, mime and dance – was an artificial but influential divide created in London by the Patent theatre rules growing out of the Elizabethan and Restoration periods, which was maintained in subsequent Jacobean and Georgian eras. In the

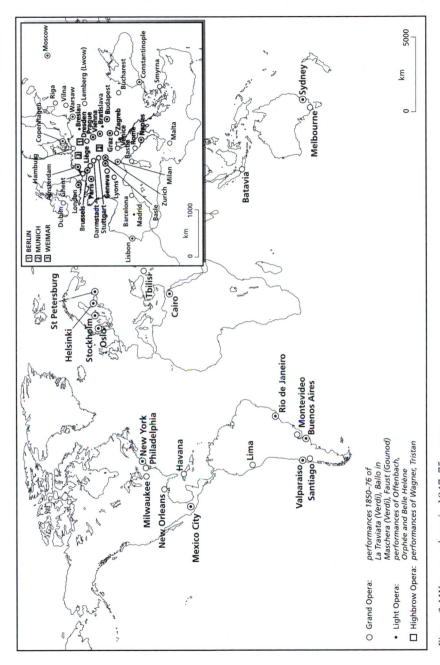

Figure 3.1 Western culture in 1847–75: opera

Source: Hobsbawm (1977: 371)

Georgian 'golden' period a building boom fuelled not only both construction and crafts trades, but also a new service economy: 'Urban renewal meant more inns and shops, coaching-houses, social centres such as *theatres, assembly rooms and concert halls*, where service employment was created, money spent and business transacted' (Porter 1982: 225, emphasis added). The effective monopoly of theatre operation and this restriction on the art form itself however continued to create endless disputes, illegal operation and battles over interpretation and licensing responsibility. When a new theatre was built, 'the Patents objected and called in the law, which was always upheld. A great many troupes of players were, over the years arrested and fined for daring to put on Shakespeare or other dramas in defiance of the Patent theatres' (Weightman 1992: 21). The other form of licence, for music and dancing, was granted by magistrates and this alternative allowed the development of the music hall since *both* licences (drama or 'entertainment') could not be held at the same time, i.e. they were mutually exclusive.

After the passing of the short (running to only five pages) Theatre Regulation Act 1843 (and in 1871 the Fair Act), which recognised the unworkable and anti-competitive nature of the previous patent system, all manner of venues could apply for a licence, although the plays they showed still had to be approved by the Lord Chamberlain's office. The key issue was alcohol – food and drink could not be served within theatres playing legitimate drama (and no smoking was permitted), whilst pubs and inns and the later gin palaces and music and dance halls of course could and did, but they could not show straight drama! Exceptions to these rules were practised whether in subtle breach or defiance of these separate licences and criteria, such as the Britannia Theatre in Hoxton, East London which had a theatre licence but allowed food, drink and smoking, whereas the West End theatres enforced this prohibition – an effective east–west divide appeared to operate, with the downmarket East End allowed to operate outside of the strict licence rules (as recently as 1911, the newly opened Palladium Theatre, West One, licensed for music and dance entertainment, was fined for presenting an excerpt from *Julius Caesar*, Weightman 1992: 29). As Weightman also observes, 'There was no reason for the theatres and music halls to develop separately, other than the vagueness of the 1843 Act and the dual licence system' (ibid.: 29). However this understates the deep antipathy of Church, puritans and their proponents, the temperance societies and rational recreationists, to popular theatre and its association – real and perceived – with degradation, prostitution, alcoholism and general bad behaviour – as well as the state's natural fear of mass gatherings: 'The City housed the Puritans, the moneymakers, the recalcitrant Commonwealths. What had the City to do with cultivated urban court life?' (Wall 1998: 152).

This divide in the road between drama and entertainment fuelled the growth of the music and variety halls, whilst straight theatre development remained largely static after the Georgian period. This growth was accelerated by the natural extension from existing places of popular music and dancing – pubs and saloon theatres, and their imported rural predecessors, the fairs and pleasure gardens. On these urban oases Weightman quotes from A. Thornton in *Don Juan in London* (1836): 'the gardens are beautiful and extensive, and contain a variety of walks, brilliantly illuminated . . . and terminate with transparent paintings, the whole disposed with so much taste and effect as to produce sensation bordering on enchantment to the visitor' (1992: 19). In 1830 there were no music halls, or *variety theatres* as they came to be termed,[2] by 1870 there were thirty-six large halls just in London: 'impresarios recognised that there was a new market in London population with a little more money and a little more leisure, and a taste for a relatively refined surroundings and architectural razzamatazz' (ibid.: 30). From their working-class audience base, the widening of their clientele and experience was soon sought, for example a picture gallery was opened in the Promenade of the Canterbury (Morton's) music hall in Lambeth: 'while providing for them the innocent and enlivening enjoyment of the music in the hall, the fine-arts gallery can be made the medium for raising in their minds ennobling and refining thoughts' (*The Builder* 1858, quoted in Chanan 1980: 157). The music hall was therefore opportunistic, the product of entrepreneur and pragmatic response to state control of cultural consumption, but it was not an original source of production, unlike the playhouses of Shakespeare's time: 'Into the music hall flowed other tributary streams: pantomime, "variety", tavern concerts, all-male singing assemblies, or "song-and-supper rooms". . . . There is about its flowering an unmistakable air of socio-economic determination' (Best 1979: 235). Another, more historic recreation was also seen in the Hippodromes and Coliseums that had little connection to the music hall, an urban hybrid, these large and technically ingenious venues took their inspiration from the circus and fairs, but their promoters effectively cleaned them up, as the bawdy music hall was to move into variety and family entertainment.

The decline and transformation of the fair and pleasure garden into the music halls, saloons and gin palaces in part reflected the changing circumstances of the working man – the urban economy meant better wages, and an active nightlife facilitated by lighting, sophistication in taste and supply of entertainment. The gin palaces, which grew in size to accommodate commuting workers on their way home, grew up along transport routes – street corners, railway stations, horse tram and bus termini, park entrances and other gateways and by the second half of the nineteenth century they vied for custom in close proximity – on the Whitechapel Road East London there were over forty-eight gin palaces and on the Strand forty-six, both in a stretch of less than a mile. Urban society also brought with it greater control, policing and morality, underwritten by a class divide: 'a social division grew between the kind of places, amusements and behaviour that were acceptable to the majority of working people, and what was acceptable to the industrious, but increasingly refined, professional classes. . . . Respectability was demanded in entertainment, just as it was in other spheres, and respectability in turn had a great deal to do with drink' (Weightman 1992: 13). The music hall attracted similar opprobrium to the playhouses, which continued into the twentieth century: 'Worthy people who had never been inside a music hall in their lives, strongly condemned them. All social evils, especially among the young, were laid at their door. (People say the same thing to-day about cinemas.)' (Willis 1948: 163–4). The rustic revels that retreated in the 1830s attracted increasing disdain and were seen as a public nuisance. For instance the Bartholomew Fair, which was held over three days in the autumn (on a site that became the Smithfield meat market), according to Hollinsghead, 'was the oddest combination of town and country ever brought together, it combined the bustle, business and attractions of a cattle-market with a congress of peripatetic show-men' (*My Lifetime*, quoted in Weightman 1992: 17). These shows played all over London with booth theatres and all kinds of menageries and freak shows. With growing concern of the authorities the scope of these touring fairs was curtailed and by 1855 Bartholomew and its satellites had closed down altogether. This decline and effective urbanisation of rural popular entertainment was mirrored in the pleasure gardens, several of which drew large numbers of participants and *flaneurs* to the gardens at Vauxhall, Ranelagh and Sadler's Wells; Cremorne Gardens, Chelsea with a range of entertainments like those of Copenhagen's Tivoli Gardens, to Crystal Palace where a gigantic garden 'included refreshment rooms, music, paintings, sculpture, tropical trees and architectural models' (Best 1979: 234). Mr Sadler had established his fairground and pleasure garden on a natural spring ('well') in Islington in 1683, on a site where Lilian Baylis and Ninette de Valois started the Vic-Wells Ballet in 1931, birthplace of the Royal Ballet and later the English National Opera, in a theatre which still bears Sadler's name, and is still the home of modern dance today. Whilst visitors took 'the waters', Sadler added musicians and a Musick House: in the words of a song of the day, 'sweet gardens and arbours of pleasure' (Senter 1998: 6). The fate of the pleasure gardens was to be similar to the fairs – sealed by the appeal of the music hall and 'dream palaces', with their wide variety of entertainment, food and drink and internal safety, on which the gardens increasingly could neither compete nor satisfy the licensing authorities: 'semi-rural places of entertainment were beginning to lose their attraction and were tolerated less and less as the bricks and mortar of the expanding city covered the fields' (Weightman 1992: 10). By 1711 Sadler's Wells's audience was described as 'vermin trained up to the gallows' and by the Inquisitor as 'a nursery of debauchery' (Senter 1998: 6) and in 1851 after many changes of style, management and illegal operation, Charles Dickens's opinion was no better: 'as ruffianly an audience as London could shake together . . . like the worst kind of fair in the worst kind of town . . . it was a bear garden, resounding with foul language, oaths, catcall shrieks, yells, blasphemy, obscenity' (ibid.: 10) and by 1876 it was turned into a skating-rink and winter 'Garden'.

Audiences in their place

Pleasure gardens tended to draw their crowds on warm, dry days and evenings, and as well as the major gardens, countless smaller ones cropped up in and around most cities: 'Wherever streets and houses lay thick, it was worth someone's while to set up a vista of Arcadia' (Best 1979: 234–5). As well as a local pedestrian population, the larger sites relied on public transport such as horse-buses and expanded rail networks, but their prime location also made them exposed to speculative building. Some pleasure gardens actually built halls on site, losing green space but maintaining their clientele in the new saloon theatres. Some of the remaining gardens were later to become urban parks, as a buffer or 'green lung' to encroaching industrial city population growth and density, but more a product of rational recreation than popular pleasure pursuit. Outdoor city entertainment also suffered, once public transport enabled quick and cheap access, from the popular growth of seaside resorts, the 'piers and promenades', and the establishment of fairs, 'winter gardens' and 'summer palaces' in less threatening and controlled environments than the inner-city. From the second quarter of the nineteenth century, the focus of urban arts and entertainment – the music halls, gin palaces, saloon theatres and subsequently the early cinemas – was increasingly on *local* provision rather than on the city-centre which, as today, is frequented by professional middle classes ('PMCs') from outside of the city: the suburbs, Home Counties being the origin of the largest attender group at West End theatres today (43 versus 40 per cent of London residents; MORI 1998). Whilst city-centre arts and entertainment zones have survived, although not unchanged, neighbourhood entertainment was not sustainable with the decline in the local music hall, pleasure garden and, later, the cinema.

The evolution of performance from court and private house, to public and open entertainment, had also brought the live arts public in closer proximity, as Best states by the mid-Victorian period: 'Theatres might be patronised by all classes; their accommodation and prices were often expressly designed to that end, and pit and gallery had long brought the mob close to the classics. Pleasure gardens were open to all who cared to pay the entrance money, and all but the lowest music halls would attract a mixture of patrons who, in their own homes and occupations would not normally meet' (1979: 221). Italian opera was of course played to a wide stratum of society, and there the pit was no more refined and respectful than in the English music hall. At the Government Select Committee held in 1866, playwright Dion Boucicault's evidence stated thus: '25 years ago the amusement-seeking public were divided into two classes: the upper classes which went to the theatres exclusively, and the lower classes which attended public houses and gardens; the music hall was the stepping stone between the two . . . the large sums of money which have been made by managers in pits and galleries of theatres has been principally due to the pits and galleries of the theatres being recruited from the music halls' (quoted in Weightman 1992: 49). As Weightman concludes, 'at this time audiences were not sharply divided . . . the division between drama and music hall imposed by the licensing laws, did not reflect a division in taste. What divided the audience along social class lines more than anything was the *behaviour* in music halls and theatres' (ibid.). Audience loyalty and frequency was also a factor, irrespective of programme, genre and playwright, and the supposed preference for high-art (neo-classical verse and tragedy, high comedy) over melodrama between upper-class and lower/middle-brow audiences is not borne out by actual attendances here, as in other European cities at this time. In Rotterdam between 1795 and 1815 the social composition of theatre audiences hardly changed, whilst changing popularity, e.g. a declining interest in opera in the late 1800s and the departure of the gallery and pit audience, reinforced the upper bourgeois presence (Gras 1999), rather than diversion to new venues such as the *salons des variétés* which attracted less than 10 per cent of the city's working population.

Rational recreation

The social and spatial divide that the early industrial city exhibited in the location and participation in public culture can be seen both historically to reflect the elite provision of arts facilities for pleasure and improvement, and the emerging control and interference in popular pastimes and gatherings by the state. State 'interference' in culture is an enduring phenomenon manifested in censorship, licensing, planning (e.g. land-use/class) and other controls, which are applied at different times and in different countries, to a greater or lesser degree (Pick 1980, 1988). In late nineteenth-century Britain, for example, theatre planning and licensing controls were extended by the London County Council's 'Theatres and Music Halls Committee' which scrutinised scripts and lyrics and sought to control music hall performances and spectator's behaviour through the notorious 'music-hall purification campaign' (Bailey 1986).[3] Human agency does not of course follow these engineered divides, witness the mixing of audiences at theatres, music halls and pleasure gardens which only urban land and income class separation reversed: 'Rich and poor, aristocracy and underworld, were never closer together than at prize-fight, the cockpit, the race-track, and the demimonde saloon and casino' (Best 1979: 221). Positive (or patronising) involvement in public culture on the other hand is evident in state provision, funding and advocacy of the arts as a social/welfare 'good'; through (arts) education and training systems; and dissemination (e.g. public service broadcasting). State involvement in monumental and civic culture has a long pedigree as discussed previously – the coming together of civic and municipal cultural provision, the control of public culture and rational recreation, is no less evident than in the Victorian era and the development of public museums, libraries and other leisure amenities that sought to reconcile these otherwise conflicting objectives. As Wilson claims, this paradox of *loisir* and *relative freedom* may however be irreconcilable, since 'leisure cannot be equated with education, medicine and shelter as a state function because there is an inherent contradiction in the planning on which the welfare state must depend on the one hand and the freedom necessarily entailed in true leisure on the other' (1988: 118).

Only in recent times has the concept of actually *planning* for the arts emerged and in comparison with other areas of amenity and city planning it still remains largely undeveloped. In Britain, for example, cultural planning is, however, not a solely modern phenomenon, in the sense of social planning and the expansion of civic cultural amenities. Many of the local and national 'flagship' and civic arts buildings and facilities are inherited from the Victorian era (and many assembly rooms and some private theatres date from the Georgian 'golden age'; Fox 1992), legislated in England in Ewart's Acts for museums in 1845 and for libraries in 1850. The Museum Act, although this did not initially cover public art galleries, gave local councils with a population of at least 10,000 powers to levy a halfpenny rate for the establishment or support of museums of arts and science, whilst the Libraries Act, Select Committee on Public Walks 1833, Towns Improvement Clauses Act 1847 and Baths and Washhouses Act 1846 all laid down statutory provision for public amenities at a local borough level. The Public Health Act 1875 also gave an impetus to public park provision by giving local authorities power to raise government loans to acquire land for recreation (Conway 1989). Prior to this, costly special legislation was required for each project. These acts also established local committees, such as for Baths and Washhouses, the Local Board of Health and later School Boards – all mechanisms for the planning and delivery of local amenities albeit within a national (central) legislative structure.

In European cities (but less so in the USA) in the third quarter of the nineteenth century, the arts fulfilled a fundamental role in meeting social demands, almost in a quasi-spiritual way, as Hobsbawm observed: 'great collective symbols of theatre, opera arose in the centres of capital cities – the focus of town planning as in Paris (1860) and Vienna (1869), visible as Cathedrals as in Dresden (1869), invariably gigantic and monumentally elaborate as in Barcelona (from 1862) or Palermo (from 1875)' (1977: 334). The industrial city and the development of public civic culture were also exported to colonial cities, which had also undergone rapid population growth. For example, Toronto, just a small town of 10,000 people when incorporated in 1834, reached

200,000 by the end of the century; Bombay was the third most populous of the Empire after London and Calcutta, in the 1880s with over 700,000 and by the 1890s over 800,000 residents. Melbourne, in the Australian State of Victoria, grew from a town of 23,000, on 'incorporation' in 1850 it had 97,000, but twenty-five years later had over 1 million people in thirty separate municipalities ('boroughs'). Termed, the 'Paris of the Antipodes' or the 'Chicago of the South', Melbourne acquired a metropolitan character as a centre for business and entertainment: 'The city centre included well-stocked and well-lit shops equal to the best in London; banks, a Theatre Royal built in 1842, where you could see Italian opera in a style worthy of the English metropolis itself' (Briggs 1990: 280). Culture and society also emulated British institutional provision, with the Society of Arts, Mechanics Institutes, Friendly Societies, The Literary Institute of the Trades Hall – the first municipal library opened in 1859 – this city even hosting a huge (in size and cost) International Exhibition in 1880/1 and again in 1888.

Museums

The development of the public museum, as perhaps the archetypal national and civic cultural institution, owes its foundation and existence to the benefactions of Crown and courtier, the proceeds of public lotteries and the philanthropic merchant and industrialist – from Sirs Hans Sloane and Henry Tate. This relationship is of course alive and well in the twentieth century – from the 'donor-memorial' foundations of Ford, Carnegie, Getty, Gulbenkian and Guggenheim, to modern Medicis, Conran, Saatchi and Thyssen – and supplemented with state lottery grants (Schuster 1994, Evans 1995a). Founders and sponsors donated sites, buildings and collections that make up the core historic fine art and scientific collections in many state and city museums, as well as specialist collections. Their locations often reflected their own roots and property (e.g. Guggenheim – New York, Venice, Salzburg), whilst today as well as wealthy individuals seeking new facilities to house their private collections for posterity, corporate ownership of arts venues and companies stretches from London's West End to cinema and theatre chains in many cities and countries. Colonial explorers also played their part in providing archives and artefacts either during their life or after their death, including those combining trade and travel such as Horniman, the tea-merchant, anthropologist and collector. (The Horniman Museum is in South East London in the Horniman Gardens, Lewisham.) Museums as manifestations of monumental and civic culture also became representations of imperial and national supremacy: 'The expansion of the commissioning of monuments in the late nineteenth and early twentieth centuries, when the plunder of colonial wars was being assimilated to European museum culture, is a statement of national identities' (Miles 1997: 63, also Coombes 1994).

The location of these museums is therefore a function of posterity and a desire for a wider audience to wonder and receive educational inspiration, as well as to serve as a celebration of their founder's and nation-state's greatness – 'a building whose primary duty is to proclaim the enlightened majesty of the monarchy' (Taylor 1998: 123) – or of man and nature's 'achievements'. The Crown had already taken a role in the training and education of artists and craftspeople through Academies of Arts, which were established in Paris, Rome and then Berlin in the seventeenth century (the Royal Academy was founded in Britain in 1768) and collections of art works, both commissioned and acquired, had built up in private palaces and aristocratic houses, whether displayed or stored: 'Buildings call for interior decoration and for contents. Open spaces such as gardens and public squares are invitations to landscape design and statuary' (ibid.: 78). As well as the aesthetic benefit of fine art and architecture to the court: 'For the kings, princes and dukes and other landed aristocracy of the age, the collection of works of art served a number of purposes. One the one hand it was an economic investment. At the same time it was a demonstration of an awareness of intellectual and aesthetic values, coupled with their sense of history' (ibid.: 81). In 1821, fifty years before the International Exhibitions which were to bequeath London its museum quarter in South Kensington (Table 3.2), the celebrated architect Karl Friedrich Schinkel was commissioned by the German king to design and build an art museum to house his constantly expanding art collections, for *public exhibition*. Although the monarch laid down his detailed specification, Schinkel's neo-classical design looked to Hellenic roots – he never envisaged architecture without sculpture and his 'great achievement is to have been the first not only to recognise and make close examination of the splendours of Greek art but also to demonstrate how the values of that art might be applied' (Alexis 1838). By the mid-eighteenth century in Europe, as Hobsbawm notes, 'Even in the most splendid monarchies [museums] belonged increasingly to "the public" rather than the court: imperial collections were now museums, operas opened their box offices. They were characteristic symbols of glory and culture' (1977: 329).

The discovery of buried cities (e.g. Pompeii) and civilisations also grabbed the

Table 3.2 Great Exhibitions and World Fairs, 1851–1939

Year	Exhibition/festival	Attendance (millions)	After-use facility
1851	The Great Exhibition, Hyde Park	6	Crystal Palace, London
1879	Centenary of Revolution	32	Eiffel Tower, Paris
1871–74 1883–86	International Exhibitions International Fisheries, Health, Inventions, Indian & Colonial		South Kensington, London Museum Quarter: Natural History Museum, Science Museum, Victoria & Albert Museum
1887–90, 1895–99	American, Italian, French, German India		Earls Court/Olympia Exhibition Halls, London, Liverpool and Glasgow
1901	Glasgow International	10	Kelvingrove Arts Museum, Glasgow
1908–14	Franco-British, Imperial, Japan–British, Coronation, Latin–British, Anglo-American	12	White City, London Sports Arena
1939/40	New York World's Fair	45	Stadium, State Pavilion

Sources: Greenhalgh (1991), Evans (1996a)

attention and popular imagination in the eighteenth century, as the Great Exhibitions were to do for industrial machinery, produce and exotica a century later. The Crystal Palace erected in Hyde Park in 1851 received an estimated 6 million visits over only six months, exhibiting works of art and industry from over the known world, and about one-quarter of visitors were Londoners. Like the advanced theatres, price discrimination separated the classes with shilling days targeted at artisans, whilst the middle classes paid five shillings on other days. Pick (1985) and Pick and Anderton (1996) see the Great Exhibition as a milestone that accelerated the separation of artistic and popular entertainment, a divide which museums and galleries have bridged more than other forms of culture, with a wider socio-economic profile of attenders than the performing or visual arts, although still a predominantly well-educated visitor, including the (school)children of the well-educated (Bourdieu and Darbel 1991, Evans 1995c). The Royal Academy exhibition in 1848 attracted around 90,000 visitors, but by the end of the 1870s almost 400,000. The pre-war Great Exhibitions, given their scale and sprawling sites, do offer a key contribution to the planning of public culture in that their inheritance provided some of the major cultural buildings and symbols of the mood of their day (see Chapter 8).

The concept of a trade exhibition celebrating national produce and culture was in fact conceived earlier in France, where between 1797 and 1849 ten national exhibitions were held in Paris, at first in the courtyard of the Louvre and then to buildings on the Place de la Concorde, before reverting to temporary structures on the Champs Elysees. Other French cities followed suit (as they have in emulating the Parisian *Grands Projets*, see Chapter 8) with similar 'mega-events' being staged in Nantes (1827), Lille (1835), Bordeaux (1835 and 1845), Toulouse (1836) and Dijon (1836). Both political and commercial events: 'they were no mere trade fairs or festive celebrations, they were outward manifestations of a nation attempting to flex economic, national, military and cultural muscles' (Greenhalgh 1991: 6). In the 1909 Golden West Exhibition in London, the promoters stated in their prospectus: 'There is no means so effective for diffusing knowledge of a country as an exhibition in the Metropolis of the World' (Willis 1948: 142) and exhibitions of art and industry had also been mounted on a smaller scale in England by the Royal Society of Arts.

Museum zones and islands

The reuse and restoration of historic buildings located in city centres, plazas and squares is a phenomenon seen in cities of both developed and developing worlds, as outlined in Chapter 2. This not surprising since their urban morphology has common roots and foundations, whether classical, Renaissance, baroque, and whether conceived by liberal or totalitarian regimes – the celebration of national glory, manifested and reinterpreted over time, remains a common act and therefore a sign of the continuity of the city which museums encapsulate both through their physical presence and their selected collections of historic significance and provenance. At a local level, municipal museums have also served, since the early 1900s, as borough archives and repositories of local history, and as reminders of both municipal and munificent benefactor – often bearing their name. The clustering of museums and galleries in 'districts', 'miles' or 'quarters' is also an architectural and urban design statement and solution to the grand axis and access considerations (e.g. public transport/interchanges); for the museum public and patron this also acts as a convenience, as it does for the curator and museum management. The Altes Museum for instance served as the first in a group of museum buildings in Berlin which needed further expansion due to the King's insatiable collecting habit. The Neues building was erected in 1843–6 by the Altes architect Schinkel to house in part a collection of Italian and Dutch paintings. Twenty years later a third gallery was established on what was a 'museum island' – the National Gallery – this time to house a collection of modern art donated by Wilhelm Wagner (to the then King). The last of the galleries to be built here and survive was the Kaiser-Friedrich-Museum (renamed in 1956 the Bode Museum). Modern, or more accurately post-modern, additions and extensions to galleries and museums have been undertaken at the National Gallery and Tate Gallery London; the Museum of Scotland (an extension to the former Royal Scottish Museum), Edinburgh and major upgrades at the Prado, Madrid, all meeting the need for larger and more modern exhibition space (larger museum exhibitions are a 'tip of the iceberg' of actual collections, many of which languish in stores, unseen), and enhancing their value to the tourism-offer and therefore visitor income potential. They also serve to reassert national as well as city pride. Whereas in Edinburgh the Royal Museum in the nineteenth century presented the 'World to Scotland', the Museum of Scotland opened in 1998 seeks to present 'Scotland to the World'. Museum quarters were created in South Kensington near an entrance to Hyde (Royal) Park, on land bought out of the profits of the 1851 Great Exhibition, creating the Geological, Natural History, Science and Victoria and Albert museums (Table 3.1), whilst parks serve as natural locations for groups of museums such as in Mexico (Chapultepec), Glasgow (Burrell, Kelvingrove and Pollock House), Hyde Park, London (Serpentine Gallery), and as new locations for modern if neglected theme-museums, such as the Museums of Modern Art, Folklore and Aeronautics in Parque do Ibrapuera, Sao Paulo (Plate 3.1). Barcelona's Montjuic Park houses the Ethnological and Archaeological museums, the Art Museum of Catalonia and Joan Miro museums, and in Park Guell, the Gaudi Museum, whilst the Parc de la Villette in Paris created a new open space on a reclaimed brown-field site and former abattoir, hosting a science museum and omnimax cinema (see Chapter 7).

Parks and libraries

Like the performance of music, drama and opera, libraries first existed in private court and aristocratic houses. However private libraries also expanded during the seventeenth century, from the collections of the diarist Samuel Pepys to the 'modest bookshelf of the yeoman's farm' (Trevelyan 1942: 279). Outside of Oxford and Cambridge universities, public libraries were rare (e.g. Hereford in the fifteenth and Leicester in the sixteenth centuries), but in 1684 a public library was built by the Rector of St Martin-in-the-Fields (Tenison, later to become Archbishop of Canterbury) in the grounds of the church above a workroom for the poor. In the late eighteenth century communities large and small established secular book clubs and for a small annual subscription (one or two guineas) had access to library collections. Such proprietary libraries were set up in Liverpool in 1768, Sheffield in 1771, Hull in 1775 and Birmingham in 1779, and the earlier Spalding Gentlemen's Society library in 1711. With the cost of books and novels being relatively expensive, serialisations were also published, paid for in instalments, as

Plate 3.1 Museums of Aeronautics and Folklore, Parque do Ibrapuera, Sao Paulo (1998)

well as abridged versions of classics, and the circulating libraries fulfilled a role as a home university library, especially for women – by 1800 there were 122 such subscription libraries in London and 268 in the provinces (Porter 1982). Following Ewart's Libraries Act 1850, public libraries multiplied although there were only around sixty by 1875. However libraries and reading rooms were also provided by local philanthropic and educational institutions, and 'plenty of working class readers used them, but, librarians reported, what they usually read was fiction, and not the heaviest kind at that' (Best 1979: 234). Nineteen British cities installed public ('free') libraries in the 1850s, eleven in the 1860s and fifty-one in the 1870s (Munford 1963 quoted in Hobsbawm 1997: 386, Pick and Anderton 1996). National libraries were also promoted, with the reading room at the British Museum constructed in the early 1850s, and between 1854 and 1875 the Bibliothèque Nationale in Paris was reconstructed (both institutions being the focus of further *Grand Projet* developments in the 1980/90s). Bennett cites the political economist William Jevons whose articulation of the public good rationale in 1883 saw the 'public ownership of cultural resources as a means of securing what he saw as "the vulgarisation of pleasures", through the *principle of the multiplication of utility*' (1998: 108): 'The main raison d'être of Free Public Libraries, as indeed of public museums, art-galleries, parks, halls, public clock . . . is the enormous increase in utility which is thereby acquired for the community at a trifling cost' (Jevons 1883: 25–9).

Book collections multiplied exponentially to meet library demand – it is estimated that there were 400 major libraries with around 17 million volumes in 1848; by 1880 almost twelve times as many and twice the number of books – Austria, Finland, Russia, Italy, Belgium, Holland and Italy multiplied the numbers of their libraries more than tenfold, Britain almost as much, Spain and Portugal nearly fourfold. The USA increased

its collections only threefold, but even here the number of library books almost quadru-
pled (Mulhall, *Libraries*, quoted in Hobsbawm 1977: 386). Hobsbawm however notes
the relatively slow development of public arts in the USA at this time, with the notable
individual exception of the influential Andrew Carnegie and the 'German/ised Jewish
middle-class'. As he remarks: 'What the arts, and notably classical music, owe to the
patronage of this small but wealthy and profoundly culture-imbued community in the
later nineteenth century is incalculable' (1977: 334). This influence of course continued into the twentieth century in the movie and music industries, a cultural milieu of a
special kind.

Public gardens and areas for walking and promenading on the other hand benefited
from existing commons and opens spaces which medieval towns possessed, and which
were used for festivals and sporting events. By the late 1700s most larger cities had commercial pleasure gardens in which concerts, dances and other recreation took place. As
urban population density and building intensified in the next century, the importance of
town squares and gardens and the need for public parks – less for active pursuits than for
rest and respite from streetlife – was recognised. With the popular success of Joseph
Paxton's pioneer park at Birkenhead, Merseyside, Manchester opened three urban parks
in 1847; Bradford opened Peel Park (jointly financed by the mayor and textile industrialist Salt); in Dundee the Baxter (People's) Park; and Bolton Heywood Park in 1866 and
several parks were opened by the Metropolitan Board of Works in London: 'The public
parks and promenades which began to be opened must have made life a little pleasan-
ter . . . in every town or city of any size, wealth and concentration, the crystallising of a
cultural apparatus providing for every level of the community. . . . The leisure patterns
of the modern industrial urban mass society now begin to take shape' (Best 1979:
219–20). Parks hosted regular band concerts that were not engaged by the local council but, of course, played with their permission. As noted above, larger parks have also
extended their natural and building heritage into live arts programmes, such as the linear

Plate 3.2 Public art at Bretton Hall, West Yorkshire (1999)

Lee Valley Regional Park (Evans and Reay 1996), Regents and Holland Park (theatres) in London, whilst local parks often serve as locations for annual community festivals and parades. Taking the environmental art association further, parks are increasingly again being used to house public art collections, extending from the traditional park sculpture and conservatory tradition (and *grand manner*, e.g. Versailles), as well as sculpture trails. In Britain there are such parks in Hampshire, Yorkshire (Bretton Hall; Plate 3.2) and Lancashire (Grizedale), and outside of Paris, the Cartier Collection, whilst botanical gardens also serve as host to historical buildings and structures as well as public art (e.g. Rio; Plate 3.3). Public parks thus serve as open museums, such as the Chapungu Sculpture Park in Harare, Zimbabwe. Its collection of Shona carvings and statues both preserves the heritage of this sculpture tradition and presents these abroad through touring exhibitions, for example in Westfalen Park, Dortmund in Germany and Kew Gardens, London, whilst Holland Park in West London hosted a specially commissioned Public Art Exhibition as part of Millennium 2000 event celebrations (Plate 3.4).

Urban reform

By the beginning of the nineteenth century the industrialisation and consequent urbanisation that created the conditions for the foundation of urban planning gave way to the problems of poverty, disease, crime and squalor, which demanded responses previously resisted by the state's prevailing *laissez-faire* philosophy (Taylor 1972). As Sudjic notes: 'The modern profession of planning got a kick start from the shock of the discovery of

Plate 3.3 Public art, 'metal origami' at the Botanical Gardens, Rio de Janeiro (1998)

Plate 3.4 Public art at Holland Park (formerly the Earl of Holland's manor house and garden, 1605), west London – part of Millennium Exhibition (2000)

an urban underclass by nineteenth-century reformers. In the wake of cholera epidemics like that of 1832 which killed 20,000 people in Paris and 5,500 in London, clergymen, commissions of inquiry, poets and journalists were all in their own ways horrified to find a parallel world out of sight of the comforts of the respectable middle class' (1993: 9). By the mid-nineteenth century there was therefore more active concern about these social problems, however it was a community rather than a state response, with private enterprise developing housing estates and utilities; voluntary groups and charities providing schools, hospitals and social ('poor') housing and self-help; and pressure groups the provision of parks and other cultural and social amenities. Furthermore this amateur tradition 'in the nineteenth century produced a plethora of musical, theatrical or artistic groups rooted in local and regional life' (Parry and Parry 1989: 17), and, as noted above, this was also paralleled by self-financing commercial entertainment. However since the 1830s the Treasury, and after the 1851 Great Exhibition, other ministries such as the Department for Science and Art had been funding some parts of the arts: museums, art galleries, libraries, as well as arts education provision through music and drama conservatoire and art schools (Best 1979). As Everitt states: 'The nineteenth century had seen the arrival of public museums and art galleries, either financed by the state, or by local government[4]. Until the 1940s, music, drama and dance and literature had had to survive in the market-place' (1992: 6).

Whilst the planning for public arts facilities in the modern town planning and spatial sense was not evident in these earlier periods, it would be misleading to present civic involvement as purely restrictive, through licensing and control, culminating in the 'rational recreationist' philosophy of the later Victorian era: 'the provision by the well-to-do of

worthwhile amusements for the people' (Weightman 1992: 97), as it would be to present arts and entertainment provision as a solely private enterprise. In the first English
Urban Renaissance of the seventeenth century, urbanisation had led to the rising demand
for social and consumer services (Jardine 1996), which provided the economic foundations for a change in the quality of urban life: 'the foundations of the Urban Renaissance
were first and foremost economic ones' (Borsay 1989: 199). The emergence of planning
on a formal or informal basis helped to create a more integrated urban design and town-
scape, which was strengthened by investment in public buildings and artefacts: 'The
provision of fashionable leisure was not a random affair, but was organized within relatively well defined temporal and spatial contexts' (ibid.: 139). This included, amongst
other concerns, the recognition of cultural services as a growing aspect of urban life: 'The
impact on towns was considerable, since they were the traditional gathering points and
service centres of society' (ibid.: 117). Outside of the patent or licensing control of
theatres in London and other cities, the popular arts and entertainments were largely
housed in public inns and coffee-houses, but from the mid-seventeenth century, public
buildings created dedicated arts and cultural venues, including town and guild halls,
market squares and assembly houses hosting dance, drama and music (Chalklin 1980).
Many such buildings surviving today still act as arts centres, civic halls and exhibition
venues. Their location and architecture expressed pride in the town and parish they represented and acted as the cultural and social centre, linked to transport and trading
systems.

From people's palace to dream palace

Later public intervention in the planning of public cultural facilities can be seen in what
has been termed the rational recreationist period (Yeo and Yeo 1981, Bailey 1987), with
the development of *People's Palaces* for example in East and North London, and in
Glasgow. In the case of London, a clear spatial approach to cultural 'deprivation' saw the
notion of bringing West End culture to the East End, with Walter Besant's vision of the
People's Palace in the Mile End Road, Whitechapel (Weiner 1989). Following the success of the Crystal Palace, which had been moved from Hyde Park to Sydenham, South
London after the Great Exhibition of 1851, North London also developed its own
People's Palace named after the Prince of Wales's wife – *Alexandra* Palace, linked by
railway to central London. The 'Palace of Delights', as Besant referred to his East End
vision, was designed to separate leisure and education facilities on either side of the
grand Queen's Hall, but this paternalistic project 'lacked a coherent objective understood and shared by the public' (ibid.: 48) and contributions for its financing fell short
of target. It was never 'owned' emotionally or literally by the 'people' (but by City
guilds and foundations and with distinctly middle-class management). Opened in 1887
by Queen Victoria, attendances reached 600,000 in the first six months with a mixture
of popular and classical shows, dances and exhibitions. Its temperate financiers however
veered towards constructive recreation and educational improvement of the working
classes, rather than popular entertainment (or the combination of the two, as the arts
centre movement was to again attempt in the twentieth century; see Chapter 4) and by
1907 the Palace became part of the University of London (the Queen's Hall forming
part of Queen Mary and Westfield College). Like the temperance society movement

that sought to develop alcohol-free venues for entertainment, these attempts to control and prescribe behaviour failed. In the 1880s for instance, coffee-taverns were set up with music and dancing licences (nine existed in London by 1892[5]) to appeal to the genteel and tempt others away from the bawdy music hall and gin palace. The National Sunday League also hired popular theatres on Sunday evenings and held concerts for free or very low cost. These 'dry' venues also failed 'because they tended to have a pious and unattractive atmosphere' (Weightman 1992: 97).

Although the People's Palaces failed to grip the imagination of the people themselves (as opposed to the moral improvers), another location for popular entertainment grew out of the industrial city and organised employees. The working men's clubs had also been the subject of temperance society and other gentrification movements, but they soon gained independence and served alcohol (the source of their self-reliance), and installed stages for music hall and other live acts. It would be hard to find their temperance roots, with the club's twentieth-century association with alcohol, blue-comedy, drag acts and a distinctly music hall entertainment atmosphere. As Weightman observes, however: 'If the temperance working men's clubs had become more like music halls, the music halls themselves had become more like theatres – tables and chairs were replaced by seating arranged as stalls, circle, gallery and so on (seeking) respectability as well as handsome profits' (1992: 99–100).

Dream palaces

The last cultural building type to emerge from the industrial era was the early cinema. Its development coincided with the heyday of the music hall and variety theatre, which did not immediately suffer from this new competition. In London the music hall continued to increase in capacity as cinema began its mass audience and building development (Table 3.3).

Table 3.3 Music hall and cinema capacity in London, 1891–1931

Year	Number of music hall seats	Number of cinema seats	Number of cinemas	Average number of seats
1891	115,000	–	–	–
1911	–	55,000	94	585
1931	142,000	344,000	258	1,333

By the 1930s weekly attendance at cinemas was estimated to have reached one-third of the population: 70 per cent of audiences being women and girls and, as Weightman posits, cinema did not take audiences away from music hall, but tapped a new audience. The last major variety hall to be built however was the Palladium in 1910 on the site of Hengler's Circus, two years later it hosted the first Royal Variety Command Performance (the music hall had received 'Royal approval' in 1901 when the King requested a selected repertoire to be performed in Sandringham in Norfolk). Ironically, this venue's annual royal performance is more associated with its televised version, and in the portentous words of Oswald Stoll the West End theatre magnate (Stoll Moss

Group) the writing was on the wall for the late music hall: 'The Cinderella of the Arts has gone to the ball' (quoted in Weightman 1992: 38). Silent movies had already been available in the 1890s, 'peep shows' (i.e. without projectors) based on the earlier lantern shows seen at fairs and pleasure gardens, and it was the showmen of the fairgrounds who first distributed these early 'films', and short movies were first shown, again ironically, in the music hall theatres. Small screenings, the 'Penny Gaffs', were held in shops or any space where chairs and a small screen could be mounted. The first custom-built cinema in London was the Biograph, Victoria in 1905 – built by an illustriously named American, George Washington Grant. New cinemas followed in the entertainment heartland of Piccadilly, e.g. the New Egyptian Hall (1907), the Electric Palace (1908), Hackney Pavilion (1913), the 1,189-seat Marble Arch Pavilion in 1914 (with tea-room attached, a far cry from the gin palace!), Stoll's Picture Theatre in 1915 – building on the site of the defunct London Opera House ('if you can't beat them, join them') – and in 1916 the first super-cinema on the site of the Pyke's Cambridge Circus (Weightman 1992). The pattern of new art(form) buildings on the site of old therefore continued. Again licensing raised its head, with the Cinematograph Act 1909 controlling cinema exhibition (under the guise of 'safety'), informal shows were held in derelict shops (Penny Gaffs, see above): 'the proprietor of the show simply disembowelled the shop, filled it with any old chairs, fitted up a screen at one end and a hissing projector at the other, and charged a penny for admission' (Willis 1948: 185). Local music halls were also converted to cinemas – The Balham Empire Music Hall (1900) by 1907 became the Balham Empire; The Palaseum in Whitechapel that opened as Fienman's Yiddish Theatre in March 1912 became a cinema within only a few weeks. In Birmingham, of six music halls, four had become cinemas by 1920. The chameleon nature of live performance venues is seen in the Islington Palace, which opened as a concert hall in 1860s, then became the home of the 'Mowhawk Minstrels', a music hall in 1902, and by 1908 it became the Blue Hall cinema. The Grand Empire Music Hall in Leicester Square built in 1882 was bought by MGM and closed down in 1925, reopening as the Empire cinema seating 3,000, equalling the Theatre Royal Drury Lane's capacity: 'Going to the cinema became far and away the most popular form of entertainment of the day. It was a social event, and the ambience of the place, the undreamed-of, centrally-heated luxury, was as much an attraction as the films' (Weightman 1992: 44). The larger cinemas could also accommodate live acts, and many old variety performers ended their working days on the stages of the new cinemas. Like the short-lived pleasure gardens in suburbia, local cinemas expanded through circuits or chains such the Odeons and Coronets, which surpassed the music hall boom in the number of new cinema buildings. A class divide was also evident, with districts supporting two or more, from the more luxurious and higher priced cinema to the basic 'fleapit'. A picture of local entertainment provision is given by Fred Hammond in the East End of the 1930s:

> I could walk to Poplar; there was a *Grand* there, a *Pavilion*, the *Hippodrome* and then right on top of me was the *Imperial* Cinema, the *Grand* Cinema and the Canning Town Cinema (the *Old Grand*). Then we could walk to Plaistow where there was a *Plaza*, the *Green Gate* Cinema, the *Bowlene*, the *Carlton*, the *Endeavour*, the East Ham Granada, the *Premier* all on top. So you see there were

eight or nine there and many more within a very short bus ride. I used to go to cinema perhaps on average four times a week, most people went two or three times.

(quoted in Weightman 1992: 129)

The heyday of popular cinema was of course short lived: after the War (in part due to a lack of supply of new films from the USA) attendance declined as it had exploded before, peaking in the late 1940s. In their detailed study of *English Life and Leisure* carried out in 1947/8, Rowntree and Lavers found that in the small Home Counties town of High Wycombe with a population of 41,000, four cinemas with a seating capacity of 4,300 attracted 24,000 people each week, with regular Saturday morning cinema clubs and even football clubs organised within leagues made up of theatres in adjacent towns (1951: 384). The influence of both supply (cinemas/exhibition) and demand in terms of the Anglo-American dominance (in production and language) had also created higher attendance rates, for example between Scandinavian countries and Britain where in the late 1940s cinema-going was twice of that in Denmark, over three times that of Norway and Sweden and twelve times that of Finland. Even in these countries, attendance habits in proportion to population revealed a higher rate in cities and towns of between five and eight times that in rural areas (ibid.). Between 1950 and 1959 cinema attendance decreased by 50 per cent in London and the South East – like the new cinematographic technology which eventually did for live music hall and variety productions, it was television that did the same for the movie. Music hall had in fact feared the impact of *radio*, which had by the 1930s a mass audience, but listeners were still curious to see their favourite stars live and weekly variety theatre attendance continued. Suburban theatres had not however survived and with the growing urban core, inner-urban and outer-/suburban growth of the city, the centre had already assumed the role as an entertainment zone, for 'going-up-West' (i.e. as a special occasion), for tourists and the social life of the well-to-do (Weightman 1992). The shifting habits and locales for arts and entertainment saw a general peak by the late 1940s, whether to cinema, football, theatre or greyhound racing, and cinemas also suffered from oversupply and, of course, the lack of any planning framework or needs assessment, since 'town planning' had no real concept of such 'leisure' activity either from an amenity or spatial perspective. The lot of those cinemas not demolished altogether was conversion to bingo halls as a basic local amenity, which did save some of these local buildings, albeit in a rundown and unloved state, others extended their lives as music venues and even hosts to new/immigrant religious groups without traditional places of worship (Plate 3.5). West End theatre however survived this technological innovation and socio-cultural change, both by going upmarket (hence their break from variety and music hall traditions) and serving a national and international audience. By the 1980s however cinema attendance was to go through a renaissance (as television viewing peaked then started to decline), not through a technological or cultural advance, but through the building of multiscreen cinemas, the *multiplex* which developed and was exported from the USA.

This decline in cinema attendance was experienced in Europe and the USA from the 1950s, albeit from widely differing attendance rates per head of population (highest in the UK and Italy, followed by Japan and the USA, and lowest in France and The Netherlands), before the recent upturn in the supply of screens. Wide variations in

Plate 3.5 Rainbow Theatre, redundant former Rank Cinema and rock venue, Finsbury Park, north London (2000)

cinema-going in the 1950s have however narrowed with less than five visits a year to a cinema per head of population in developed countries now the norm. Historically therefore the level of cinema-going is still a fraction of its peak as demonstrated for example in Britain in Tables 3.4 and 3.5 and this current resurgence (now peaking again) was therefore never on the scale of the mass audiences seen in the 1930s when little competition existed. Today commercial cinema combines screen choice and ancillary facilities and screen advertising, which together generate at least as much income as entrance tickets, and in consequence cinema attendance has doubled since the mid-1980s.

In contrast, The Netherlands, which had an increase of over 10 per cent in the number of cinemas over twenty years, has also seen attendances halve over this same period (Table 3.6).

The multiplex, at first an out-of-town phenomenon, then a town centre opportunity, has also raised particular planning issues and problems, as discussed in Chapters 4 and 5. Different countries sit on various points in this continuum of the rise and fall and resurgence of film-going and cinema exhibition, which also depends not only on distribution, but also on programming, substitution and cultural taste. In the case of New Zealand, for example, like Britain, post-War cinema attendances declined, but not until the early 1960s when television and then video first began to impact, but then a dramatic decline from a peak of 40 million in 1960 to 12 million in 1972 occurred. The recent upturn in cinema-going in this case has been attributed to pricing strategies, not the multiplex, and also to the growth and success of national cinema production and home-based films such as *The Piano* (Ministry of Cultural Affairs 1995: 100).

Table 3.4 Fall and rise of film attendance in Britain, 1933–99 (million admissions)

Year	1933	1940	1950	1960	1970	1980	1990	1995	1996	1997	1998	1999
Admissions	903	1,027	1,395	500	193	101	97	114	123	139	135	140

Sources: British Film Institute, LIRC (2000)

Table 3.5 UK cinema screens and multiplexes, 1985–1999

Year	1985	1989	1990	1991	1992	1993	1994	1995	1996	1997	1998	1999
Number of screens	1,284	1,550	1,673	1,777	1,845	1,890	1,969	2,003	2,166	2,383	2,564	2,758
Number of multiplexes	1	29	41	57	64	70	76	83	85	142	167	186
Multiplex screens	10	285	387	510	562	625	638	732	742	1,222		1,710
Percentage of total	< 1	18	23	29	31	34	36	37	34	51		62

Sources: *Screen Digest*, Febuary and September 1994, Evans *et al*. (1997), LIRC (2000), BFI (2000)

Table 3.6 Cinemas and attendance in The Netherlands, 1970–94

Year	1970	1975	1980	1985	1990	1991	1992	1993	1994
Number of visits (000s)	1,863	2,083	1,982	1,060	983	990	904	1,041	1,042
Number of cinemas		403							445

Source: SCP (1996: 367–77)

The Hollywood-multiplex entertainment duopoly is however showing signs of saturation and over-supply, with attendance peaking in Europe and even declining in the USA itself. Whilst the number of screens increased by 22 per cent to a total of 37,000 between 1997 and 2000, cinema attendance in the USA increased by only 3 per cent, with a 7 and 10 per cent decrease in annual revenue and attendance respectively in 2000 over the previous year. Their over reliance on a youth audience and lack of diversity has also meant that demographic change impacts disproportionately on cinema attendance, as the declining under-twenty-five population takes effect in Europe and North America. These mono-cultural facilities are also less flexible than arts centres or even traditional theatres and are harder to 're-purpose' in retail complexes. Responses by cinema exhibitors have been price increases, which risk further depressing audience numbers, and disposal of these now-surplus venues, with the effect of blighting the areas where they have often been the main or anchor development in mixed-use and leisure developments.

Conclusion

As this chapter has examined, what the early industrial city and emerging secular, urban society exhibited in cultural provision, was the rise of entrepreneurial entertainment on the one hand, and the development of social welfare cultural amenity on the other, with a hierarchy of arts facility maintaining its city centre and 'cultural capital' role. State intervention through licensing and other controls has been an irresistible feature of urban society, whilst the popular entertainment which had begun, once free of court and private house, with a more catholic taste and mix of users, gradually divided, for both commercial and social control reasons, underpinned by mercantilism, the new bourgeoisie and the rise of the middle classes. The mutual reliance between culture and commerce that supported crafts and artisans also suffered from industrial and mechanical (re)production (photography being one of the first to compete with artistic creation) and import trade. According to Crimp, Malraux perhaps fatally admitted photography within his *Museum without Walls* (1978): 'But once photography itself enters, an object among others, heterogeneity is re-established at the heart of the museum; its pretentions of knowledge are doomed. Even photography cannot hypostatize style from a photograph' (1985: 51). This mechanisation and devaluation of traditional crafts and artefacts of course stimulated the arts and crafts movement itself: 'whose anti-industrialist, implicitly anti-capitalist roots can be traced through William Morris's designing firm of 1860 and the pre-Raphaelite painters of the 1850s' (Hobsbawm 1977: 332). This relationship was also weakened as home life took over from collective consumption, epitomised in the Victorian era when home-based entertainment and hobbies became established (and home comforts gained appeal over the saloon theatre and gin palace), and as wartime austerity further dampened cultural consumption outside of the cinema, before the next home-based distraction took over – television.

A recurrent pattern in the location and development of buildings and sites for public culture has been the reuse of sites and buildings for subsequent cultural and amenity use, as Borsay confirms: 'During the early phase of its development, polite urban culture had to make do with facilities that already existed rather than enjoying the benefits of

Plate 3.6 Old Hampstead Town (Vestry) Hall (1878, grade II* listed), former borough planning office, converted InterChange Studios arts centre, north London (2000)

purpose-built accommodation' (1989: 144). Whether due to sentimental or sound marketing reasons (maintaining the goodwill and memory of previous operation), or the prosaic locational advantage from existing sites, transport routes, interchanges and axes, it has been preferred in many cases to site arts buildings where such activity previously took place, whether in the open fair or pleasure garden, circus or arena, or in the case of the early Exhibitions which begat prime monuments and arts and recreation complexes. Even new-build cultural facilities such as libraries or assembly rooms (e.g. for dancing) were often added to existing civic buildings and this is not necessarily an imposed, bureaucratic convenience. For example in the Australian city of Wagga Wagga (population of 60,000), New South Wales, a new civic centre incorporating arts gallery and library was the subject of both an architectural competition and community consultation exercise. This encouraged the council to build the centre on the site where the historic Council Chambers stood, rather than elsewhere in the city (Guppy 1997: 46). Former town halls now serve as homes to many arts centres – from Battersea to Hampstead in London (Plate 3.6) as seen in Chapter 4, indirectly ensuring a rare vestige of municipal culture and heritage in the local landscape.

Planning for new cultural facilities in new places was not of course a social planning consideration until town planning proper had been established and responses to urbanisation set in motion – manifested in the Garden Cities, suburbs, satellites and New Towns (and in socialist city planning regimes). Post-colonial development also looked to updated versions of Renaissance urban planning, adapted to already hybrid multicultural populations such as in South America and in resurgent national movements in countries such as Finland, and in Cuba where for instance:

> The *Plano del Proyecto de la Habana* mostly designed between 1925 and 1926 established a framework of extraordinary magnitude in twentieth-century urban

history and contrasts sharply with Havana's appearance since the late nineteenth century. Since the beginning the plan embraced all aspects of urbanism, from a regional scope to detailed design of public furniture.

(Lejeune 1996: 165)

In Britain, the first comprehensive national legislation for the development and control of land-use (Town & Country Planning Act 1947) succeeded by only one year the foundation of the Arts Council of Great Britain, which was emulated in British Commonwealth countries and paralleled elsewhere in Western Europe, notably France and Sweden. The beginnings of modern town planning and arts policy, and the development and distribution of arts amenities through the civic venue and arts centre are therefore examined in depth in Chapter 4. This is then followed by a detailed critique of the emerging planning 'norms' and standards of amenity and recreation provision, and the atypical position and treatment of the arts and culture within these approaches to urban and new town facility planning.

Notes

1 The role of the aristocracy was of far greater importance to the dissemination and distribution of the arts than the Crown where performance and the fine arts were concentrated in the 'court'. The English aristocracy 'had not one centre but hundreds, scattered all over the country in "gentlemen's seats" and provincial towns, each of them a focus of learning and taste' (Trevelyan 1967: 414).
2 The term 'music hall' was rarely used, 'palace of varieties' never. People referred to the hall by name. They went to the Tiv (Tivoli) or Mo (old Mogul Tavern – to become the Middlesex on Drury Lane) and did not generalise the whole as 'music hall' (Willis 1948, Weightman 1992).
3 The LCC's puritanical image in this period has been exaggerated and developed from real concerns in health, education and worker protection, which included 'cleaning-up' various 'acts' – winding up of the Metropolitan Board of Works (also known as the 'Board of Perks' for its corruption); ridding parks and open spaces of vagrants, antisocial behaviour and gambling, as well as legislation on public health, child and worker protection (e.g. the Shop Hours Act 1892). Indeed, the moves towards licensing and control of leisure and recreation did not go far enough for the Nonconformist 'lower middle classes' whose growth had put the Progressives out of power and elected the first Conservative group to the LCC in 1907. The licensing of places of public entertainment was one of the vital issues for the 1889 LCC election and where 'creating a civic culture, at once humane for the deserving and punitive for the corrupt or dissolute, still seemed a worthy endeavour to many in the metropolis' (Pennybacker 1989: 148).
4 Lottery proceeds also paid for the purchase of the original collections of the British Museum (Wilson 1989, Evans 1995a: 225).
5 One of the largest, the Royal Victoria Hall became the Old Vic, Waterloo (South Bank) after which Lilian Baylis took over in 1912, showing between 1914 and 1923 all of Shakespeare's plays. The theatre later housed the first National Theatre company in 1963, under Laurence Olivier before its move to new premises adjoining the South Bank Arts complex in the 1970s. Bought by the Canadian Mirvish family in the 1980s, who had become famous impresarios in Toronto, it was sold in the 1990s finding neither commercial nor subsidised programming success, in part due to its location, isolated from other venues or entertainment centres and facilities on the 'wrong' side of the river.

4 Amenity planning and the arts centre

Introduction

The conditions that led to the recognition of culture as an aspect of amenity and social welfare provision, and the growth of public participation in national and local cultural activities, can be linked fundamentally to the growth of urban and city populations – in density and industrial conurbations. This was (and is still today) also concurrent with the need for reinforcement of national identity and culture, which rises and falls in scope and intensity as either are threatened, whether from without, e.g. war, economic competition, new technology, or from within, e.g. political, social change movements and creative milieu. Whilst *force majeure* incidents, notably great fires and natural disasters, had provided the opportunity for the rebuilding, planning and cultural renewal of major city areas, the destruction caused by modern warfare combined with the need for reconstruction of both the social and economic fabric, and also opened up the foundations for late urbanised society and therefore systematic town planning and consideration of the arts as an element in social welfare provision. As Rasmussen pondered: 'I often wonder if there would have been any progress in London planning if there had not been a war. . . . The war period became the third phase of the great comprehensive plans for the entire London Region' (1937/1982: 427) and for example in Germany, ten years after the War, one hundred theatres were built or reconstructed. The effect of public and private transport technology and provision should also not be underestimated in opening up recreational opportunities and cultural consumption beyond largely pedestrian limitations, as train, bus, tram and the motor car extended the travel horizon to a widening social group between the 1830s and 1930s.

Urban(e) and rural

Urbanisation, which threatened the removal of what was perceived as the essential rural character, had however been resisted and demonised by seventeenth- and eighteenth-century Puritans in Britain and America and also by Rousseau in France, manifested for instance in theatre regulation and control of popular entertainment throughout this period, as discussed in Chapter 3. The destruction of the values of 'traditional' English society was of course placed at the door of the modern nineteenth-century city, against which A. W. N. Pugin, John Ruskin and William Morris reacted, with a harking back to the aesthetic and vernacular harmony of the countryside and rural life. As Dr John Fell had remarked over two centuries before in 1680: 'I will tell you why my Lady Hatton is very happy. She is removed from the infectious conversation of the Town, where the precious time and estate designed for the purposes of charity is to be wasted on impertinent and uncharitable visits' (quoted in Wainwright 1993: 1). The great cities of the earth were in Ruskin's view 'Loathsome centres of fornication and covetousness' (1880, quoted in Hall 1996) and as Froude observed in *Oceana* in 1886, 'The tendency of people in the later stages of civilization to gather into towns is an old story. Horace had seen in Rome what we are now witnessing in England – the fields deserted. The people crowding into cities. He noted the growing degeneracy. He foretold the inevitable consequences' (quoted in Briggs 1990: 59). From another perspective, Elizabeth Wilson maintains that urban life in the 1800s was projected as undesirable arguing that: 'C19th planning reports, government papers and journalism created an interpretation of urban experience as a new version of Hell' (1991: 108). Wilson's feminist critique of the town planning movement as 'an organised campaign to exclude women and children, along with other disruptive elements, the working class, poor, and minorities – from this infernal urban space altogether' (ibid.) targets the planners' hegemony, including Abercrombie's ambitious plans for post-Second World War London. In Wilson's opinion, 'there was a whiff of authoritarianism about his solution' (1991: 14). However the position of women in the city is also seen to present both opportunities, as well as threats, as 'a place of freedom and opportunity economically and socially, but as a potentially dangerous place sexually' (Greed 1994: 102). Women were also largely absent and excluded from the opportunities provided by the earlier rational recreation movement (and coffee-houses), with its focus on (male) sport, outdoor pursuits and clubs.

The view of the city as a place of moral depravity, chaos and disorder, threatening the natural order of the countryside, applies distinctly Old Testament language and sentiments, whilst the panacea of the 'rural idyll' and 'natural order' belies human intervention in the countryside itself, from intensive agriculture, control of landed estates, landscaping and the feudal systems from which the city promised the only escape – one fundamental reason for their continued expansion as the economic mix with social refugees from the derogatory, 'suffocating' small town, and from provincial and village life. Raymond Williams in *Culture and Society* traces this intellectual, holistic tradition, 'which interrelated aesthetic, moral and social judgements' (1958: 137). Increasingly, however, urban life was contrasted with rural life, which was perceived as 'uncivilised', whilst city life was perceived as 'urbane' and 'cultured'. Meller maintains that 'what happened in each large city . . . was part of the national response to the challenge of civilisation' (1976: 7) and Geddes also: 'The central and significant fact about the city is that it functions as the specialised organ of social transmission' (quoted in Mumford 1940: 198), whilst Mumford himself likened big cities to museums where 'every variety of human function, every experiment in human association, every technological process, every mode of architecture and planning can be found within its crowded area' (1961: 640). The urban 'condition' and notions of civilisation and citizenship therefore present a balancing act that has existed since urbanisation (*urban* – 'other than rural') began. As Cheshire was later to suggest, major cities were also predicted to serve a role as cultural centres, and less as the location for industrial production, and 'to be much closer to those they had before the industrial revolution –

as commercial and administrative centres, as cultural centres in the broadest sense of cultural, and as providers of higher level services and urban amenities' (LPAC 1991: 7).

From the 1840s, the institutionalised rational recreation movements had also sought to meet the need for both social reform and control and the relief of poor health and 'poverty' – including moral and educational – and to underpin the notion of civilised nationhood (Bailey 1987, Yeo and Yeo 1981). The Chartist movement and early trades unions were concerned for the recreation of working-class groups and what they saw as the risk from excess capitalism and 'shiny barbarism' (Haywood *et al.* 1989). These responses were one of the foundations of Foucault's 'governmentality' thesis – power and control over the threat from urbanisation (1994: 62). The enlightened paternalism of government also provided a response in the form of legislation for public museums, libraries and recreation facilities as already discussed, however the industrial age also witnessed the first phases of industrial capitalism in leisure and transport that served these emerging mass markets, such as music halls (Bailey 1986, Weightman 1992) and the railways. The advent of the railways enabled early tour operators such as Thomas Cook in Britain and Fred Harvey in the South West USA to exploit the demand for cultural tourism and serve the previously exclusive resorts, which became accessible to the working classes in the industrial conurbations. This distinctly urban recreational goal allied to national culture, productivity and pride has, however, been perennially hampered by the tensions between urban society seen as an essentially dehumanising and amoral state in contrast to the rural idyll and aspirations of what came to be the Arts & Crafts and Garden City movements. The reality was that in major capital and regional cities urbanisation was here to stay, and indeed provided the only real sophistication in cultural consumption and taste – creating a real urban and rural spatial and spiritual divide. It would be wrong however to ignore the resurgence of localism in the mid-nineteenth century, as Harris maintains: 'much of the cultural and intellectual life in early Victorian Britain flourished, not in the metropolis, but in the provinces and Scotland' – the source of 'patrons of high and low culture, popular media . . . and social laboratories of social reform' (1994: 18). Much of the legislation that gave local and parish government powers to finance public libraries, art galleries and educational institutions was privately sponsored, however as Harris goes on to point out: 'Yet, all this dynamism and variety notwithstanding, provincial communities in the 1900s were less overwhelmingly dominant in society and culture than they had been a generation before, and the late Victorian period saw a subterranean shift in the balance of social life away from the locality to the metropolis and the nation' (ibid.: 19).

Agenda 21

Today, as in the 1840s and 1940s, this meta-view of cities as either the causes of, or solutions to, the problems of urban environmental and social disaster features highly in late twentieth-century sustainable development agendas, articulated through Rio[1] (i.e. Agenda 21), and subsequent global environmental summits, such as Habitat II (Istanbul 1996), which 'can be viewed as an attempt to extend an ethically, socially and culturally, reformed modern project into the future' (Knutsson 1998: 30). Writing on a review of city and urban parks, this anti-urban sentiment resurfaced: 'I believe the cultural role assigned to cities is greatly exaggerated . . . modern cities are environmental disaster zones' (Nicholson Lord 1994). However, the city-as-solution model is also coming from an unlikely source, the Green Movement: 'For inside the problem of cities lies the solution. The city – always the place of greatest dynamism and creativity – may also present the greatest opportunity for a greener future' (Baird 1999: 8). Land-use, distribution and workable habitats, it is argued, benefit from planning – a statement of the obvious but one that reflects the weakness of development and facility planning against less-consensual realities. As Ward and Dubos claim: 'There is no single policy that deals more adequately with full resource use, an abatement of pollution, and even the search for more labour-intensive activities than a *planned and purposive strategy* for human settlements' (1972: 180 emphases added). As well as the core environmental and physical impacts considered by Agenda 21 such as principles of sustainability with regard to climate, biodiversity and forests, the notion of environmental rights and objectives was also enshrined by over 150 signatory countries in a section of Agenda 21 relating to social and economic dimensions, which also focuses on the strengthening of local economies, changing consumption patterns (locally and globally) and also on strengthening the role of local communities in their environment and *provision of amenities*. Indeed Chapter 28 of Agenda 21, which calls for local authorities to develop a Local Agenda 21 (LA21) plan, requires both consultation and consensus involving all sectors of the community and the setting up of mechanisms for community involvement: 'LA21 is the new agenda of sustainable development. It is succeeding because I suspect, it embodies many things that people, individual people, believe deep down in themselves. It is a planned, democratic process involving the whole community. It is about improving the quality of life for everyone – but within constraints set by the natural environment' (Prince of Wales, quoted in Harman *et al.* 1996: 41). With a target for LA21 plans to be in place by 1996, however, this voluntary policy initiative and example of 'glocalisation' has not had the impact or degree of community involvement for which advocates had hoped (Leslie and Muir 1996), with the USA as one of world's largest consumers and polluters, in particular resisting its implementation (and the Bush presidency reneging on the Kyoto treaty). 'Culture' was also absent from this environmental agenda, as it has been from definitions of and planning legislation for 'amenity', below (even Ruskin had made the point that it was futile to have 'Art' until there was clean air and water), and again the professional and environmental bias has limited the democratisation of this global venture (Bohrer and Evans 2000; Worpole *et al.* 2000).

Town planning and amenity

Town planning, in attempting to assimilate diverse pressures and interests ranging from utopian reform to design and practical administration, has had a complex history (Foley 1973). Firstly, urban population and density varies, notwithstanding global convergence towards a broadly *urban* state, in Britain for instance considerably more so than the USA and most other West European countries due to the duration and depth of its industrial revolution, subsequent congestion and limited land availability. This is also the result of precise choices in planning policies and to cultural preferences (e.g. low-storey houses with gardens versus apartment living). Distribution as well as density are also factors in planning that differ between Europe and North America, where 'American people live in several megapolises sprawling over once productive land and requiring enormous expenditures for utilities and commuting. The effort to take advantage of the city while still having a bit of land of one's own, the ideal of suburbia, is losing its charm' (Daly and Cobb 1989: 264). This contrasts with Continental Europe with the drift to towns and cities at the expense of rural, agricultural areas, and the magnet of mega-cities in developing countries in Central and South America and South East Asia, which have outstripped the population size and land area of the first-generation world cities of London, New York and Paris (Friedmann 1986, King 1990).

The extent to which theoretical foundations have influenced town and environmental planning and its professional and practical implementation suggests that the view of amenity and the reluctance to plan for the arts relates, at least in part, to the theory underlying town planning and its formation for instance in Britain, as distinct from say American city planning or Continental regional planning where, for example, in France the *aménagement du territoire* literally means 'management of the territory'. In London, however, Bell observes that 'After the mid-fourteenth century, urban planning virtually collapsed, and for the next three centuries remained a dormant force . . . with the exception of a small portion of Stuart London, there was little planned extension of existing settlements' (1972: 68). Even by the seventeenth century, the situation had not much improved: 'While Renaissance Europe forged ahead with sophisticated urban schemes, England remained rooted in the dark ages of planning' (Borsay 1989: 87). By the nineteenth century, other countries had developed formalised planning systems, including city and regional planning laws in Spanish Latin American colonies from the sixteenth century onwards,[2] beginning with Puebla, Mexico in the 1530s with planning statutes closely resembling recent British town planning (e.g. the Town & Country Planning Act (TCPA) 1947) and in new town developments at La Plata, Argentina from the late 1800s, as well as Haussmann's mid-nineteenth-century Paris. Other cross-cultural influences on British planning were those of Prussian land policy and urban design and the later North American influence in the field of environmental impact assessment, outdoor recreation and national parks (Rydin 1993). The consideration of arts and culture as an aspect of 'amenity' had however been universally absent from town planning. From this it might be concluded that the higher support for and legislative protection of the arts and culture by some European countries owes something to their historical approach to land-use and strategic planning and to the position and role of the planner: 'Although the [British] Arts Council have developed some guidelines . . . there is little to compare with the French, Dutch or Scandinavian policies of seeking spatial equity in arts provision' (Burtenshaw *et al.* 1991: 180).

The ideology of planning also provides a philosophical basis for the activity itself, as outlined in Chapter 1, it indicates the main goals and approaches and provides a basic operational rationale. In so far as town planning is a governmental function, its ideological base provides a broad means for winning over and maintaining the allegiance of politicians, officers and the community, i.e. *consensus*, a particular feature of British public policy and planning, including arts policy from Britain's near revolutionary political changes of the 1830s – until the 1980s largely apolitical and with a low profile in governmental and ministerial terms (Hewison 1995, Pick 1980, 1988). The British tradition has also relied on greater trust given to public officials, both elected and appointed, in the protection of the 'public interest', than is the case in America. As Glass states: 'British land-use planning has *a prioristic* and utopian origins. They are the idea of nineteenth century reformers. . . . Since then, society has become more complex and the prospect of social change far more ambiguous, and yet the old ideas have been maintained, have become fixed prejudices' (1973: 55). It is to these ideological 'super-egos' of planning thought that is attributed the perceived anti-urban bias in British town planning. This historical and utopian position provides a clue to the limitations of amenity in urban planning: 'Amenity is one of the key concepts in British town planning, yet nowhere in the legislation is it *defined* (Cullingworth 1979: 157 emphases added). Paradoxically, whilst not defined, 'amenity' is also claimed to be one of the most reliable concepts in British town planning! Amenity in this context has been defined as 'a quality of pleasantness in the physical environment [which] ranges from an essentially negative restriction against nuisances to a notion of visual delight'[3] and as Foley observed: 'one sometimes gets the feeling that the British have quite self-consciously sought to protect themselves against the pragmatic inventiveness of their own designs' (1973: 81).

The distinctive influence of the Garden City movement also evolved into a broader, decentralising new town movement, predating town planning proper. Its influence on the Greater London Development Plan (1969), for instance, was profound. Indeed the Garden City Association formed in 1899 became the Town and Country Planning Association, which was formed in 1914. Town planning had come to be distinguished from regional planning and also from countryside (rural) planning: there had been virtually no recognition until recently of metropolitan or conurban planning. The preoccupation with the separation of town and country, of the new town and decentralisation solutions and consequently with the green belt and recreational objectives has clearly influenced and limited the positive approaches to urban planning seen elsewhere in Europe and North America (Burtenshaw *et al.* 1991). Despite town planning's normative role as an extension of social policy, this foundation and formation in the post-War reconstruction and settlement period has limited the development and scope of amenity and its extension to the cultural sphere, particularly in the urban environment. This has persisted in the subsequent periods of urban development: the periods of technocratic planning epitomised in the high-rise, high-density and new road building of the 1950s and 1960s, and its counter-movement, the 'flight from modernism' and the city (ibid.: 37–41). The contemporary association of economic development and 'boosterism' with physical planning at the local and city level, although linked from the late 1970s through urban and regional economic policy initiatives, has not generally been reflected in central government economic and employment policy for which responsibility lay with separate departments. As Glass observes: 'the various aspects of planning were separated, Economic Planning was split up into various branches, and physical planning was set apart. . . . While one Ministry [dealt] with a major aspect of economic planning – location of industry, another is entrusted with town planning – no longer including the word planning in its title' (1973: 51).

In the UK, the planning of towns and cities as a professional discipline and statutory function, and one which therefore dictates land-use designation and the control of building, is a largely twentieth-century development, notwithstanding the Crown and major landowners who exercised a similar role through land and property ownership and leasehold terms over the previous two centuries. Whilst the historical perspective on state involvement in arts provision and town planning and its place in urban cultural society is of relevance, particularly in view of the inheritance and notion of civic provision and 'public good' noted previously, I will concentrate on the period from which town planning and arts policy was formalised, in Britain manifested in the TCPA 1947[4] concurrent with the foundation of the Arts Council of Great Britain in 1946. In the formulation of a national arts policy, the ACGB's founding father, Maynard Keynes, argued: 'How satisfactory it would be if different parts of the country would walk their several ways as they once did and learn to develop something different from their neighbours and characteristic of themselves. Nothing can be more damaging than the excessive prestige of metropolitan standards and fashions' (1945, quoted in Pick 1991: 108). Preliminary steps in the pre-Welfare State period (*c.*1890–1939) had also established land-use planning and recreational uses, notably the TCPA 1909, 'which itself marked a significant stage in the state's willingness to intervene in spatial development' (Travis and Veal, in Henry 1993: 13). This was followed by the Physical Recreation Training Act 1937 and the Green Belt Act 1938, each of which had specific recreational objectives. In London, the most important planning milestone was represented by the County of London Plan and Greater London Development Plan (Abercrombie and Forshaw 1943, Abercrombie 1944), following a long period of *ad hoc* and *laissez-faire* development. Precursors to the 1947 Act in 1909 and 1919 sought to establish the ground rules for town planning, in terms of the concern to improve social conditions and housing, and as a response to urbanisation and overcrowding, which was directed at the new town and urban fringe policies, not least in Abercrombie's decentralising Greater London Plan (1944). The statutory requirement to produce development plans, defining future land-use and to control new development in the light of the approved plan, was to be a key feature of the 1947 Act. This also gave planning authorities powers to deal with specific problems of amenity including the preservation of trees and woodlands and of buildings of special historic or architectural interest. However, arts and cultural provision were not considered alongside other amenity considerations, such as open space and recreational land. According to the Arts Council, in one of its periodic but short-lived strategic reviews, urban amenity planning and city cultural life had thus been separated:

> In the nineteenth and twentieth centuries, British urban planning and the arts parted company. The earlier view of the city as a work of art, a planned series of aesthetic experiences, was lost. The city came to be conceived as a functional unit, with the emphasis more on efficiency and economic prosperity than on quality of life or the cultural aspirations of its citizens.
>
> (1993a: 110)

This link between the arts and the urban renaissance had however been recognised and advocated earlier (see below) – an attempt by the national arts agency to reclaim both legitimacy and protection for the subsidised arts at a time of hardening free-market policies and a commercially driven development and planning regime:

> There is little awareness nationally of the important role which the arts are playing in revitalising depressed urban areas. The Arts Council has launched the 'Urban Renaissance' project to inform those involved in redevelopment – policy-makers, property developers and inner city agencies – on the ways in which the arts can stimulate economic and social regeneration.
>
> (Rees-Mogg, quoted in Arts Council 1986a: 1)

State arts agencies – policy and plans

Political systems and administrative structures – notably the point they occupy on the continuum between central control and national priority on the one extreme, and on the other, local democracy and control of public resources (e.g. land, tax revenues) – are likely to directly influence national attitudes towards the arts and public provision, and related land-use planning and public investment. Comparative studies of arts policy have sought to analyse state rationales for intervention; the mechanisms for carrying out arts policy, and irresistibly, contrast the levels of public funding of individual art forms between countries and the overall level of arts support in terms of their Gross Domestic Product (GDP) (Schuster 1995, Zimmer and Toepler 1996, Feist *et al.* 1998). Such comparisons are the stuff of calls for increased funding and cultural provision (as resourcing 'norms'), however they suffer from the pitfall of the cross-cultural study (Schuster 1996, Aitchison 1993). In this case, variations in definition for example of 'arts', 'culture', 'heritage'; departmental responsibilities and the degree of policy integration ('joined up government'); and the historic evolution of cultural development, facilities and participation as discussed earlier, together weaken any quantitative comparison and produce reductive conclusions. This includes the strong regional basis of cultural provision in the German *länder*, Italian regional and structural planning, compared with French cultural policy (Looseley 1997) which remains largely centralised despite much investment in decentralisation since 1993, and the British London-centrism. These differences undermine simple comparisons of funding distribution and provision levels, and ignore differing historic, urban development, arts participation and production activity (e.g. cultural preferences, strengths, habits, etc.). North American and Japanese comparisons, further stretch the *like-for-like* test, particularly regarding the extent of private and corporate patronage versus public provision and subsidy (Stewart 1987, Hillmand-Chartrand and McCaughey 1989, Schuster 1995). In Britain *London-centrism* was evident even at the time arts policy purported to be pursuing a re-distributive agenda (see above). As Arts Council Chairmen Lord Keynes and later Goodman successively made clear, the Council's business was 'to make London a great artistic metropolis, a place to visit and wonder at' (Keynes 1945 quoted in Pick 1991: 108), while though 'Goodman's Arts Council [had] two arms . . . the jewels, its power houses, centres of excellence, benchmark of quality, the other groped for local initiatives, fumbled for distinctive and different "somethings" in each locality' (ibid.: 49).

However, one important outcome from a more regionally based political system with greater devolution and regional independence from the administrative centre-capital, has been the tendency towards higher levels of cultural facility provision, more widely distributed (i.e. regional cities), than in more centrist states such as Britain, France and Greece, where the capital dominates in higher level provision, such as opera houses, theatres and cultural production. In Federal Germany, unlike London, Paris and New York where professional theatre is predominantly concentrated, seventy-six communities maintained a public theatre in 1970 – twenty-one state, 102 municipal theatres (with 88,000 seats) in addition to twenty-seven private theatres, forty travelling and eighteen small theatre companies. This enthusiasm for the stage has seventeenth-century roots (see Chapter 3) and this regional strength in cultural facilities was underpinned by the Basic Law (*Grundgesetz*) of 1949, which made cultural affairs the sole responsibility of the separate states (*länder*), and subsequently a *Permanent Conference of Cultural Ministers* was set up as coordinating body. With the adoption of the Treaty of European Union in Maastricht 1993, the Federal State acquired, albeit very limited, direct powers in relation to culture. Participation in the lyric arts is higher in Germany than in other European countries, not only in part due to the level of supply and proximity to populations, but also through a system of attendance rings (*Besucherring*). These are local communal organisations acting as season-ticket holders such as the *Popular Stage* (*Volksbuhne*) with 450,000 members in over one hundred local branches. Each member attended an average of ten performances annually, representing over 20 per cent of all theatre audiences in Germany.

However, in the absence of such regional cultural promotion, subsidy and facility development, even where urban city population distribution *is* more evenly spread such as in The Netherlands, cultural provision and consumption is still skewed towards the cultural capital, Amsterdam, and the three other major cities (Table 4.1). Here these cities attract considerably above-average and higher usage rates than in smaller towns and municipalities.

Table 4.1 Share of visits (%) in people's own place of residence (twelve years of age and over) in The Netherlands, 1995

	Dance halls/ discos	Cinemas	Museums/ exhibitions	Theatre productions
Four major cities†	82	91	70	77
Other towns‡	66	77	40	79
Smaller municipalities	33	27	9	29
All Netherlands	43	51	25	46

Source: adapted from SCP (1996: 377)

†Amsterdam, The Hague, Rotterdam, Utrecht
‡less than 100,000 inhabitants

The cultural and political renaissance seen in European and other city–regions has been one feature of the late twentieth century, both as a response to this imbalance, this centrism, and also to the effects of globalisation which further threaten to accelerate the decline in regional and ethnic identity through cultural convergence and acculturation. Furthermore, within this regional movement, the tension between cultural capital – fulfilling its role as cosmopolitan and international city – and the regional governments' notion of 'identity' often involves a rewriting of history and a mono-cultural image, irrespective of diversity, plural artistic expression and aspirations and the desire for cultural interaction and universality. This in one sense reflects the oppositional notions of what are national and local cultures (Williams 1961) which are bound up with the perspectives of real and imagined communities (Anderson 1991), the country and the city (Williams 1975), and what may be considered as the 'ideology of the small', for instance as portrayed by the separatist *Lega Nord* (Northern League) in Italy (Albertazzi 1999).

Cultural policy and planning: a case of Britain and France

Two cases of national cultural policy that directly influenced arts planning and facility development, emerging from the Second World War reconstruction, were France (and also under the French Front Populaire Government of the 1930s) and Britain, and attention to their respective formation is therefore paid. These benefit from in-depth review and analysis over the last twenty-five years, most recently in Looseley (1997), Wangermée (1991), Hewison (1995), Pick (1991), Marwick (1991) and also in a comparative European context (Ellmeier and Rasky 1998).

In addition to the nineteenth century enabling legislation for the establishment and support of a range of public amenities and national cultural institutions (Chapter 3), the British Arts Council's formation had been preceded by several other national cultural institutions in the pre-War period – in 1933 the British Film Institute (BFI) was formed to protect the national film industry against the dominance of Hollywood and in 1934 the British Council was created initially as a response to the propaganda machines of Italy and Germany (Hewison 1995). The Standing Commission on Museums and Galleries was established by Treasury Minute in 1931 following a Royal Commission on museums, and the British Broadcasting Corporation (BBC) was incorporated by Royal Charter in 1927 following five years of broadcast monopoly.

Notwithstanding earlier piecemeal public involvement and civic building for culture, largely undertaken by local authorities and often involving public subscription or private patronage, in Britain the creation of the first Arts Council (ACGB) out of the wartime 'Council for the Encouragement of Music and the Arts' (CEMA) in 1946 marks the beginning of a formalised if 'arms length' central government promotion of certain arts activities and forms, and the development of arts policies, which have come to influence the development of new and existing arts facilities. As Keynes promised at the Arts Council's inception in 1945 the vision was 'to decentralise and disperse the dramatic and musical and artistic life of the country' (Pick 1991: 108). This model of a state 'arms length' agency was also adapted in Canada and Australia to fund and promote national and professional arts practice, and other social democratic states such as the Swedish National Board for Cultural Affairs and the Finnish Culture Foundation in 1937 had created similar models predating the UK Arts Council. The major premise on which public subsidy was based can also be traced to the 'right' of access to culture, as specified in the 1948 Universal Declaration of Human Rights: 'Everyone has the right freely to participate in the cultural life of the community, to enjoy the arts . . .' (quoted in Shaw, Arts Council 1983: 7) and the national Arts Council therefore formed part of this post-War settlement: 'so accepting for the first time the contemporary and performing arts alongside museums and art galleries [government funded since the eighteenth century] as a permanent national responsibility' (Hewison 1995: 29).

However, as with the response to demands for social need and change a century earlier, it was private and voluntary action that both preceded and prompted formal state involvement in arts policy. In the 1930s' depression the Pilgrim Trust charity, endowed by the American Harkness Foundation in 1930, had supported the touring of art exhibitions and the appointment of music and drama organisers and local museum education officers and loan services to areas of particular deprivation. With outbreak of the War in 1939 and the curtailment of most professional and amateur artistic activity, the Board of Education wished 'to show publicly and unmistakably that the Government cares about the cultural life of the country' (Leventhal 1990: 293). Through Lord Macmillan, who was both the government Minister for Information and Chairman of the Pilgrim Trust, a pump-priming grant was given by the Trust to a newly formed Council for the 'Encouragement of Music and the Arts' (CEMA – see below), which was to continue the wartime touring of the arts. Subsequently matched by funding from the Treasury, CEMA was the institutional and cultural model to be used in 1945 as the basis for the Arts Council of Great Britain. The alternative model rejected by this choice was the 'Entertainment National Services Association' (ENSA), which had been formed in 1938 to entertain the troops in anticipation of the outbreak of war, and which was staffed and organised by the commercial entertainment industry. (In Pick's view this was an unfair move since 'CEMA . . . was over-praised by about the same proportion as the more earthy activities of ENSA have been undervalued'; 1991: 23.) By turning their backs on ENSA in the formation of the Arts Council, the government effectively prescribed and separated 'high-art' from popular culture. Thereafter public subsidy was almost exclusively directed at the former, leaving the promotion and development of popular and amateur arts to the commercial, independent and voluntary sectors, from West End theatres, cinemas, to publishing, pop music, and the amateur and folk arts. As a later Secretary-General of the Arts Council observed: 'almost from the beginnings, an ideological conflict underpinned the theory and practice of public funding of the arts. Serious efforts were made to encourage a holistic approach to cultural policy – but gradually the interest of the public as audience, reader or spectator overtook that of the public as doer, maker or participant' (Everitt 1992: 6).

In France, most cultural affairs were grouped under a single ministry in 1959 under Secretary of State André Malraux (Deputy Minister-level), after an earlier move by the short-lived Front Populaire Government in 1936 ('Ministry of Leisure' under Leo Lagrange), and although not called the 'Minister of Culture' at the time, Malraux decreed 'to make the major works of humanity, starting with those of France accessible to the greatest number, to provide the widest possible audience for the French cultural heritage and to encourage the creation of works of art and of the mind' (Wangermée 1991: 7). In fact, several key areas of the cultural services remained separate, as in the UK, with education, universities and science, communication and broadcasting, *Aménagement Territoriale*, defence and foreign affairs ministries all having substantial cultural roles and together accounting for more than half of all central government cultural expenditure, a situation unchanged today. Malraux's vision was one of cultural diffusion (today 'access') and the notion of cultural democratisation to be achieved through *action culturelle* or *development culturel*, with the state intervening to stimulate both artistic expression and wider public participation in the arts and heritage (Cook 1993). In 1963 the Regional Committees for Cultural Affairs (CRACs) were established, with their programme including a ten-year plan for musical development in providing or supporting music and dance conservatoires, orchestras and opera houses, in order to offer and popularise music in provincial cities (UNESCO 1970). A key plank of Malraux's distributive policies was the promotion of *Maisons de la Culture* as *polyvalent* centres for arts activities at the community level. This was very similar to Jennie Lee's *Policy for the Arts* in Britain a year later (1965), which set up the *Housing the Arts* fund as a key part of its own strategy (see below). France also introduced a new law in

1964 that provided subsidy for private developers agreeing to include studios in new buildings and to let them to artists at a moderate rent (UNESCO 1970), a forerunner to 'percent for art' and public art agreements in the UK and USA (Shaw 1990a, b). In 1967 the Centre National d'Art Contemporain (CNAC) for contemporary visual artists was established, organising exhibitions, research and information on living art and the commissioning and purchase of art works. Marc Chagall for example was commissioned to paint ceilings at the Opera de Paris. Despite new provision and support (although largely for the same 'high-arts' production), audiences did not significantly increase, and after the Paris riots in May 1968, Malraux's policy was criticised for its elitism and middle-class art promotion (Cahiers Français 1993). In 1971 Pompidou's government appointed Jacques Duhamel as Secretary for Cultural Affairs, whose policy was: 'creative liberty, no cultural dirigism, incitement and coordination' (ibid.: 1993). An increased department budget also expanded into architecture with conservation of a higher number of buildings, not limited to restoration of a small number of 'edifices'.

In Britain, the new Labour Government elected with a small majority in 1964 transferred responsibilities for arts, libraries and museums funding from the Treasury to a new Arts and Libraries Office of the Department of Education and Science (DES), though most 'heritage' responsibilities stayed with successive Public Works and Environment Ministries until the 1992 formation of the Department of National Heritage, and the National History Museum remained with the DES, funded as a research institute until this date. The Labour Party came into office in 1964 with no official programme in the field of culture and it did not feature in election debates. However the appointment of Jennie Lee as the first Minister for the Arts in 1965 brought forth the milestone government White Paper: *A Policy for the Arts: The First Steps* (Lee 1965). This was accompanied by a 30 per cent increase of the Arts Council's revenue grant, followed in 1967 by a revision of its Royal Charter. References in the Charter's original 1946 Objects to the 'Fine Arts' and the 'improvement of standards' were removed and the amended Charter reflected the renewed distributive and access aims in taking the arts to the people. The 1965 White Paper repeatedly referred to the maintenance of 'artistic standards', 'high points' and 'excellence'. However, in order to accommodate these promotional and distributive aims, the *Housing the Arts* fund was created by the Arts Council initially allocated £250,000 in 1965/6 and by 1986/7 £661,500. This provided capital grants for the development of arts facilities, and when combined with this newly elected Labour administration's policy of wider distribution and access to the arts (mainly the same 'high' arts as in France), an arts centre movement emerged (Lane 1978, Forster 1983, Hutchison and Forrester 1987). Arts centres were envisaged as 'Centres where light entertainment and cultural projects can be enjoyed', and made more inviting 'to provide additional amenities (restaurants, lecture rooms) at existing centres' (Lee 1965). By 1970, £1 million had been contributed towards new arts centres, valued at £5 million in total (i.e. the Housing the Arts fund had provided an average of 20 per cent of the capital cost), supporting over 125 centres, in addition to twenty-six regional film theatres.

Arts centres and *Maisons de la Culture*

Writing in the same year that the Arts Council issued *The Glory of the Garden: A Strategy for the Development of the Arts in England*, Stark documented the rise and proliferation of arts centres in England, observing that 'This phenomenal growth is in no sense the result of central, regional or local planning by any one agency, least of all the Arts Council. It is, and has been *unplanned*' (1984: 126 emphases added). Because of this, arts centres had certain characteristics, distinguishing them from other arts facilities:

1   Their unplanned status meant that there was never enough food for them on the table [i.e. funding].
2   They are architectural opportunists; [over 80 per cent of arts centres were housed in second-hand buildings, from churches, drill halls to town halls, over fifty per cent of urban centres were in buildings over a century old (Hutchison and Forrester, 1987)].
3   They are economic and efficient [multi-use/purpose].
4   They are masters of disguise – in terms of their programme, purpose, attracting a wide mix of funding, in addition to 'arts' funding – inner city programmes, unemployment training, education, youth and community services.

Source: Stark (1984: 126–7)

Earlier 'models' of arts centres could be seen in the eighteenth-century coffee-houses, nineteenth-century working men's clubs, mechanic's institutes, in the socialist people's palaces ('clubs'), and more recently the 'little theatre' movement of mainly amateur dramatic societies, between the First and Second World Wars. Three seminal post-War arts centre projects were associated with key individuals: Joan Littlewood at the Theatre Royal, Stratford East; John English at the Midlands Arts Centre, Birmingham; and Jim Haynes at Drury Lane *Arts Lab* (short for 'Arts Laboratory', where experimental art and participatory work was the focus, in non-institutional settings). Other arts centre and community arts projects were identified with local communities, and were sometimes neighbourhood based: '… buildings, a programme which combines presentation with workshops, a serious commitment to more than one art form, and a policy of working with particular communities' (LAAC 1984: 3). By 1969 there were 180 projects claiming to be *arts labs* (White 1969) and from a survey conducted in 1970 there were over sixty designated arts centres in the UK, excluding single-use venues such as theatres and purely amateur and community dual-use centres. In a national survey in 1986, 242 arts centres were identified in the UK (Arts Council 1989), although ten years later in a further study (MacKeith 1996) only 129 were listed, this net loss being put down not only to closures, but also to a narrowing in the establishment's definition of 'arts' activity (which excluded wholly non-professional, youth/amateur work).

The 1960s were therefore a boom-time for arts centre development – a combination of 'post-war idealism' and ''60s revolt' (Lane 1978) – in Europe, Australia and North America, where by 1970 over forty-four arts centres were identified, with a further fifty-five planned in the USA alone (Tables 4.2 and 4.3). A combination of local aspirations

Table 4.2 Arts centres opened in the 1960s/1970s

Year opened	Location	Facilities
1965	Canberra, Australia	main theatre seating 1,200 (symphony concerts, opera, ballet, plays and conferences), playhouse seating 312, art gallery, club room, balcony room/restaurant
n/a	Victoria, Australia	town hall (1,000 seats), art gallery (two gallery spaces, exhibition hall, small theatre, 200), lounge/supper room (400)
1966	Victoria, Australia	Mildura Arts Centre, theatre seating 400, art gallery, history museum
1968–	Melbourne, Australia	Victorian arts centre, exhibition area, library, theatre court, sculpture garden. Phase II: concert hall (2,500), playhouse (750), lecture/experimental theatre (350–1,000)
1968	Calgary, Canada	established in 1945, reconverted factory in 1960 with theatre, art gallery and classrooms
1969	Ottawa, Canada	National Arts Centre, opera house/concert hall seating 2,372, theatre seating 800, studio (288), film projection, salon
1963	Montreal, Canada	Place des Arts, theatre seating 2,983, L'Edifice des Theatres added in 1967, two theatres seat 832 and 1,290 (vertically integrated)
1959	Vancouver, Canada	QE Theatre and Playhouse, main hall seats 2,880, rep theatre (650), recital room (175), exhibition areas
1964	Usti nad Labem, Czechoslovakia	cinema, two theatres, exhibition hall, three lecture rooms, twelve club rooms, library, studios, darkrooms, ballet hall, rehearsal rooms, social hall (seats 750)
1958	Berlin, Germany	Academy of Arts, exhibition areas, studio theatre seating 600, used for plays, concerts, dance, films and lectures, reception room, studios
1969	Wairarapa, New Zealand	gallery, reception foyer/exhibition space, studio, film projection
1955	Warsaw, Poland	Palace of Culture and Science, three theatres seating 750 (drama), 120 (rehearsal), 540 (experimental and classics), state theatre Lalka (280), congress hall (3,200), three cinemas, technical museum, Youth Palace with workshops, studios, library. (Also sports/PE, University, Polish UNESCO committee)
1964	Skovde, Sweden	library, disco, art gallery, lecture theatre, meditation room, theatre (seats 509), dance and congress hall (for up to 1,100) and café

Table 4.2 (continued)

Year opened	Location	Facilities
1967	Zurich, Switzerland	Centre Le Corbusier (his 'last work'), exhibition gallery, lecture hall, cinema
1968	Basildon, UK	auditorium seating 476, used as theatre, cinema, exhibition area, studios for pottery, sculpture, darkrooms, restaurant (80 seats)
1967	Swindon, UK	theatre, for plays and film, club room, two meeting rooms (40, 20)
1965	Boston, UK	theatre seating 219, concert hall and film, two rooms and foyer exhibition area, licensed club
(1717)	Liverpool, UK	Bluecoat Chamber, concert hall seating 399, gallery, studio, printing press, music studios for arts and crafts, film theatre, small studio theatre
1964	Bristol, UK	multipurpose theatre, drama, film, poetry, two exhibition galleries, coffee bar, restaurant
1965	Camden, London, UK	exhibition area, gallery, restaurant, activity/teaching area
1966	Shrewsbury, UK	experimental theatre, club room, recital/lecture rooms, bar
1960	Somerset, UK	theatre seating 242, lecture/demo room, club, exhibition space
1969	Brighton, UK	Gardner Arts centre (University of Sussex), gallery, four music studios, three visual arts studios, auditorium seating 500
1970	Hull, UK	theatre seating 150, bar, exhibition foyer
1964–	Birmingham, UK	Midlands Arts Centre, two theatres (seating 200 and 100), art gallery, cinema, exhibition area, studios and workshops, bar
1973	Crouch End, London, UK	Mountview, two theatres, small hall, cinema, six rehearsal studios, two restaurant/bars (performing arts school)
1971	Peterborough, UK	theatre seating 395, exhibition area, bar/restaurant, activity rooms
1955	Weymouth, UK	main hall seating 200, art room, members room, exhibition area, coffee bar
1964	Portland, Oregon, USA	Albina Arts centre, gallery, meeting and classrooms, library, workshops for pottery, photography, rehearsal rooms

Year	Location	Description
1958	New Jersey, USA	exhibition area, three visual arts studios, dance studio
1972	Birmingham, Alabama, USA	Civic Center, exhibition hall, indoor coliseum, theatre seating 800, concert hall seating 3,000
1968	Troy, New York, USA	chapel and cultural centre, hall for drama, films, dance, concerts adjoins baptistery, conference room and chapel
1967	New Jersey, USA	Fair Lawn Arts Center, former municipal library and firehouse, meeting, exhibition rooms, theatre seating 150
1971	Fort Wayne, Indiana, USA	music halls seating 2,500, school of performing arts, theatre (500), art school, museum
1968	New Jersey, USA	amphitheatre seating 5,000, lawn area > 3,000, nature trails, Monmouth Museum for art, nature and science
1970	Washington, DC, USA	Kennedy Center for the Performing Arts, opera house seating 3,000, concert hall (2,761), Eisenhower theatre (1,142), cinema/experimental theatre (500), gallery
1970	Midland, Michigan, USA	main theatre/hall seating 1,539, film projection, two rehearsal rooms, four practice rooms, music library, instrument room, workshop, little theatre (399), workshops
1964	St Paul, Minnesota, USA	two theatres seating 650 and 300, projection/lectures, art museum, science museum, classrooms
1961	Zagreb, Yugoslavia	theatre hall seating 500 for plays, cinema, concerts, small hall (200), lectures, exhibitions, library, reading room, restaurant
1974	Warwick, UK	Warwick University Arts Centre, theatre seating 1,200, Gallery, catering, studios
1969	Lancaster, UK	Nuffield Theatre, University of Lancaster, 350 seats
1969	Kent, UK	Gulbenkian Theatre, University of Kent, 342 seats
1969	Southampton, UK	Nuffield Theatre, University of Southampton, 468 seats

Sources: White (1969), Schouvaloff (1970), Hutchison and Forrester (1987)

Note: no arts centres were notified for Denmark or Greece

Table 4.3 Arts centres in the USA and UK by 1970

Community arts councils in the USA operating or planning an arts centre in 1970	UK arts centres listed by the Department for Education (DES) and regional arts associations (by county)
Alabama	Cornwall (3)
Alaska	Cumberland (3)
Arkansas – Little Rock, Pine Bluff	Derbyshire
California – Carmel, Fullerton, Oakland, Pomona	Devon (4)
	Dorset (2)
Connecticut	Durham (2)
Florida (2)	Essex (2)
Georgia	Gloucestershire
Illinois (3)	Hertfordshire (2)
Iowa	Kent
Maryland (4)	Lancashire
Massachusetts (2)	London (5)
Michigan	Norfolk
Montana	Northumberland
New Hampshire	Nottinghamshire (2)
New York (6)	Shropshire
North Carolina (7)	Somerset
Ohio (3)	Staffordshire (3)
Oregon	Warwickshire (3)
Pennsylvania	Wiltshire (2)
South Carolina	Worcestershire (3)
South Dakota	Yorkshire (2)
Texas	Scotland (2)
Utah	Wales (3)
Virginia (3)	
Washington (3)	Total = 49 + 40 arts lab projects
Wisconsin (2)	By 1994/5 sixty-four arts centres were
Virgin Islands (2)	supported through the regional/national arts
	funding systems, representing about 50 per
Total = 55	cent of all 'arts centres' in England

Sources: White (1969), Schouvaloff (1970), Hutchison and Forrester (1987)

and strengthening of the power of local administrations, together with growing affluence (leisure-time and spending) and artistic freedom and experimentation, together fuelled the local arts centre movement in many countries. The map of provision at both local and city levels (the latter also benefiting from major performing arts centre projects, e.g. Kennedy and Lincoln Centers in the USA; South Bank and Barbican centres in London), therefore changed dramatically in this short period, just as the Elizabethan theatre and later the music hall had done in London. Arts centres did not tend to replace traditional and historic theatre and museum buildings, but created a more contemporary and less institutional setting for arts participation and spectating, but one that was also outside of the music and dance hall tradition, and which therefore defined their class base.

The multi-use and multi-form aspect that identifies an arts centre as distinct from a single-use facility, reflects its *physical* nature (i.e. design, layout), while its accessibility in the widest sense, e.g. new audiences, defines its *location*: 'A *Maison de la Culture* is defined in terms of the kind of audience that constitutes it' (Malraux 1966). The scope and scale of arts resources were to be capable of creating a *synergy* between differing levels of ability and experience, art forms and opportunity: amateur and professional; youth and adult; diverse cultures (multicultural); mixed-arts, crafts and media; local, regional and national networks and so on. Figure 4.1 gives an indication of the thought processes with which the early arts centre designers were engaging.

Arts centres thus sought to break down barriers between passive consumption ('audience') and active participation, and between art forms and practice and therefore create links in the 'production chain' (see Chapter 6) – between rehearsal and performance; workshop and display/exhibition; production processes (e.g. crafts and designer-making, audiovisual media); print/media and communications – including new technology such as graphic design, interactive video and digital imaging (Evans and Shaw 1992: 7). Technology, knowledge and scale were therefore also important in the multi-use and mixed-arts centre. A summary of ticketed events at arts centres in 1994/5 in England (Table 4.4) provides an idea of the attendances at various productions with the 'live arts' representing 61 per cent of programmes. On average, non-professional productions represented over one-third of arts centre performances of which over 50 per cent was drama and 18 per cent contemporary dance/mime. Drama and opera/musical theatre also had the highest number of non-professional performances.

Increasingly, technical and management skills and resources (including fundraising, production, marketing) were required for the individual and small-scale to develop. Some of the best arts centre models have been those where professional arts and resources mix with new artists and companies, including youth, non/unemployed, and in community education (Macdonald 1986), i.e. acting as a seed-bed and showcase for new work and offering scope for collaborative and outreach work with residents/local communities, schools and touring productions – the modern equivalent of the 'strolling

Table 4.4 Number of paid attendances at arts centres by art form

Art form/production	Total	Percentage	Median
Dance	203,643	5	737
Cabaret/comedy	105,899	3	929
Drama	787,159	20	3,238
Music	1,270,121	33	3,457
Live art/interdisciplinary work	12,023	†	300
Literature	22,052	1	239
Cinema	687,001	18	3,053
Paid exhibitions	786,749	20	5,000
Total	3,874,647	100	17,981
Free exhibitions	91,258	–	14,379

Sources: O'Brien (1997), Arts Centres, Statistical Appendix

†negligible percentage

WORD GAME
take two words

ARTS CENTRE

civic hall   public hall   community   centre
PEOPLE   PLACE   **COMMUNION**

P A S T I M E S

plays   theatre
films   cinema
ballets   theatre
exhibitions   gallery
lectures   theatre
concerts   concert hall
recitals   recital room
operas   opera house   RECREATION

(PERFORMER / SPECTATO R)

MAISON DE LA CULTURE

culture   POP   popular   eating
         music dance   drinking
       folk   song
         beat rhythm

talking   DIALOGUE   Plato
          relaxation   Tolstoy
watching   looking   listening   doing   Craig
                         Gropius
            SILENCE   Aymé
                         Weidlé
         S O U N D   Gimpel

laughing   clapping   shouting   cheering

leisure   C O M M U N I C A T I O N
les voix du silence   nothing

       MAGIC

AUDIENCE
       appreciation   education   enjoyment   fun
PARTICIPATION   fun palace

PALAIS de DANSE
discothèque

                   palace   of   culture
             PLACE   civilization   art

ARTS
arts lab   workshop   forum   arena   ἀγορά
                     school   ring   wrestling
environment   gymnasium   CIRCUS   gymnastics
         focus   centre
                          theatre   performing arts
                  performance   arts centre
       concentration
ENVIRONMENT

ENVIROTHEATRE

Figure 4.1 Word game take two words: arts centre (Schouvaloff 1970: 83)

players'. Adult and community education facilities both historically (see Chapter 3) and even in their more beleaguered state (falling between the stools of leisure, youth, community and education services), have had a tradition in providing the sole cultural activity for a local area, as Rowntree and Lavers found in the late 1940s: 'praise is also due to those in charge of the recreational institutes, most of which are established in urban deserts, for nothing flourishes but seemingly endless rows of mean streets, cheap cinemas, public houses and poor shops . . . the recreational institutes are true oases in these deserts and are carrying out a truly civilizing task' (1951: 321). Advances in design and building technology also made possible the multipurpose space or hall, which served dedicated sporting, performance and exhibition requirements, including surfaces, layout, staging and mixed/multimedia work, offering opportunities for a mix of users – sports and arts, not normally attending the same venue (a combination long recognised and practised in Eastern Europe). Such multi-use is not limited to new and rediscovered venues, however – village, church, school/college and town halls have long provided a focus for community activity, from badminton to ballet, classes to choral societies. In rural areas, the village hall has often provided the only venue, and this has been recognised by regional arts associations which cover remoter areas and host touring networks of small-scale companies and productions.

What the arts centre offered in planning terms was the end to the separation of single-use venues and between professional and community-based activity, which influenced the operation, design and layout of large prestigious arts centres, not only in the combination of performing spaces and resources, but also in their relationship with arts in education and community development – now an accepted part of the overall programme of any major arts venue organisation. For example, 'American cultural institutions have distinguished themselves as public educators. To a large extent this dedication to public education has increasingly preoccupied our orchestras, symphonies and more recently our arts centers' (Adams 1970: 206).

In France, for instance, Malraux's policy of decentralisation saw the *Maison de la Culture* as a new instrument, a place of cultural creation with a specialisation in a particular art form (e.g. theatre, music, ballet, art, film). They were financed 50:50 by central and local government and in 1964 the first *Maison* was inaugurated in Bourges (Table 4.5). The importance of these new buildings was evidenced by architectural competitions held for their design, in contrast to the *Centres d'Action Culturelles* (CACs) located in second-hand buildings like the majority of arts centres in the UK. Malraux foresaw at least one *Maison* per *Département* based on population level, but the lack of financial resources and reservation from some municipalities (the *Maisons* were perceived as a 'third power') meant that this vision was never realised. In 1989 the last *Maison de la Culture* 'in action' closed.

The southern French city of Montpellier provides a useful example of the development of cultural policy through *Maisons* from the late 1970s (Table 4.6). Montpellier, had like other regional cities, a growing population – from 90,000 in 1950 to 210,000 by 1990 – in this case due to resettling Parisians, returnees from Algeria and industrial relocation by the US multinational IBM. A growing (and 'imported') middle and professional class and university student body provided a base for expanded cultural consumption and activities, whilst administrative devolution cast the city as regional capital of Languedoc in 1984. The significance of regional capital status is evidenced by the

Table 4.5 Maisons de la Culture opened in France, 1963–71

Year opened	Location	Population	Facilities
1963	Bourges	65,000	two theatres seating 949 and 376, exhibition room, library, recording studios, rehearsal rooms, coffee bar, television room
1963	Paris, 20th *arrondissement*	200,000	theatre seating 1,000, rehearsal room
1965	Amiens	110,000	two theatres seating 1,100 and 300, disco, library, exhibition space, three meeting rooms, coffee bar, television room
1966	Firminy	26,000	two conference halls, library, exhibition space, meeting rooms, television room, disco auditorium for music seating 150, studios, bar
1968	Grenoble	163,000	two theatres seating 1,300 and 325, library, disco, exhibition hall, three meeting rooms, café, television room
1968	Rennes	158,000	two theatres seating 1,200 and 35–550, cinema, exhibition gallery, disco, café, children's room
1969	Nevers	42,000	theatre seating 1,000, small auditorium, disco, library, multipurpose rooms
1969	Rheims	165,000	two theatres, exhibition space, café
1971	Le Havre	185,000	n/a
1971	Angers	123,000	two auditoria, exhibition room, gallery, bar, disco

Sources: Ministère des Affaires Culturelles, Paris, Schouvaloff (1970)

proportion of the regional cultural budget that was allocated to Montpellier, from a negligible sum in the mid-1970s, this jumped to 28 per cent in 1977 and by 1980 had reached 36 per cent. The leftward shift in French politics also saw the introduction of cultural policies and actions to attract middle-class voters: 'this cultural theme was a common feature of the political rhetoric used by the left to attract voters throughout almost every town in France in the 1977 municipal elections' (Negrier 1993: 137). After these elections the new left-wing city government 'endorsed the construction of an urban cultural strategy . . . inspired by three objectives: the decentralisation of cultural activities and facilities; cultural animation, and the democratisation of cultural policy-making. The first of these was symbolised by the setting up of the *Maisons pour tous*' (ibid.) – multipurpose institutions of which fifteen were built. These were designed, like the earlier culture-houses and communist 'people's palaces', to house under one roof neighbourhood-based cultural, sporting and leisure activities, but serving smaller populations in adapted rather than new buildings. Like the *Maisons de la Culture* the re/building programme was not sustained, in some respects not only an indication of sufficient provision, but also of a shift in policy objectives and whilst sounding good rhetorically, the policy objective of 'cultural democratisation' in this city as in others was effectively buried. The 1980s' cultural strategy moved towards the higher profile projects – a new theatre, concert hall, conference centre and mass spectacles, with the traditional cultural sector declining in resource allocation – falling from 17 per cent of 'flagship' arts spending in 1986 to only 3 cent in 1990: 'The city's new cultural policy was closely related to the imperative of inter-urban competition' (ibid.: 143).

Table 4.6 Montpellier City Council's expenditure on *Maisons pour tous* as a percentage of its total capital expenditure on culture (FF in thousands)

Year	Spending on Maisons pour tous	Percentage of all capital expenditure on culture
1978	700	56
1979	2,200	76
1980	2,551	58
1981	4,666	56
1982	2,990	18
1983	3,790	25
1984	2,192	15
1985	885	3

Source: Montpellier City Council in Negrier (1993: 139)

Without pursuing a technical argument over what qualifies as and constitutes an 'arts centre', and more importantly what is *not* covered by the official definition – 'a regular base for substantial programmes . . . in more than one art form, with professional input . . . primarily used for arts activities' – the emergence of local arts development did change the focus and role of many arts centres, with the rediscovery of the *animateur* and outreach role and the resource nature of many organisations – a growing networking function (NAAC, in Hutchison and Forrester 1987). In view of the changing functional nature of arts centres, as well as many outward-looking museums and galleries, that reflected local needs and existing provision (commercial and public theatres, galleries, studios, etc.), the essence of the arts centre ethos was thus: 'it lies equally in the spirit of participation that tends to inform and shape the activities at most centres' (ibid.: 216). The dynamic and non-intimidating form of the arts centre provides further opportunities in arts planning, beyond the preoccupation with designated arts space. With hindsight, whilst many of these centres developed unique arts production and acted as a reflection of their times, particularly in attracting young artists and playwrights, they generally did not achieve or sustain a cultural democratic role (staffed by the well-educated, pursuing an 'alternative' culture/lifestyle), and few if any, could claim a place in the cultural democratic process:

> arts centres have slipped from a dominant position in the cultural chain – they've lost part of their role as value givers within the subsidised world. . . . The vortex of commercial popular culture and the consumer, commodity and technological revolutions has taken so much of what the arts centre world values most – smallness, locality, the love and sense of community.
>
> (Wallace 1993: 2)

Whilst arts centres therefore met political goals and the aspirations of willing community and professional arts practitioners (whose own motivations partially coincided through the community arts/avant-garde movements), they specifically fulfilled a role at a local level as an extension of amenity provision which local authorities, district and borough councils effectively adopted. The local orientation of these community arts groups was particularly attractive to left-wing borough councillors and many were consequently grant-aided by local authorities (Davies and Selwood 1999). This has meant that the national pursuit of cultural democracy and distribution that dried up in both Britain and France (particularly the provision of capital grants) was however overtaken by the provision of arts centres and community arts provision at a local level as a natural component of the portfolio of recreation and leisure facilities, alongside sports, parks, libraries and local museums (Lane 1978). Despite the early visions of arts planning and the national support for arts centre development therefore, the primary role in policy, provision and planning for local amenity has rested largely with local authorities and this had been the case in many countries since municipalism established itself:

> For over a century the Town Hall and Civic Centre, local library and art gallery, the Mechanics Institute and School of Arts, the local parks, gardens and rotundas have been part of our landscape. They have enabled Australians from all walks of life to attend dances, plays, art and craft classes, to view floral and art exhibitions, to hear bands and concerts, to become educated, to attend public meetings and more. Local government provided the facilities and spaces.
>
> (Australia Council 1991: 45)

Prior to the 1940s, however, in Britain local authorities had no general power to fund the arts (in contrast with their long-standing library and museum powers), although some obtained Local Acts of Parliament authorising specific projects. The Emergency Powers Act 1939 was widely used by local authorities to permit provision and funding of dancing and entertainment during the War, including joint initiatives with touring CEMA productions (see above) and from 1943 the annual *Holidays at Home* – weeks of summer arts events. In face of impending loss of very popular public facilities – especially social and recreational dancing – with the end of the Emergency Powers Act provisions, the Local Government Act 1948 gave local authorities limited powers to continue to support and provide dancing and entertainment. As a consequence of successive reviews of local government, powers to fund, promote and develop arts provision were specifically permitted under the Local Government Act 1972 (section 145; Marshall 1974), although such provision was not mandatory (i.e. no minimum levels of per capita funding or facility provision were set). This measure removed the sixpenny-in-the pound maximum rate, imposed under the 1948 Local Government Act. Roy Shaw observed that 'thirty years later few English authorities have achieved an annual expenditure of even half [that] amount' (Arts Council 1978: 10). The 1963 national survey of Municipal Entertainment in England and Wales showed that net council spending on the arts, culture and entertainment was only the equivalent of one (old) penny rate – the pre-1948 maximum. Of this expenditure, 43 per cent was on the upkeep of buildings, and the most popular areas of spending were on band concerts, art exhibitions,

children's entertainment, ballroom dances and orchestral concerts, all much more popular than theatres, which ranked only fourteenth (Mulgan and Worpole 1986). Unlike
sports and recreation, however, arts facilities tended still to reflect a mixed economy and
independent model, as distinct from directly owned municipal provision of sports centres, parks and libraries. This distinction is important in planning terms, since arts
centres tended not to be located on predetermined sites or in new buildings. As a
national survey of arts centres in the UK revealed, whilst most centres had begun operating in the late 1960s/70s, 47 per cent occupied buildings that were over a century
old – 45 per cent of professional and 62 per cent of voluntary-run centres in urban areas.
Ten years later another national survey found that 60 per cent of arts centres were either
leased or rented from the local authority and over 8 per cent occupied second-hand and
'non-customised' buildings (Tables 4.7 and 4.8) (MacKeith 1996).

Table 4.7 Age of arts centre premises (%)

Age (years)	All centres	Urban professional	Urban voluntary
< 10	11	19	–
10–49	18	13	5
50–99	24	23	33
> 100	47	45	62

The previous occupation of these centres also echoes the reuse of cultural buildings
through the ages – further indication of the preference for and symbolic value ascribed
to places of civic, communal and cultural activity, a symbolic importance that new
buildings and sites often never really attain (Table 4.8).

Table 4.8 Previous use of arts centre buildings (%)

	All centres	Urban professional
Church/chapel	26	13
School	18	16
Factory/warehouse	23	42
Residence	22	18
Town hall	10	11

Source: Hutchison and Forester (1987: 13–14)

The generic arts centre can also be categorised by three separate scales of facility. A
fourth general-purpose large-scale venue is the stadium or arena, most associated with
major sporting and exhibition events, but also used for large concerts such as stadia-rock
(e.g. Toronto's SkyDome, Wembley Arena):

1   Flagship Arts Centre complex (e.g. USA – Lincoln, Kennedy; UK – South
    Bank, Barbican).
2   Middle-Scale Professional (300-plus seat main theatre, touring production
    base).

3   Community arts centres – small-scale, 'studio' theatre, resource-based, multi-arts, dual-use, e.g. community/village and sports 'halls', clubs, etc.

The promotion of new centres (Point 1) is evident in the downtown, architectural statements which are familiar and still emerging in post-industrial cities, whilst the traditional large proscenium arch theatre struggles to survive financially and culturally whether commercial or civic venue. The small- and medium-scale with exceptions (fringe/alternative and innovative venues/producers) have also floundered in the problematic zone between arts-as-amenity, social arts provision and experimentation/agitprop, and the economic imperatives for arts investment as part of regeneration, gentrification and visitor-based socio-economic development. The extent to which these levels of provision and performance –Points 1–3 – provide a link between one another and an opportunity for upward and downward movement, is therefore a particular issue in cultural planning and programming, and in the treatment of community facilities within planning standards. The approaches adopted and resisted in prescribed planning and scale hierarchies of cultural facility are therefore discussed further.

As is evident in this and further so in the next chapter, the absence of a definition or parameters for the notion of *amenity* in town planning, particularly today in the contemporary urban cultural sense, has served to limit the equitable distribution and treatment of cultural activities when compared with other recreational facilities and the conservation of heritage in its various forms. This has also reinforced the urban–rural/city–suburban conflict and sentiment that has associated much arts and entertainment as problematic in social and amenity terms. The arts centre and its equivalent in the *maisons* and *casas*, the village halls, community and education institutes which serve, often in a second-hand and compensatory way to meet local cultural provision, has offered a benign solution to this position, and to the growing political adoption of distributory models for arts provision as distinct from the centre-dominated, state monuments and prescriptive socialist solutions to collective cultural activity and exchange. Their expansion and adoption throughout the world attests to this, however this movement has been caught in the decline in public service provision and local amenity generally, a political preference – economies of scale, profile, economic impact benefits – for larger but fewer flagship and municipal centres. This has been a response to commercial development and the out-of-town leisure–retail and downtown mixed-use phenomenon, as well as changing technology and cultural consumption habits, fed by a segmented culture and entertainment industry and subcultures which exist at least initially outside of the legitimate sites and locations for culture. Much arts and recreation is however supply led and location is therefore a significant factor in participation. More responsive forms of planning and greater attention to the relationship between the supply and demand for cultural facilities looks to existing models of recreation planning and the greater integration of cultural factors in the urban planning process. These Chapter 5 considers, drawing on a range of spatial, demand and consultative planning mechanisms, in particular the advantages of conducting cultural mapping exercises as the basis for cultural amenity planning.

Notes

1 In 1992 the United Nations Conference on Environment and Development, the 'Earth Summit', was held in Rio de Janeiro. The conference was the culmination of initiatives that can be traced back to the UN's Stockholm conference of 1972 on the Human Environment and the Bruntland Report of 1987. However the Rio meeting adopted a more holistic view, manifested in the title 'Environment and Development'. Some international agreements were reached of which none perhaps is more comprehensive and wide-ranging in scope than that of Agenda 21. By far the greatest proportion of this Agenda related to matters that require locally based action – thereby reflecting and reinforcing the maxim 'think global – act local' – namely Local Agenda 21 (Leslie and Muir 1996: iii).

2 The layout of Latin American cities is still largely present in their historic quarters, dating from the fifteenth and sixteenth centuries. This practice codified in the 'Leyes de Indias' produced what some consider as the only true Renaissance cities ever built. Another interpretation traces the origins of this practice to the military settlements established in Spain in the fifteenth century to consolidate the territory gained from the Arabs in the 'Reconquista' wars, a practice that owes its origins to the Roman 'castrum' (Rojas 1998: 2). In contrast, Portuguese colonial city layout inherited the traditional organic pattern of streets and public spaces evolving as settlements grew according to the natural topography as well as defence needs.

3 See Chapter 10 'Amenity', in *Town and Country Planning, 1943–1951*, Cmnd 8294, London: HMSO: 138-54.

4 Although preceded by the 1909 and 1919 Housing and Town Planning Acts and the 1925 Town Planning and 1932 Town and Country Planning Acts, the 1947 Act was 'The crowning piece of comprehensive planning legislation heralded by the 1944 White Paper *The Control of Land Use*, which imposed a compulsory planning duty on all local authorities for the first time' (Rydin 1993: 26).

5 Planning for the arts
Models and standards of provision

As the 'unplanned' but responsive arts centre experience suggests, the notion of developing and promoting measurable standards for arts amenities, either within the statutory planning process or as part of wider cultural policies, has not gained wide acceptance by either planning or arts policy practitioners. However, some attempts at population-based and/or comparative provision levels of arts facilities have in the past been proposed within the arts sector at national and more successfully at regional levels particularly in the context of distributive arts planning. The contrast between planning for sports and open spaces (e.g. parks, playing fields, promenades) and the arts also highlights the differing treatment and resourcing of these two arguably key elements of public 'leisure' provision. Whilst there is a clear relationship between the influence of the supply-led nature of much sport and recreation provision on participation and the impact of planning norms, this is in contrast to the less homogeneous arts (versus libraries) where difficulties in defining and accepting arts planning standards has prevented the adoption of any systematic norms of arts facility provision. In consequence there tends to be significantly more local sports facilities (e.g. swimming pools and pitches) in total and more evenly distributed than equivalent arts amenities, and higher participation rates as a result, i.e. activity is location and supply led.

Amenity and land-use planning cannot however take place in a vacuum if competing needs, present and future, are to be met and 'Pareto' losses minimised.[1] This was a basic tenet of the modern town planning movement that responded to public concern about the uncontrolled development of the nineteenth and early twentieth centuries. The direct result in urbanised Britain was the passing of the Town and Country Planning Act 1947. This made universal the preparation of local plans by local authorities in the literal sense, i.e. maps marked with existing, proposed and permitted changes in land-use, such as zones for new housing or industry, together with systems of development control by means of planning application and permission.[2] Other provisions included the listing and special protection of buildings of historic or architectural interest, and facilitating the improvement in economic, environmental and community amenities. Furthermore, an analysis of cultural policy in relation to planning for arts provision and cultural facilities is enlightening on the grounds that plans are, or should be, policy led, or at least influenced by policy objectives.

Several years before taking up appointment as the second Secretary of the Arts Council (1950–63), W. E. Williams had published 'Are we building a new culture?' (1943) in which he foresaw a Great Britain 'covered with a national grid of cultural centres' (quoted in Pick 1991: 22), drawing, although not overtly, on the socialist 'community centre' concept promoted before the War, for example by the *Populaire Front* in France and in the Soviet Union (May 1931). Whilst planning guidelines and standards have never been mandatory for arts and cultural facilities, some early attempts to quantify arts provision were made. In 1943 Williams expounded some notion of arts planning based on the concept of a *National Grid of Arts Centres*, with echoes of the socialist planning models which had informed the development of communes and cities of the Eastern Bloc in the previous decade:

> instead of our present dispersal of the public library down one street, the art gallery (if any) down another, the workingmen's club somewhere else . . . let us plan the Civic centres where men and women may satisfy the whole range of educational and cultural interests between keeping fit and cultural argument. Let us so unify our popular culture that in every considerable town we may have a centre where people may listen to good music, look at a painting, study . . . join in a debate.
>
> (quoted in Pick 1991: 23)

The emerging Arts Council in 1945 had also produced a pamphlet and touring exhibition called *Plans for an Arts Centre*, which 'designed to show how the arts can be accommodated in a medium size town . . . a town where it is not economically possible to run a separate theatre, art gallery and hall for concerts'. The special role for the flexible arts centre, as opposed to a single-activity building (e.g. theatre, gallery), was also later recognised by the Council of Europe in its 'Symposium of the Council for Cultural Co-operation', entitled *Facilities for Cultural Democracy* (Janne 1970), and later adapted by national and regional arts associations, including the National Association of Arts Centres (Hutchison and Forrester 1987). In 1959, the Arts Council produced a survey, *Housing the Arts in Great Britain*, that listed eight general rules on the needs of regions, cities and towns expressed entirely in terms of physical facilities for arts performance or exhibition:

1 A region with 10 million inhabitants should have one permanent professional opera company.
2 A region with 5 million inhabitants should have one permanent symphony orchestra.
3 Towns of more than 150,000 or more should have one theatre large enough to house major touring productions including opera and ballet.
4 Towns of 100,000 or more should have one permanent repertory company, with its own theatre.
5 Towns of 75,000 or more should have one hall suitable for large symphony and choral concerts.
6 Towns of 50,000 or more should have one museum and/or art gallery, and one professionally staffed Arts Centre (in use all year).
7 Towns of 20,000 or more should have one Arts Centre which may be part of another establishment; one Music Club or Arts Society, presenting regular series of professional events; one amateur orchestra (on a scale of at least one

for every 60,000 inhabitants); facilities for showing regular touring exhibitions.
8   Towns of 10,000 or more should have an amateur dramatic society, a Choral Society, and an Amateur Art Society or Club (each on a scale of at least one for every 30,000 inhabitants).

Source: Arts Council (1959)

However, when in office during the 1950s, Williams's egalitarian tendencies were subordinated to the Arts Council's Royal Charter objective of pursuing excellence in the professional arts. Mary Glasgow, Arts Council Secretary-General between 1946 and 1950, had already commented thus: 'An actual conflict developed between what may be called the amateur and the professional point of view' (Glasgow and Evans 1949: 47). Although supporting professional artists working in education settings and for a period the funding of the National Federation of Music Societies (NFMS, subsequently devolved to the regional arts associations), the professional bias dominated for some time to come: '[Williams] argued forcibly for the need to concentrate on raising standards, believing that too great an emphasis on spreading would lead to the diffusion of mediocrity . . . in 1975, the climate of opinion had changed, and Williams' views seemed like "elitism"' (Shaw, quoted in Arts Council 1983: 7). Unlike the New Town movement, therefore, the idea of arts planning and the ideals of arts centres or similar facilities existing as a service to a community and as a place where amateurs and professionals would work together disappeared from Arts Council thought until the late 1960s/early 1970s. John Pick also rejects this 'relish for planning' as one 'which might please a Soviet Planner' (1991: 23) and which also confirmed the Arts Council's London-centric view of arts provision: 'There has never been anything which demonstrated more plainly the Arts Council's London mind (the figures are subdivisions of the scale of provision in the capital) or its narrow view of what constitutes the arts' (ibid.: 55–6). Whilst standards were not officially adopted, the tendency to plan has not disappeared, creating an 'uneasy history in the arts: British arts funding bodies have always had an ambivalent relationship with the idea of planning. One underlying approach has been the very top-down approach involving the production of apparently prescriptive documents' (Stark 1994: 12).

In his 1989 review of the structure of arts funding for the Office of Arts and Libraries (OAL), Richard Wilding recalled Keynes's BBC broadcast at the formation of the Arts Council in 1945, namely 'the artist walks where the breath of the spirit blows him' (quoted in Arts Council 1984: iii), stating, rather naively: 'Art is resistant to bureaucratic planning. It may crop up anywhere. The wind bloweth where it listeth and we must keep our ears cocked if we are to hear the sound thereof' (ibid.: 17). Peter Hall, writing in the *Financial Times* in the early 1970s, put this more sharply: 'The [Arts] Council did not try to plan art into existence – always a barren and schematic procedure. The policy was to watch out for creativity wherever it occurred and then encourage it with a mite of subsidy. The process was organic' (quoted in Stark 1994: 12). The tendency to prescribe provision has, however, been replicated regionally. At the formation of the Southern Arts Association in 1970 its document *The Arts in the South* contained an application of the Arts Council's 1959 'rules' based on a quantitative hierarchy of facility across the cities, large and small towns of this mixed region of coastal/port, rural and urban areas:

(i)   Permanent repertory theatres with their own buildings in Portsmouth, Reading and Southampton.

(ii)   A large touring theatre in Southampton.

(iii) Large concert halls in Havant, Poole, Reading and Swindon.

(iv) Museum/Art galleries in Havant, Gosport, Fareham, Crawley and Poole.

(v)   Portsmouth, Southampton, Bournemouth, Reading, Poole, Havant, Worthing, Gosport, Fareham and Crawley should have professionally staffed arts centres, in use all year round.

(vi) Maidenhead, Eastleigh, Farnborough, Aldershot, Salisbury, Basingstoke, Bognor Regis, Winchester, Christchurch, Windsor, Horsham, Newbury, Andover, Ryde, and Chichester should have an arts centre which may be part of another establishment.

(vii) Portsmouth, Havant, Fareham, Crawley, Eastleigh, Farnborough, Aldershot, Lymington, Bognor Regis, Christchurch and Ryde should have a music club or local arts society presenting a regular series of professional events.

(viii) Poole, Havant, Fareham, Farnborough, Basingstoke, Bognor Regis, Newbury, Andover and Ryde should have facilities for showing regular touring art exhibitions.

The report also concluded: 'It is clear from the present low level of public investment in the arts at both local and national level that *something needs to be done*' (emphasis added). Whilst this 1970s' vision was more of 'shopping list' a than an investment strategy, it does closely resemble the current distribution of subsidised provision in the Southern Region, and was therefore influential in both district and county council arts funding through John Lane's *Every Town Should Have One* sentiment (1978) based on a regional map of existing provision. This facility-led approach mirrored the leisure planning seen in sports and recreation: these arts 'plans' focused on physical buildings, with little consideration of the 'soft' infrastructure of artistic creativity, education or specific art forms, particularly those that are not building-based such as literature and broadcast media. Over twenty years on from the Southern Arts Association proposals, the successor agency for the region, the Southern Arts Board, reaffirmed the seemingly contradictory approach, mirroring the national position: 'Southern Arts is not in business to plan the arts, but to plan for the arts, to create a climate of opinion, a strategy and a framework for support in which the arts can flourish and develop' (1990: 1). For example, within this region the County of Hampshire Arts Department has been structured around two *arts centre* officers, following a spatial mapping approach to provision rather than an art form or cultural-need approach. Provision in this sense has therefore tended to be homogeneous, mirroring sport and other recreation amenities.

The *local* arts plan has therefore been the prime mechanism by which regional arts bodies have attempted to influence local arts and develop a network of provision and thereby the adoption of arts policies, effectively taking a comparative approach, despite the absence of any agreed arts planning norms. A conclusion here is that it was the regional arts associations that increasingly sought to take on the role of planners, rather than the planning or arts and recreation departments (and the planning and leisure professions) themselves. The preparation of arts plans, informed by audits of borough facilities, provision and participation profiles, has been a feature of regional and borough arts development since the mid-1970s, for example in Germany where there has been integration with other municipal policy areas (Bianchini 1994), and the late 1970s/early 1980s in Britain where regional arts agencies acted as the catalyst to the encouragement of boroughs and districts in an arts audit exercise. These are normally followed by an arts policy and plan to promote certain art forms, to fill 'gaps' in provision and target specific locations and catchment areas, as well as to signal an increase in local authority arts funding, often in partnership with regional arts and other funders and sponsors. The weakness of such arts audits and plans has, however, been their top-down approach, and one that does not engage with the wider artistic, cultural community or take sufficient account of the urban social and economic context within which local and regional artistic and cultural activity exists. The audit of arts facilities may also be seen as an obvious, even a superfluous task, however a move towards a cultural policy and plan would require as much emphasis on cultural activity in non-arts and informal settings and locations, including education, youth and community, religious and amateur centres, as well as commercial leisure, entertainment and work spaces. The profile of urban arts centres nationally provided in Hutchison and Forrester (1987) and similarly for London (Forrester 1985), largely located in converted and second- or third-use buildings, provides some indication of this as discussed in Chapter 4, particularly when barriers to participation are most pronounced in institutionalised or elitist arts venues (Dobson and West 1988). In the city situation this was seen to be an opportunity where 'Experience has shown that there are still buildings with hitherto unknown, previous Arts Culture and Entertainment [ACE] uses which can be relatively easily brought back into active ACE use' (LPAC 1990b: 3).

Cultural audit and mapping

The importance of a comprehensive 'map' of existing, redundant and prospective arts facilities and participation (and non-usage) is therefore an essential prerequisite to planning for the arts and its supporting infrastructure. At a strategic level such assessments also need to be underpinned by research and facility audits:

> Ideally Unitary Development Plans should be supported by research into the demand for and supply of Arts, Culture and Entertainment facilities. This is important not only to assess the level of provision needed within the Borough and its effective location, but also to ensure that it is reconciled with that in neighbouring Boroughs.
>
> (ibid.: 3)

However, as Landry points out: 'If the audit is undertaken in a narrowly focused and unchallenging way it may be useless' (2000: 169). In practice, selectivity and preconceptions as to who constitutes a community and what culture is being audited can create a biased view of cultural activity (e.g. Tower Hamlets and Stepney, East London; Landry 1997a, b, Evans *et al.* 1999). From a more recent Australian perspective on cultural planning, a *Community Cultural Assessment* approach has been promoted that brings together these audit, mapping and consultative phases of local area planning (Guppy 1997: 14–15):

- Use demographic (e.g. census) data to identify relevant characteristics of the local population.
- Examine the cultural and social needs of different groups within the population.
- Categorise and list and/or map the area's cultural resources, including facilities, activities, people, organisations, valued places and landscapes, previous cultural projects, community services/facilities, economic activities and information.
- Identify plans for new or expanded cultural resources.
- Consider the relationships that exist between the area's various cultural resources.
- Identify barriers of access to cultural development activities by different population groups.
- Examine the actual or potential leadership and support roles in cultural development of civic, social, educational, religious, business and other organisations.
- Overview strengths and weaknesses in community cultural activity.
- Evaluate existing facilities/programs and needs for new or expanded ones.
- Evaluate the outcomes and appropriateness of previous cultural projects and activities.
- Consider relationships between cultural development and other areas of activity (e.g. tourism, employment).
- Cultural assessment needs to be a consultative and participatory process involving all interested groups within the local and artistic community. Community arts and other cultural activities can be used to stimulate ('tasters') interest.

This last point is fundamental since such a seemingly comprehensive but potentially mechanistic process carried out by 'experts', officials, and arts bureaucrats and special interests alone is where notions of 'arts planning' and town planning generally have failed in the past. As Landry suggests, 'planning should be more consultative and participatory as the discipline is too technocratic and incomprehensible to citizens because it expresses itself in a form that has little meaning in terms of day to day experience' (2000: 268). This consultative approach is therefore one with which wider city and area planning also needs to engage more successfully and more often. The frequency with which such exercises are undertaken will obviously depend on the dynamic nature and change factors within an area or community, changing aspirations and opportunities, but where a town or city planning cycle (political term, town plan) is the norm (e.g. every five years), undertaking cultural planning concurrently enables greater integration within land-use and environmental planning reviews (see Toronto *City Plan* below).

Planning methods and techniques

In the absence of any consideration of the planning requirements of the arts in town planning proper (as opposed to arts planning guidelines), and with no workable definition of 'amenity' in town planning legislation nor any real place for the 'arts' in leisure planning guidelines (Sillitoe 1969), a review of planning approaches to broader recreation and related amenity provision may offer possible applications to arts and cultural provision, as well as reasons for their differing treatment. The link between leisure and social problems and change had in any case followed a more normative approach to planning for leisure from the 1970s (Cullingworth 1979: 190), whilst at the regional level a degree of integration was promoted by government environment departments, including 'issues of structural importance to the area and their inter-relationships, e.g. employment, housing, transport and conservation, recreation and tourism' (DoE 1974: Circular 98). In 1973 (TCPA) local plans were made mandatory for boroughs and districts, and in the era of the first build-up of post-War youth and structural unemployment, studies into 'linkages' were undertaken, notably *Recreation Deprivation in Inner Urban Areas* (DoE 1977a) and *Leisure and Quality of Life Experiments* (DoE 1977b). As with leisure and recreation planning legislation and guidelines (Sillitoe 1969), arts and cultural activities and provision were largely absent from these policy- and profession-led developments, and research and literature in leisure and recreation planning itself was more developed both in North America, particularly for outdoor recreation (Walsh 1986), and in the UK, largely for open space, Green Belt and physical recreation (e.g. PR Acts; Sports Council 1968, Veal 1982).

The following planning approaches have therefore evolved from a largely quantitative, normative (rational recreation) and participatory philosophy, and when combined with a spatial application they draw from human geography, epitomised in planning for New Towns and the garden suburbs of London's sprawl (Howard 1902, Veal 1975).

Normative approach – the use of standards

A 'standard' in planning for leisure amenities is normally a prescribed level of provision of facilities or services, usually related to the level of population served. The value of such standards to planners is their neutrality and simplicity, and ease of understanding when communicating with politicians and local communities. As Judy Hillman noted on the much-disputed 1969 Greater London Development Plan, 'elected politicians are still politicians and few believe that an electorate is able to cope with intelligent discussion of the alternatives and difficulties ahead. So the picture was painted bright' (1971: 10). Examples of standards used in the UK include those given in Table 5.1.

As universal norms, such numerical standards can avoid duplication of provision, for example at local and regional levels. They also obviate local authorities having actually to assess *need* – a politically controversial, costly and sensitive exercise with the risk of raising false expectations and irreconcilable needs. Standards are also attractive because they can be presented in terms of public equity aims, since the standards are applied everywhere (or can be used to justify provision on equal terms), and a certain equality in provision per capita is ensured. This also simplifies resource allocation processes, particularly government grant-aid and local authority investment and public spending criteria. Evaluation of provision is also more easily measured by this quantitative approach, and a minimum, if 'conservative', level of provision is a likely outcome. Despite long-standing planning norms for open space provision for instance, actual quantitative provision has seldom reached these minimum standards. The French standard of 25 square metres per inhabitant compares actual provision of only 10 square

Table 5.1 Recreation facility planning standards in the UK

Facility	Planning norm/standard	Source
Playing fields	6 acres per 1,000 population	NPFA (1971)
Allotments	0.5 acre per 1,000 population	Thorpe Committee (1969)
District indoor sports centres	one per 40,000–90,000 population, more than one for each additional 50,000 population	Sports Council (1972)
Local indoor sports centres	23 square metres per 1,000 population	Sports Council (1977)
Indoor swimming pools	65 square metres per 1,000 population	Sports Council (1978)
Golf courses	one by nine-hole unit per 18,000 population	Sports Council (1972)
Libraries	one branch library per 15,000 population, maximum distance to nearest library in urban areas: 1 mile. Book purchases: 250 per annum/ 1,000 population	Ministry of Education (1959)
Children's play	1.5 acres per 1,000 population	NPFA (1971)
Open space	hierarchy – local accessibility standards	GLDP (GLC 1967)

metres in Paris and 18 square metres in Lille; the UK and NPFA standards of 16–18 square metres compares with only 10 square metres in London, with Rome (9 square metres), Berlin (13) and Ottawa (14) all falling short of these urban parks standards (Baud-Bovy and Lawson 1998, Bohrer and Evans 2000). It is therefore not surprising that in Europe (e.g. Sweden and the UK; Worpole *et al.* 1999, 2000) and North America the tendency is to move away from quantitative standards of recreational facility provision in preference for more responsive planning and integration of recreational amenities in development and design guidance and in regional plans.

There are also fundamental drawbacks in the use of such standardised criteria for provision, in that they rely on a hegemonic assessment of what is the *right type* of provision, and the *right level*, in short, by whom and how are standards determined? Local conditions also need to be reflected in the quantitative approach – these will vary between and amongst communities and groups from demographic (particularly age, given the wide group variation in certain arts activities), socio-economic, cultural to spatial. Standards also tend not to be dynamic or reflect socio-cultural change, including lifestyle, mobility and cultural trends, and diversity. A quantitative approach also ignores quality issues, provision and processes – a high-quality facility may 'compensate' for lower capacity – including design, ambience, location and age of facility – key factors in access and in maintaining participation levels (Craig 1991). The degree of substitution within provision also needs to be considered – if an area has no cinema, will theatre attendance be higher? What is the impact of non-arts leisure activity and facilities in terms of attendance and participation, home and out-of-home? The lack of qualitative, spatial (e.g. transport travel time, i.e. Clawson effects, physical access; Clawson and Knetch 1986) and local issues and preferences, and most importantly assessment of community needs and the requirements of artists and cultural intermediaries (e.g. *animateurs*) is generally overlooked in such standards. Standards may also fail to reflect historic or community factors such as provision or access under a legacy or 'trust' (e.g. heritage building/site, collection, public park), as well as ethnic and religious observances.

Moreover, a purely quantitative arts planning approach is not easily accepted (see Pick, Amis, Rees-Mogg, Eckardt), although in longer-term planning a hierarchy of provision may need to consider the equitable spread of arts facility-types and spaces for a given community. One of the difficulties in setting arts norms has been the heterogeneous nature of arts experience, programming and development (unlike, say, a swimming pool, museum or even a library), the mixture of local and touring work, and the need to experiment and take risks, leading to a dynamic range of provision and programming. Arts provision also sits alongside commercial arts and entertainment as well as multi-use venues (e.g. halls), but is distinguished by having insufficient accessible arts resources and, as noted above, is predominantly housed in second-hand or temporary buildings. The arts, outside of the palaces of culture and museum 'zones' or 'quarters', as discussed above, are therefore often the poor cousin in municipal leisure provision (Hutchison and Forrester 1987) and, it could be claimed, despite these drawbacks, that this directly results from their exclusion from amenity and planning standards.

Gross demand or comparative approach

This method, associated with a Maslow 'hierarchy of needs' approach to human development and psychology (1954) and with the assessment of under/non-participation, takes the overall level of participation for each selected cultural activity, as derived from national and regional surveys, and compares and applies these to a local area population. Participation and consumption rates are typically obtained from national census, household, family expenditure and other regular studies undertaken by government and other agencies, in addition to market and other consumer surveys. In addition to social surveys – e.g. in Britain the General Household Survey (GHS), Family Expenditure Survey (FES) and in The Netherlands the Social and Cultural Planning (SCP) survey – the Arts Council has since the 1980s commissioned audience surveys for both a range of art forms by social class, age, gender and region, and more general public opinion surveys on attitudes to and participation in cultural activities and events – whether subsidised or not. Figure 5.1 indicates attendance rates by socio-economic groups in comparison with the national/regional population distribution for a range of arts activities, based on this annual survey (BMRB; Evans *et al.* 2000). The higher income groups AB (professional/managerial) and C1 (manual skilled) are consistently higher attenders than lower ones, even in the more 'egalitarian' activities of cinema, jazz and pop/rock. Notably, women outweigh men in arts attendance with the exception of cinema and pop/rock, significantly so in ballet and contemporary dance audiences – from 63 to 69 per cent compared with 31 to 37 per cent. This profile of legitimated cultural consumption persists in most Western societies and is perhaps the most problematic challenge for cultural policy and planning, in some respects undermining their public/merit good status altogether:

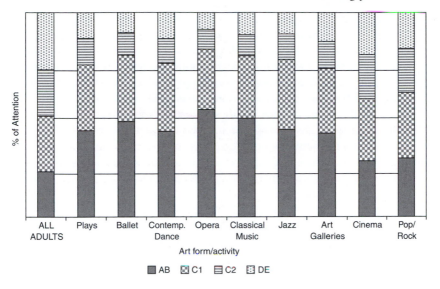

Figure 5.1 Profile of arts attenders in Britain by social grade, 1996

Source: BMRB International – Target Group Index (BMRB 1996)

Note: percentages are based on the sample attending each performance type. For example: 9.7 per cent of adult attenders at opera are in social grade C2 (while 22.5 per cent of all adults are in this grade)

> With the exception of a few art forms, e.g. murals, monumental sculptures, broadcasting, a majority of artistic activities lack properties attributed to the so-called 'public goods' . . . a large number of audience surveys show that these are made up largely of well-to-do persons. . . . Various studies however nevertheless show that government involvement in supporting the arts is welcome by the citizens.
>
> (Knutsson 1998: 26)

In practice, this 'comparative' approach draws largely on planning for recreation provision, for instance as put forward in 1968 by the Sports Council in *Planning for Sport* (Veal 1982). The approach can be taken in stages, the initial simplified version takes an overall level of participation for a particular activity as derived from a national or regional survey, and then applies this participation rate and profile to a local community, such as a local or district authority area. For example, the UK *Social Trends* (ONS 1999) indicated that 22 per cent of the adult population visited a museum or gallery in 1997/8 in the previous three months, 17 per cent a theatre and 34 per cent a cinema. A district with a population aged sixteen and over of 200,000 would therefore be expected to contain some 44,000 regular museum and gallery and 34,000 theatre visitors and 68,000 cinema-goers. From this usage norm, a calculation of the number and size of spaces required to meet this demand could be estimated taking into account frequency and population profile (e.g. families/children), and the result measured against actual provision in the area concerned. Profile is crucial of course since average participation rates conflate variation between age and other groups. In theatre-going this is relatively evenly spread in terms of age groups – between 14 per cent (16–24 year olds) and 18 per cent (25–34 and 45–59 year olds), in museums 26 per cent of 35–59 year olds visit at least quarterly, but only 18 per cent of 25–34 and over-60 year olds, whilst cinema is youth-dominated with 65 per cent of under twenty-fours participating, declining to only 11 per cent of those in their sixties. This method is more readily acceptable again, with more homogeneous provision, such as parks, play and certain sports facilities (see the norms above), but less so in the case of arts and cultural facilities. A flexible, multi-use arts centre may lend itself to such a quantitative demand approach, possibly dedicated arts venues such as theatres, cinemas, but in the nature of museum collections, perhaps less so. (The definition and valuation of what constitutes a local museum or art collection is complex, dependent upon historical and patronage influences, but collections are now less fixed or rooted in origin or 'place' – see Chapter 8.)

Comparisons with national or regional participation rates for a range of leisure and cultural activities can also be combined with a catchment area defined by access/transport (see below). This might be adjusted for differing distances from a venue or facility depending on user group (e.g. junior school, youth, adults) and transport mode (e.g. walking, car, bus, train). Investment in accessible public transport – price, frequency, safety – and dedicated transport services linking cultural and recreational venues have proven to increase usage and widen participation rates amongst different user groups (e.g. Jubilee Line Extension, London; CELTS 2000), whilst transport connections and links can also serve to regenerate areas and help support the viability of local as well as higher level facilities (e.g. Sheffield 'SYS', Lawless and Gore 1999; Bay Area Rapid Transit, San Francisco, Cervero and Landis 1997). Here this method can take a single point, i.e. a proposed new centre or venue, and measure a catchment in terms of distance travelled/journey times in relation to existing facilities using a comprehensive mapping approach, see above. This can also apply existing participation rates in the population catchment area based on national surveys (General Household Survey, Target Group Index, see above) indicating whether local participation is above or below regional/national averages, and if below this might indicate under-provision, if significantly above, over-supply might be created by additional facilities.

Examples of planning that have actually been based on existing participation data are however severely limited. For example, a review of cultural statistics in the European Union (1995) revealed a patchwork of data on cultural consumption and participation, with little or no comprehensive information of the supply of arts facilities, their spatial distribution or planning standards. A UNESCO initiative has also attempted to establish a comparable basis for cultural statistics and data, although problems with cross-national and cultural definitions will limit any results in the arts to high-level aggregates and simplistic categorisation. Improvement in the quality of cultural data – economic, participation, consumption and production – is therefore required, for instance in Italy: 'The main problem seems to be that the research on cultural statistics does not have high priority', and 'there is no comprehensive and uniform system of cultural statistics in Germany yet. . . . It can be said that the data generally in the areas of art and culture is unsatisfactory' (EU 1995). In the UK, the annual Arts Council Target Group Index (TGI) survey does provide a regional analysis of participation (or rather *attendance*) across a range of art forms and activities. Although this method does afford some simplified comparison, it again does not allow for regional/local and cultural differences, particularly variances in the *supply* of arts facilities, or their quality, artistic and otherwise, and no account of 'externalities' – socio-economic and other variables affecting demand and attendance.

The comparative approach however gains some credence in the absence of national cultural planning, particularly where the 'centre' dominates in the quality and quantity of arts facilities. In Greece, for example, a quantitative analysis of the *Geographical Distribution of the Cultural Spaces* (Deffner 1993) compared each region with the national whole, in terms of artistic and educational cultural spaces. This took this simple formula:

$$\frac{\text{Number of spaces in each region X/population of region X}}{\text{Total number of spaces in Greece/total population of Greece}}$$

Thus the resulting quotient for each region is 1.00 when the particular region has the same concentration of spaces as the whole of Greece. Whilst the capital, Athens, hosts nearly 24 per cent of the country's designated cultural spaces, its high population density (33 per cent of the total population living in the basin of Athens; Population Reference Bureau 1995) means that on this calculation the city has a 'below-average' proportion of cultural provision. In practice, of course, the inner core area of the city hosts the majority of facilities and even taking the larger metropolitan area, the national ratio of artistic spaces for Athens is 0.94. This per capita formula also ignores spatial, qualitative and cultural diversity issues, and the typology used is limited (*artistic*: theatres, music and dance spaces, cinema clubs; *educational*: museums, galleries, libraries and cultural centres; Deffner 1992a, b). Cultural planning in Greece also needs to be seen in the context of what Deffner terms 'a crisis in cultural spaces, a phenomenon which is connected to the side-stepping both of open and public spaces' (1993: 8). The reasons he gives for this crisis include the application of functionalism in space, in connection with rationalism; the commodification of space and time; the privatisation of space; and critically the dominance of the private use car.

A more sophisticated application of the comparative method breaks down population groups into age, gender and, if appropriate, other socio-cultural groups in terms of participation rates. Whilst the gross demand approach affords a comparison in terms of participation trends over time and a shorthand benchmark for local authorities seeking to justify investment on the grounds of 'under-provision', like expert norms, no spatial or access considerations are taken into account, notably public transport provision. A full-blown application of this method may however effectively model an area and incorporate such determinants of demand, using multiple regression and related econometric analysis methods. The major flaw in this approach, however, is its reliance on participation, as synonymous with 'demand', which leisure planning has come to adopt (Burton 1971, Wilkinson 1973, Field and MacGregor 1987). *Demand* in this sense is in fact *consumption*, and as such ignores unmet need, representing latent, excess or unrealised demand. This arises from a complex array of factors, in addition to the obvious supply-led nature of demand and non-availability of arts facilities in the first place, such as lack of information ('marketing'), location–time–price interactions, and less tangible determinants, such as education, skills (*cultural capital*) and substitution effects such as competition from similar activities – home video for cinema, compact disc

(CD) for live concert. As Ellison wrote on the joint Arts Council/BBC report into orchestral provision: 'Orchestras "lack audiences not fans". . . . Average classical music lover prefers to listen at home. . . . Can new means be found to lure them from their couches?' (1994: 3). In response today CDs are now being sent in the post pre-concert for premiers of new works as part of the ticket price (the cost of CD production is less than the paper ticket), and CDs are being sold two hours after a Boulez concert as a souvenir of the performance, whilst CD-ROM tickets for rock concerts let you order the T-shirt, programme, list gig dates, travel arrangements, band images and lyrics.

Barriers to participation

The policy of stimulating demand/usage and the resourcing of arts (and sports) development by local and regional arts authorities has from the 1970s been closely allied to arts centres and community arts projects as discussed in Chapter 4, and local arts development agencies and *animateurs*, and since the early 1980s education and outreach officers and programmes at theatres, museums and galleries. As Jacobs observed: 'Art's role in our society will not be effectively established until it permeates our social systems and is not thought of as just something inside the doors of a museum. Working outside the institution – in other sites, with everyday means, with daily issues – is a start in shifting the ideological position of art in our culture' (1995). Emphasis on inducing demand from non-users of arts facilities has concentrated the minds of arts venues and agencies, with the dual but potentially conflicting aims of new audience development and income generation to offset reductions in public subsidy, and the widening of audience profiles – including young people (future audiences), under-represented groups (race, gender, class) – and thereby meeting socio-cultural and public good objectives. Evidence in countries that have maintained longitudinal studies of arts participation and consump-

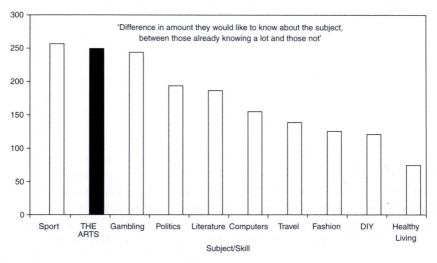

Figure 5.2 Knowledge gap between arts participants and 'non-users'
Source: Darton 1985: 20

tion reaffirm that the higher-income and socio-economic groups are disproportionately high 'beneficiaries' from public culture (Figure 5.1) and that therefore efforts at distributing the professional arts through physical dissemination and networks of facilities alone has largely failed to reach those under-participants and 'non-users'. Barriers to participation are deep-seated, as the work of Bourdieu and others has revealed, and Figure 5.2 gives an indication of the 'knowledge gap' between arts participants and non-users in comparison with other 'leisure pursuits' (Darton 1985: 19), suggesting that cultural capital is harder to acquire than other types of social capital and skills.

In this survey people were asked how much they knew about various subjects and also how much they would like to know in the future. Activities to the left of the chart are those where people already participating are enthusiastic about developing their interest further, but those not participating have little interest in doing so in the future – the knowledge and perception gap: 'on the whole the arts have failed to have an image of being fun, even to those who know a lot about them, and therefore appear inaccessible to those who do not have any expertise in arts-related matters' (ibid.: 19). Five years later, these barriers were still significant according to a follow-up survey (Table 5.2).

Table 5.2 Factors encouraging out-of-home leisure activity

Would you spend more time in away-from-home activities if?	Percentage citing yes	
	1990	*1985*
Cultural facilities were more to your liking: (comfort, convenience, atmosphere)	65	55
Facilities catered for the whole family	58	56
Streets were safer	54	54
Better parking facilities were available	43	29
Facilities for a variety of activities were in one small area	44	41
Better public transport	36	28
Better baby-sitting facilities	24	20

Sources: Henley Centre for Forecasting/PSC Surveys (1985, 1990: Chart 2 in Stewart 1990)

Environmental factors continue to rank and grow in importance, and the design and planning of the public realm and access to arts and cultural facilities provides some clues as to how cultural planning might approach user needs and support services. However, the paternalistic view of *users* (and critically 'non-users') of cultural and other community facilities has been long established in the essentially normative provision of civic amenities, and in the design and planning professions themselves. Here *users* signify *occupiers, inhabitants,* even *clients* – 'those who would not normally be expected to contribute to formulating the architect's brief' (Forty 2000: 312) – but all suggesting a powerless even disadvantaged role for the faceless 'user' and the problematic, ungrateful 'non-user'. Lefebvre had recognised this tension in *The Production of Space*: 'The word "user" [usager] . . . has something vague – and vaguely suspect – about it. "User of what?" one tends to wonder. . . . The user's space is *lived* – not represented (or conceived)' (1974: 362). As Forty has also observed: 'the decline of interest in the "user" and "user needs" corresponded to the decline in public-sector commissions in the 1980s. Perhaps another reason for dissatisfaction with the "user" has been that it is such an unsatisfactory way of characterizing the relationship people have with works of architecture: one would not talk about "using" a work of sculpture' (2000: 314). At the same time, 'use-value' was preferred over 'exchange-value' as a more emancipatory relationship between user and provider/designer and the more functional determinism which underscored the earlier socialist and post-war welfare boom in public and recreational facility building from the 1950s onward (see Chapter 4). The resurgence of public commissions of cultural buildings and facilities, together with an emerging community architecture and planning movement, has seen a return to the notion and primacy of the 'user' and the values of urban design quality (CABE 2001) and the public realm in attracting and maintaining attendance at cultural venues. Today, the dominant political ideology in Europe and North America has now taken this concept even further (Le Grand 1998) as an exercise in governmentality (Foucault 1991). Socially included/excluded now substitute for user/non-user (in the USA, *ghettoised*) but the connotations are social citizenship-based, with policies to combat exclusion and encourage inclusion adopted in cultural (DCMS 2000) (see Chapter 9) as well as other social spheres. In an increasingly commodified and privatised public realm, the environment, the concern for safety, ease of access, as well as the image and comfort that community and cultural facilities need to project, continue to rank as key factors in participation. How far private control of public spaces (e.g. CCTV, gates/guards) meets these essential access needs seems doubtful however, particularly for those without cars; or with children or families; and for the majority for whom familiarity, local proximity (Tables 5.4 and 5.5) and a non-intimidating relationship is a basic prerequisite for engagement in the first place.

   Time (or rather lack of it) is often cited as the prime reason for non-attendance, even above cost/finance, and therefore convenience and proximity to residential and work place as well as efficient and reliable transport access are predeterminants of capturing and maintaining audiences for the arts (as local sports, parks and library standards verify). Factors that act as barriers to participation in out-of-home activities and those which encourage activity also vary amongst different groups and over the life-cycle. Figure 5.3 shows the results of a study undertaken for UK art centres (Darton 1985) from which there are clear issues that affect potential participants according to their income, transport (e.g. car ownership), gender and age (where street safety is of greater concern).

   Consumption 'at home' does not of course imply subsequent participation in public arts activities, although this is a measure of latent demand for cultural provision if some of the barriers – real and perceived – can be adequately overcome and communicated to non-participants, as Table 5.3 suggests. Variations by region are also an indication of the relative difference in the supply of arts provision, such as the lower non-attender rates for the capital than less urbanised and endowed regions of the country.

   Linking home-based viewing habits with potential interest in a live or out-of-home equivalent activity may of course be simplistic, however the extent to which cultural consumption has gone indoors is evident for instance in the case of France where the decline in cinema attendance was in inverse relation to the growing number of films shown on television (Wangermée 1991). Cable film channels and videos also shorten the time-lag between first cinema showing and availability on television.

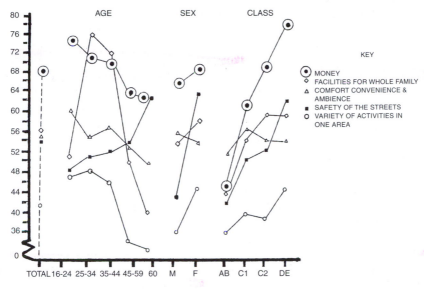

Figure 5.3 Issues encouraging people to stay at home

Source: Darton (1985: 13)

Table 5.3 Potential interest among non-attenders, by country and English region (percentage that does not currently attend these events but like to watch them on television)

	All adults	Greater London	Wales	North	England	Scotland
Plays	19.7	17.0	23.4	20.4	20.7	19.9
Ballet	7.0	7.0	6.3	5.9	7.1	5.8
Contemporary dance	4.7	6.0	3.8	2.9	4.8	4.0
Opera	6.2	6.3	7.0	5.5	6.3	3.9
Classical music	8.6	7.6	11.6	7.0	8.7	7.2
Jazz	5.4	6.4	4.7	4.4	5.5	4.6
Art galleries/exhibitions	2.8	2.9	2.4	2.8	2.9	3.0

Sources: adapted from BMRB International – Target Group Index 1995 and 1996 in Evans *et al.* (1997, 2000)

Note: percentages are based on the sample of all adults shown at the head of each column. For example, 5.8 per cent of adults in Scotland do not attend the ballet, but like to watch it on television compared with 7.0 per cent in the UK as a whole

Spatial approach and hierarchy of provision

The spatial approach to recreational planning not surprisingly draws mostly from outdoor recreation planning in North America (Clawson and Knetsch 1966, Walsh 1986), and also in the UK (Burton 1971, Wilkinson 1973, Henry 1980, Veal 1982, 1983). As discussed above, the arts centre and associated community arts movements from the 1960s had promoted the advantages of a wide range and scale of arts activities, facilities and low usage/entry costs, but over twenty years on, in the *National Inquiry into the Arts and the Community*, it was still noted that 'most of the population is still not in easy reach of such a facility. . . . The most common complaints concerning arts centres are the lack of them in many areas, the unsuitability of some for their purposes, and low public profile' (Brinson 1992: 68). Physical proximity is therefore an obvious but under-considered factor in amenity and in cultural planning in particular. As a graphic illustration of this, an assessment of facility needs and the associated distance relationships is shown in Figure 5.4 based on an English New Town plan. This shows the different spatial expectations between various community and recreational amenities in this case and also demonstrates the weakness of the simplistic spatial and other quantitative 'standards' approaches, namely their reliance on existing participation and levels of provision which place a theatre's catchment as 14 miles (for the car-owner), compared with 8 miles for a pool and squash court. Such models are therefore self-fulfilling in reinforcing hierarchies, rates of participation and access, which are dictated by predetermined levels of cultural provision, as well as programming (e.g. variety, participation) in traditional venues.

A more realistic but strategic approach to arts planning is therefore to consider catchment areas for out-of-home cultural provision, recognising that there are generally identifiable areas from which most users come. This method therefore relates the neighbourhood (ward, parish, estate) and 'centres' to population size, profile and accessibility, in particular the relationship of the location of arts facilities to work/home residencies and transport, both public and private. From empirical evidence (again audience- and participation-derived), theatres generally have larger catchment areas than swimming pools (and among pools, large, new pools a wider catchment area than small, older facilities); audiences travel further to modern dance and comedy; a new multiplex has a larger catchment than a high street cinema, and so on. The commercial planning of leisure and entertainment facilities (and emulated by the public sector) is increasingly predicated on the car-borne attender, and therefore the new multiplex cinema or leisure park operator will require a catchment population of 250,000–500,000, based on a *drive-time parameter* of one to two hours maximum (Grant 1990). The implications for cultural planning and provision are considerable and problematic: 'The concept of the working town, where work, housing and recreation are integrated, is in danger of being pulled apart by the centrifugal forces of out-of-town shopping centres, green field private housing, car-based leisure provision, and retirement villages' (Worpole 1992: 21). Planning here is dictated by transport links and the necessity of adequate car parking, along the lines of US arena developments, where the planning norm is one car parking space for each five seats. This is demonstrated in a new twelve-screen multiplex opened in the outer London Borough of Enfield, where 1,100 free car parking spaces were advertised (*Hollywood Comes to the Lee Valley*, London Borough of Enfield 1993). This new facility was expected to kill-off existing local cinemas in the borough altogether and this has succeeded here as in other town centres – over one hundred cinema screens closed in 1999, mainly small, local venues, whilst multiplexes continue to grow (LIRC 2000) (see Chapter 2). Those town centre venues that survive do so in a rundown state (Plates 5.1 and 5.2), whilst alternative uses are further limited by a general decline in retail activity due to the impact of new retail and leisure developments, whether fringe/out-of-town or city centre based.

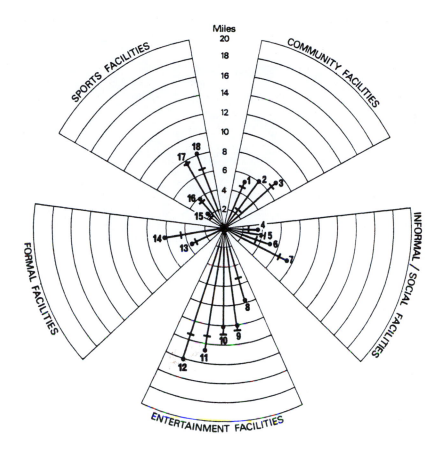

Miles
20
18
16
14
12
10

SPORTS FACILITIES

COMMUNITY FACILITIES

INFORMAL / SOCIAL FACILITIES

FORMAL FACILITIES

ENTERTAINMENT FACILITIES

1	Community Centre			10	Dance Hall
2	Church Hall			11	Discotheque
3	Village Hall			12	Theatre
4	Old Folks Facilities			13	Playschool
5	Youth Club			14	Library
6	Café	●——	Opinions of those with access to vehicle	15	Children's Play Area
7	Restaurant			16	Playing Field
8	Bingo Hall	⊥——	Opinions of those without access to vehicle	17	Squash Court
9	Cinema			18	Swimming Pool

Figure 5.4 Facilities needed: distance relationships
Sources: TRRU (1979) in Veal (1982: 29)

Plate 5.1 Cinema in decline, Milan (1998)

Plate 5.2 Cinema in decline, Rio de Janeiro (1998)

This catchment area approach can however also be applied for a range of local and regional amenities from arts, recreation to community resources, establishing a *hierarchy of need*, and identifying four potentially linked levels of provision: neighbourhood; local; city/borough-wide; and strategic (Figure 5.5).

The *hierarchy* scheme for example was applied in the planning of arts centre provision in the city of Portsmouth (Evans and Shaw 1992). Portsmouth has a population of

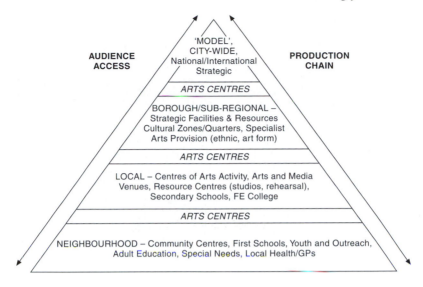

Figure 5.5 Hierarchy of arts provision and the pyramid of opportunity

Sources: Evans and Shaw (1992), after Veal (1982)

about 200,000 (1991 Census) and is one of the most densely populated urban locali-
ties in England outside of parts of Inner London. Following a policy initiative for all
council services, a *pyramid of opportunity* goal was set to represent the development and
access to cultural amenities in the city (Portsmouth City Council 1991), as Stark subsequently maintained: 'the importance of such facilities and their place in the *pyramid
of opportunity*' (1994: 16). This sought to make links in the production chain between
levels and quality of arts and cultural provision (see arts centres *form and function*
above), from neighbourhood to strategic centres: 'the appropriate structure for community based performance and visual arts development is perceived as a pro-active
network model . . . this has utilised a series of centres, with both specialist functions and
neighbourhood commitment, operating within an overall network on a regional basis
to enable a sharing of skills, resources and mutual support' (Portsmouth City Council
1991: 2). As part of this process, the current catchment and impact of existing arts provision was measured, using a facility planning chart (Figure 5.6).

Considering cultural facilities in terms of their strategic importance within a complex
city–region environment, requires certain criteria to be set (LPAC 1990a). Such an
approach also takes into account existing and potential centres and resources, where arts
activities – local, resident and touring – can take place, including 'non-arts' locations
such as schools, colleges and adult education, community and youth centres, as well as
sports and leisure centres, parks and museums. An effective mapping of an area in
terms of provision and participation and related access and spatial relationships is then
developed, at its most sophisticated drawing on user and audience surveys of existing
facilities as well as non-user and attitudinal research. This can identify gaps in the quantity, typology and quality of provision, and related infrastructure, such as transport,
zoning, parking, street lighting and pedestrianisation.

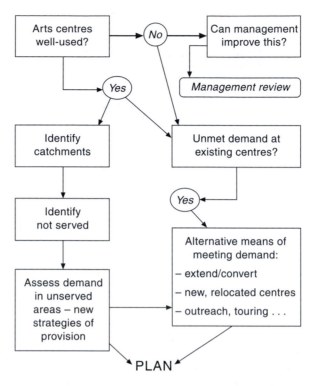

Figure 5.6 Arts facility planning
Source: Evans and Shaw (1992: 12)

Participation rates would also take into account frequency of use, within catchment areas, producing a penetration rate (PR) for a given population area:

$$PR = \text{frequency} \times \text{percentage of the population participating}$$

An annual participation rate in terms of art gallery visits of 10,000 per year may be made up of 10,000 individuals visiting once per year, or 2,500 people visiting four times, or in practice a mixture in between the two – a fundamental difference in terms of social and cultural objectives, and also in marketing and demand assessment. (Similar marked differentials can be recognised if classified operationally, e.g. length of visit: variety, facility needs, spending, etc.) For example, frequency of participation as distinct from the proportion of a population that participates (at least once a year) can reveal important differences between leisure activities, as the local area survey in Table 5.4 confirmed.

From this survey, whilst sports participation is three times the level of live arts, the penetration rate is only twice as high – a smaller population group attends sport more frequently than a higher population group which attends arts activities less frequently. The influence of supply (e.g. swimming pool, sports hall versus theatre, arts centre) and

Table 5.4 Participation in leisure pursuits for the residents of Portsmouth (three wards)

Activity	Average number of times participated	Proportion of residents who participated (%)	Penetration rate
Live arts	5.29	77	4.1
Sport	17.75	53	9.4
Exhibitions	4.3	78	3.4
Cinema	2.26	46	1.0
Other (museum, heritage, countryside, park)	7.88	90	7.1

Source: adapted from Vaughan (1992: 3)

travel distance to each facility are therefore key factors in comparative participation rates, as the survey later confirmed (see below). Simple attendance and participation rates can also be qualified by the timing, duration and ultimately the 'quality' of experience ('outcomes', aesthetic, audience and expert criticism, etc.; Evans 2000b).

Needs and community development approach

A criticism of the normative approach to arts and recreation planning has also been its tendency towards paternalism and minimum standards, rather than towards responding to the dynamic needs and demands of actual (and future) communities and participants. The view was expressed by some senior officers that there was a 'lack of demand' (i.e. expressed need) from Portsmouth residents for additional arts facilities and that this should be a test of any new initiatives and investment. Whilst this may seem reasonable, expressed demand generally arises from the more articulate and knowledgeable, seldom representative of a wider/larger group, and a need will normally arise only where a lack of provision is felt. For this, individuals and groups will normally have had some expo- sure to arts experiences. As the Portsmouth study (Evans and Shaw 1992, Vaughan 1992) and other research (Henley 1988, 1997) has indicated, there are also perceptual barriers to arts participation, e.g. education, skills, language, as well as the obvious finan- cial and access constraints. The importance of arts in education – in Bourdieu's terms, 'school and upbringing' – becomes paramount, requiring an integrated approach to arts development at community, school and in adult participation as well as in higher levels of provision.

Bourdieu's surveys of theatre, museum and gallery attendance and predeterminants of demand laid the basis for the importance of education, parental influence and over- all cultural capital in adult participation and evolution of 'taste'. With hindsight, his rigorous sociological analysis and emphasis on the tenets of structure (e.g. education) and agency (e.g. upbringing) ignored environmental determinants on cultural partici- pation, non-institutional (legitimate and popular) cultural activity and relied too heavily on national differences in education systems and the influence of cumulative national cultural capital. In particular, his discounting of cross-cultural visitation does not stand up to today's multicultural and cultural tourism flows to the very art institutions he chose to survey in the 1960s and 1970s (Bourdieu and Darbel 1991, Evans 1998a).

Furthermore, the special role of arts centres in arts in education and in encouraging adult participation has been argued from North American research into child exposure to the performing arts (Morrison and West 1986, Dobson and West 1988). This research on adult participation and attendance and childhood 'exposure' to the arts made a strong, positive linkage between participatory experience in *informal* settings (youth, community arts centre; Forrester 1985, Macdonald 1986), as opposed to formal and passive attendance at school, theatre trip or museum visit. This was echoed in Harland and Kinder's useful review of attitudes and barriers to youth participation which concluded that there was 'very little evidence on the school's contribution to encouraging applied and independent engagement with cultural venues . . . schools could help turn young people on to the arts, but they could also turn them off' (1999: 36–7). They also recommended that initial engagement should be more entertainment and experiential rather than educationally motivated ('school work') and that 'change through dramatic conversions experienced at single arts events (Hargreaves 1983) is less common than sustained support from significant others who mediate the arts over a period of time' (ibid.) – a sentiment shared with Williams and Braden in the 1970s. This correlation was high, irrespective of economic and educational backgrounds – the supposed determinants of cultural capital. For example, the arts centre and planning exercise carried out in the city of Portsmouth asked householders in three wards what changes in facility provision would encourage them to participate in arts activities, with the following results (Table 5.5).

Table 5.5 Portsmouth household survey (three wards)

	Agree (%)	Disagree (%)	No view (%)
If there were more arts facilities in this area I would go more often?	37	31	32
School buildings should be used more for arts activities (e.g. plays and music)?	68	9	23
Existing local facilities could be used more for arts activities?	63	5	32
Adult education centres could be more widely used for arts activities?	68	5	27

Source: adapted from Vaughan (1992: 22)

Notably it was not a *quantitative* increase in activity and facilities that was wanted but an increase in the use of *existing*, familiar community venues for arts activity – neighbourhood provision is therefore important in arts planning. However the pattern of cultural development has seen the growth of higher-scale facilities in city centres and suburban towns at the expense of local provision and access (Evans 1993a, 1999b). Arts centre and community provision should not therefore be seen solely as compensatory for the disadvantaged, but a key link in the arts production chain and infrastructure (see Chapter 6). The supply-led nature of arts activity is a continuing influence on attendances and the relative 'popularity' of one art form over another, for instance, dance which occurs in classical (ballet), modern (contemporary) and a host of popular forms and practices. However, in Britain, once a less self-consciously dancing and singing nation, despite the development of national and regional dance agencies and new venues, audience 'resistance' to modern dance persists (Table 5.2). Val Bourne, founding Director of the annual Dance Umbrella festival, cites the lack of large dance spaces, pointing out: 'Statistics say that [modern] dance audiences have dwindled but the fact is there's simply less for them to see' (Mackrell 1995: 20). Ironically, dance as a whole, from ballroom to bhangra, is the most *popular* cultural activity and according to Michael Argyle's survey in *The Sources of Joy* (1995) the most *joyful* pursuit, placing it first on a list of popular activities, followed by voluntary and charity work, and only then sport. The Arts Council of England also estimated (Mackrell 1995) that 5.5 million people – 10 per cent of the British population – regularly engages with dance as artists, audience or participants, and dance regularly attracts larger television audiences than opera (Table 5.3).

Many arts and cultural activities may therefore be 'hidden', undertaken within institutions and by amateur, youth groups, ethnic minorities, religious and other communities, 'privately'. It may be felt by some of these groups that public or official arts resources are 'not for them'. Some, however, may welcome support and facilities to develop and to participate/demonstrate more publicly, for example Vigar writes from the second-generation Cypriot perspective: 'For some practitioners it is important to challenge received notions of "Cypriotness" from within the community, so that a new identity can evolve which takes into account a broader, more universal reference of British Cypriots . . . clearly there is gap in provision here' (1991: 16–17). Furthermore, community development using the arts as an element of social action and empowerment has been closely allied to the growing community arts movement in Europe and North America from the late 1960s and 1970s (Kelly 1984, Braden 1977) where: 'Community artists or groups typically provided an alternative to the traditional buildings-based arts of theatre, arts centre or art gallery. Their low overheads created opportunities to be innovative; they worked with communities . . . to develop community, outreach and education programmes, including theatre-in-education and artists' residencies' (Davies and Selwood 1999: 71). European-wide initiatives originating in the early 1970s (before the UK's membership of the EEC), also focused on both arts development and notions of cultural democracy (see Chapter 7). Distinctions exist between the distributive and network approaches used by all political parties, to a greater or lesser extent, which became facility led (arts centres *maisons* and catchments, see above) and a more cultural democratic stance which would require a less paternalistic and more community-led approach to arts and cultural amenity and subsidy. Williams had argued that 'to achieve cultural growth, varying elements must be equally available and that new and unfamiliar things must be offered steadily over a long period to make a general change' (1961: 365). Su Braden, writing in the mid-1970s on artists in residencies, a popular approach to arts in education and community settings (visual artists, poets, etc.), challenges this perspective: 'to take a particular art form and expose a community to it in the hope that it will become less mysterious and more relevant was confused and wrong . . . it will succeed only when art is seen as a *part* of culture not the whole if it' (quoted in Patten 2000: 42). By definition, such change would require the end to the dominance of central arts councils and agencies and 'high-arts' and heritage preferences in resource allocation and planning, and therefore such an approach is generally rejected by the state and elite cultural hegemonies, as a serious alternative to national arts policy and interests. The degree of consultation in the formation and assessment of arts and plan-ning policies is, however, a measure of local authority and arts agency interest in the communities they (purport to) serve. The greater respect for and return to a plan-led system requiring community consultation in local area plan formulation (Healey *et al.* 1988, 1997, Nicholson 1992) suggests that the community development approach will need to be given greater attention, particularly if moves towards subsidiarity in decision-making and resource allocation are to turn from policy into practice. This extends, but is by no means limited, to the drafting, review and evaluation of land-use and environ-mental plans and both community and cultural involvement in their interpretation.

Arts Plan for London

At a city–region level, for instance, the *Arts Plan for London: 1990–95* (GLA 1990a) produced by the regional arts agency provides a particular example which drew on a spa-tial and physical plan, where over- and under-provision were matched with population concentration and growth, and with target group and other community interests. The *Arts Plan* was therefore structured around issues rather than art forms, placing stress on the needs of the user/consumer. This was based on research undertaken as part of the plan formulation (GLA 1990c: Appendix 2: 5), which concluded that:

1   The typical arts user tends to be white, middle class and middle-aged, although demand was increasing.
2   Non-use of the arts was highest amongst the working class, those on low or no incomes and people from ethnic minorities.
3   There was significant latent demand for arts activities amongst both users and non-users.
4   There are major physical and perceptual barriers which prevent people from attending arts events or participating in arts activities.
5   The views of the consumer are rarely sought by arts providers either in devel-oping existing arts provision, or in determining what new provision should exist.

Rather than adopt the traditional 'art form' based analysis such as drama, dance, music, visual arts, etc. therefore, the London *Arts Plan* used the approach developed by other regions and counties in a more norms-led distribution of cultural facility provision, and the promotion of access (equity – cultural and social, new audiences). During the same period a similar approach was adopted in Toronto in its *City Plan*, which was even more closely considered as part of the mainstream metropolitan area plan, with specific input from the Toronto Arts Council which was made up of practising artists (TAC 1992a, Evans 1996c). In London, the regional plan departed from the traditional art form/facility system and the regional arts board in consequence altered its own admin-istrative structures along the lines of a strategic view of arts provision in the capital: 'structured around an examination of issues rather than art form or geography . . . [which] places stress on the needs of the consumer with equal weight of the arts providers' (GLA 1990c: 5). This also divided London into subregions, and taking a comparative and consumption ('gross demand') analysis it targeted those areas such as

North East and Outer London boroughs that had less cultural facilities and arts resources than other London areas. Issue-based strategies included the soft infrastructure of marketing ('effectiveness'), as well as economic development and urban regeneration. The focus was particularly on arts in education and training, the disabled and the concentration of development funds on under-provided areas of the capital, notably Outer and East London: 'moving away from art form based strategies and guidelines into issue based strategies and function based guidelines' (GLA 1990d: 5). This led to collaboration between arts agencies and officers within planning, architecture, and other private and public agencies in developing and advocating specific arts planning policies, particularly in the context of new urban planning strategies. Characteristically they aimed to develop a close relationship with the regional (land-use) planning body (LPAC). This was in contrast to the period of the last city administration (Greater London Council 1981–6) where little or no collaboration existed between the regional arts association and the more proactive and populist GLC arts committees (Bianchini 1987, 1989). Despite political allegiances, the London boroughs generally supported the abolition of the GLC: 'a metropolitan authority tends to have too little power to be effective, and too much to be acceptable' (Young 1984: 5).

The collective term 'Arts, Culture and Entertainment' (ACE) was coined at this time and used throughout subsequent policy and planning guidelines. Unlike the various definitions emerging from the 1980s around the cultural and creative industries (see Chapter 6), this approach was inclusive and considered both space and linkages between cultural process and flows. This phrase now appears in most borough land-use plans (Appendices I and II) and serves as a useful compromise between the left- and right-wing positions and disquiet with the use of the terms 'culture', the 'high-arts' and 'popular entertainment'. In the planning context ACE has thus been defined as:

> A complex range of creative, enlivening and recreational activities; ranging from fine arts to ice shows, publishing to the theatre, photography to steel bands. They may be actively creative or passively responsive. They contribute to the intellectual, artistic and social quality of life of those living, working or visiting London. Some require specifically allocated spaces or facilities, others take place in shared buildings or in public spaces. They may be public or private, non-profit making or commercial or professional, be independent entities in their own right or form part of other activities. They are heavily inter-linked and interdependent with other activities, including sport and recreation; and manufacturing, business and service industries – filming, television, advertising, fashion, retailing, catering, publishing.
>
> (LPAC 1990a: 4)

This definition also recognised the convergence not least in a *world city*, between arts practice and consumption; the popularisation of high-arts (classical music, opera) and authentication of popular culture (e.g. jazz, ethnic arts), from 'pop classics', stadia opera to 'classical jazz' and world music – a global cultural fusion and interaction seen most vividly in youth culture, fashion and music. In Hannerz's words: 'there is now a world culture. It is marked by an organisation of diversity rather than the replication of uniformity. It is created through the increasing interconnectedness of varied local cultures, as well as through the development of cultures without a clear anchorage in one territory' (quoted in King 1991: 16). In London the lack of arts and cultural input to amenity and environmental planning was acknowledged by the London Planning Advisory Committee, which in consequence proposed *Criteria for Defining Strategically Important Arts, Cultural or Entertainment (ACE) Facilities* (LPAC 1990a). These developed a hierarchy of arts facilities, as outlined below. ACE facilities have complex functions – those that are physically small may have a more extensive role than their size implies, while some of the larger facilities cater for predominantly local audiences. Catchment area is therefore the main criterion for identifying those that are 'strategic'. Conventionally, it is usually assumed that facilities that draw visitors from more than one borough are *strategic*. However, local circumstances may require a more flexible definition. The following guidelines may assist in this assessment. It is suggested that if a facility falls within one or more of the following definitions it can be considered strategic:

1   A facility which draws a significant proportion of its visitors either from: abroad, the rest of the country; the rest of the SE region, London as a whole (these can be considered 'higher level' facilities) or from more than one Borough.
2   A facility which provides a service for areas where there is a concentration of workers, a significant proportion of whom travel there from outside the Borough.
3   A facility which is unique to this sector of London (a sector being defined here as a Borough and its neighbours).
4   A type of facility with special amenities, e.g. access for the handicapped, and which is unique to this sector.
5   A type of facility which caters for specific groups, e.g. cultural minorities and which is unique to this sector.
6   A facility which is or will be used by a significant number of visitors to London.
7   A type of facility able to accommodate special events which is unique to this sector.
8   A type of facility with particular historic associations which is unique to this sector.

(LPAC 1990a: Appendix 2)

Strategic provision here, whilst catchment-oriented, also recognises the importance of special needs, physical and cultural, and reflects the issue-based approach adopted by the regional *Arts Plan* (GLA 1990b). Furthermore a cultural planning approach would, for example, stress the overriding importance of the arts to the quality of a borough's environment and economy, in the broadest sense, where this was seen to meet local, regional and even international potential: 'In appropriate locations, to sustain and encourage the provision of arts facilities, to address the diverse needs of local communities and London's visitors, enhance the environment, widen and improve employment prospects and support the borough's contribution to London's role as a regional, national and international centre' (LPAC 1990b: 2). Whilst the above *Strategic Planning Policies* dealt with the principles and structure, a further guidance paper was issued: 'Partly because this is a relatively novel issue for many planners, the translation of these strategic policies to the local level could be assisted by the preparation of model policies for incorporation, either in whole or in part in UDPs' (ibid.: 1). The detailed inclusion in the borough plan itself provided key policy statements and wording which could be used directly or adapted by boroughs in their UDPs. This was a shrewd attempt by the arts advocates and sympathetic planners to talk planners' language and make it as easy as possible for them to recommend the adoption of arts planning policies in their UDPs, through the dissemination of these standard policy statements. The extent to which this novel set of cultural guidelines was adopted and interpreted is evaluated in a survey of the thirty-three local authorities in terms of the treatment of arts and cultural issues and measures in borough plans (Appendix I). This is followed by an extract of *Space for the Arts* (GLA 1991) that listed the mechanisms by which model planning policies could be implemented in practice.

A conclusion from this analysis of borough cultural policy and planning coordination, whilst not revealing a clear pattern, certainly indicated a considerable move towards valuing the input of the arts to town planning and to urban regeneration and policy development generally. This is in contrast with the situation a decade before when leisure planning was largely limited to sports and recreation provision and participation (Veal 1982, Stark 1994). The borough plans of the late 1970s and early 1980s contained little or no reference to arts and cultural facilities (except for policies aimed at the safeguarding of cinema, Steele 1983a, and this is still a specific concern given the decline in traditional town centre cinemas), and certainly no consideration of their contribution in urban and economic development. This is not surprising, since such consideration was neither encouraged nor prescribed through government planning rules and guidelines. From an analysis of borough-wide and local area plans between 1976 (GLDP) and the mid-1980s, arts, culture and entertainment facilities were simply listed in terms of facilities and in a minority of cases, an assessment of usage, demand and need in relation to current private and public provision. This took the form of a norms-led assessment of provision, for example in one borough: 'the need for a new central library, lack of community centres and play facilities' (Ealing Borough Plan 1982) and a degree of demand determination in another: 'demand for facilities is seen in terms of time and money; demand for activities by workers, residents, visitors' (Camden Planning Survey 1975, quoted in Steele 1983a: 29). Existing provision is matched against such latent and expressed demand, to reveal deficiencies in provision and distribution. The basis of needs assessment and norms during this period had been the regional *Recreation Study* (GLC 1975) which modelled demand, participant profiles and supply of recreation facilities. These were dominated by sports, play and municipal amenities (libraries), and drew on the use of population/facility standards and comparatives in planning for local amenities, as already discussed.

This traditional town and amenity planning approach was the basis of the 'shopping list' of arts and cultural facility 'need' that still persists over fifteen years later – a mix of expressed need (local interest groups), local councillor support and/or resistance, and intervention by entrepreneurs – social or commercial. The former can be identified by the community and arts centre movements that provided much of the impetus to growth in local arts provision and animation during the 1970s, as outlined in Chapter 4. As also concluded above, this was not the result of arts or borough planning, while the pursuit of minimum standards in provision has also largely failed to gain support (largely due to resource implications). Whilst boroughs may have been willing to look to arts and urban regeneration initiatives as part of specific development sites, the adoption of borough-wide policies is still resisted and restricted by narrowly focused, 'territorial' planning departments and officers, unwilling to cooperate or share with other departments and officers. For example, 'percent for art' and planning-gain policies were not included in several borough plans, which tended to refer only to the dual-use and safeguarding of existing arts facilities and the designation of a cultural quarter as part of their town centre strategy – it was clear that there had been little involvement of councils' own arts officers and no involvement from artists/arts groups in the plan preparation process itself (e.g. artist studios/workspaces, see Chapter 6).

The opportunity and imperative for local areas to adopt a plan-led approach also arose from the introduction of a new British National Lottery in 1995 as it has in lotteries which have significantly changed the basis of arts funding and distribution in countries such as Ireland, Australia, New Zealand, Canada and several US states (Shuster 1994, Evans 1995). In Britain, for instance, 75 per cent of Lottery fund applications were expected to emanate from local boroughs (BID 1994). This was borne out by the Voluntary Arts Network (VAN 1994) which carried out a survey of 270 local authorities of which 71 per cent were planning to pursue a capital scheme or lottery application for a local arts facility over the next three years. Exhibition (40 per cent) and performance facilities (38 per cent) were the most sought after, followed by rehearsal and meetings spaces (29 per cent). This enthusiasm for capital arts investment should be seen in the context of a decline and virtual standstill in local government spending over the previous fifteen years (Evans and Smeding 1997). Lottery funds were therefore primarily used to meet years of under-investment and lack of maintenance of existing facilities, i.e. substitution of public finance with lottery funds (Evans 1995), rather than the creation of new arts facilities. At the same time nearly 60 per cent of these authorities were in the process of establishing a local arts development plan or had agreed one in the previous three years and a further 10 per cent expected to agree such a plan in the following year.

Conclusion

Planning for the arts has been a prime tool in the pursuit of political ideology through cultural democracy and dissemination (at the extreme, a form of propaganda or compensation; Pick 1988), seen in the early enthusiasm for greater distribution and participation in national culture and recreation. At a national level these political programmes have not been sustained (as they have not, of course, in the Eastern Bloc), and the dominant hegemony of centralised national arts has in effect been the 'default' position of cultural provision, even where replicated at a regional/city level. This is no more evident than in the case of the *Grands Projets* (see Chapter 8), the largely unchanging profile of high-arts attenders, and the disproportionate annual allocation of resources to the major arts institutions (Evans *et al.* 2000).

Although *national* planning models have not tended to gain acceptance, at a regional and local level the desire to pursue more equitable distribution of facilities is persistent, even where no national or formal planning standards of provision exist. On the other hand, the acceptance of the more passive and homogeneous recreation amenities in terms of quantitative and comparative levels of provision has ensured greater integration and consideration within environmental planning and the planning process itself, notably protection ('presumption') against loss of amenity, e.g. green space, sports pitches, heritage. Higher levels of sports and recreation provision and participation than in the arts are strong arguments for the adoption of planning standards of arts facility provision. Quantitative norms do not however easily transfer to the variable nature of arts provision, particularly the single-use venue and facilities that are unique or rooted in the local or vernacular (e.g. museums). Here the hierarchy principle goes some way to developing a plan for levels of arts facilities which can provide a national network (e.g. for touring productions and exhibitions), on a regional and subregional basis. The *pyramid of opportunity* would seek to ensure that local and other levels of provision and activity can be linked to higher scales of facility (e.g. amateur to professional, small- to medium-scale, youth to adult, etc.). The multi-use and multi-art form centre does however offer a more universal facility model, and the expansion of local and larger-scale arts centres in many countries reflects this demand and potential, but this model also needs to be flexible and responsive to changing cultural tastes and forms – including the intercultural, local production and media technology.

Political associations and problems of definition when considering cultural aspirations in planning terms – the very nature of cultural expression and dissemination – have together confused attempts at arts planning and have repeatedly been used to support *anti-planning* arguments. The failure of redistributive policies to impact significantly on cultural participation and consumption habits also suggests that both a more sophisticated approach and, in some senses, one that is less politically driven is called for. This also means that a less hierarchical assessment of what makes up the arts, for example, moving towards a 'zero-base budget', would deal more equitably with the actuality of what people do and aspire to culturally: 'From the mid-1960s to the early 1980s, the aim of arts and cultural policy has been to make the contemporary expressions of high-art forms universally available through subsidy: both hip-hop and heritage, on the other hand were market-led' (Edgar 1991: 21). Whilst community planning generally entails greater bottom-up and community involvement in plan formulation, and consultation on development, cultural planning naturally would form part of such a process approach. The shifting locales for cultural consumption, the relationship between public arts and commercial entertainment, and the clear environmental factors that play as significant a part in cultural attitudes as those of cultural capital point to a greater consideration of the arts within town planning itself. This was recognised in the first attempt at developing arts or 'ACE' planning guidance in borough development and land-use plans, as detailed in the comparative survey in Appendix I. The attention to comprehensive cultural mapping and profiling of cultural and related activity (users, non-users, barriers, etc.) is also consistent with the inventory and trend analysis exercises undertaken in multi-step planning (So and Getzels 1988), however it is at the forecasting and scenario-making stage that planning agencies must involve the community in the evaluation and cost–benefit analysis of the possible outcomes in terms of the local environment, public amenities and responses to the development process in all its forms – economic, land-use and design.

Whilst planning for arts and cultural facilities has at best formed an aspect of amenity planning (and in the case of heritage, specific conservation consideration), as cities and national economies developed significant service economies and post-industrialisation spreads, economic planning imperatives have begun to look to cultural activity as both a commodity and production type. Planning for the arts has therefore also widened its scope to form part of economic development strategies, particularly linked to tourism and the cultural industries. Although this has raised the profile of arts provision and practice and related urban design, this has increasingly been in economic and employment rather than in amenity and environmental terms. The emergence and recognition of a cultural economy in the post-industrial town and city, and issues of planning arising from this, are now therefore discussed. As Scott observes: 'At the dawn of the twenty-first century, a very marked convergence between the spheres of cultural and economic development seems to be occurring' and as he goes on to warn: 'a deepening tension is evident between culture as something that is narrowly place-bound and culture as a pattern of non-place globalized events and experiences' (2000: 2–3).

Notes

1   The idea of Pareto-efficiency is used in modern welfare economics and is named after the economist Vilfredo Pareto, whose *Manual D'Economie Politique* was published in 1909. An allocation or land-use is Pareto-efficient for a given set of consumer tastes, impacts, benefits or resources if it is impossible to move to another which would make some people better off and nobody worse off. Winners and *losers* arising from a development would therefore be inefficient and be a Pareto loss (Begg *et al.* 1994).
2   Development control – the process through which a planning authority (e.g. borough or district council) determines whether a proposal for development should be granted planning permission taking into account material considerations such as any relevant development plans for the area.

6 The cultural economy

From arts amenity to cultural industry

Introduction

As previous chapters have suggested, the emergence and adoption of a social welfare rationale for public arts and cultural provision which built on the notions of civic culture and national glorification and the support of an industrial, urbanised workforce, cannot be entirely divorced from continuing and changing forms of popular culture, commercial entertainment and trade in cultural goods and services. The pattern of state control over cultural expression and popular pleasure has seen both a response in legitimated forms and places of cultural consumption, and the entrepreneurial efforts of impresarios, avant-garde/alternative art movements and a growing commercial entertainment world already exhibiting signs of globalisation. As Scott and others have noted: 'From their earliest origins, cities have exhibited a conspicuous capacity both to generate culture in the form of art, ideas, styles and ways of life, and to induce high levels of economic innovation and growth' (2000: 2). Whether the 'cultural is embedded in the economic', or vice versa: 'It is becoming more and more difficult to determine where the cultural economy begins and the rest of the capitalist economic order ends, for just as culture is increasingly subject to commodification, so one of the prevalent features of contemporary capitalism is its tendency to infuse an ever widening range of outputs with aesthetic and semiotic content' (ibid.: x). This is not however a contemporary phenomenon, as globalisation and cultural imperial processes and effective hegemonies have proven in earlier cosmopolitan societies.

The very instruments that fed the industrial revolution and manufacturing also made possible the mass production of cultural goods, notably print and published material, textiles and furnishings, and also later photography, film – silent and the 'talkies' – and recorded music. Indeed the post-War cultural policy-makers that emerged in Britain and France were well aware of the threat of cultural imperialism represented by Hollywood. André Malraux, in a newspaper interview in 1945, stated, rather Baudrillard-esquely, that 'European Culture did not exist', predicting that an American-led form was in gestation which he called 'La culture de l'Atlantique'. For ten years from the Liberation in 1944, French cinemas were required to show 50 per cent of English language films (i.e. Hollywood) ostensibly as part of the de-Nazification and democratisation process. This threat of course has driven French policy towards francophone culture, notably film and music, ever since, personified in the 1970s and 1980s through Culture Minister Jack Lang's anti-American stance and protective legislation. In 1981 there was a call for this quota to be increased to 60 per cent. Today French film protectionism is administered through the National Cinematography Centre, which redirects tax totalling over £250 million a year from television sets, cinema tickets and video sales to new films, including made-for-television (mostly to established film directors/producers) as well as to cinemas where attendance in France declined by 20 per cent between 1977 and 1987 (Wangermée 1991).

Britain also, in the short-lived honeymoon period of regional cultural development, saw a culture of difference as important in resisting the encroachment of the US movie, in the words of Keynes in 1945: 'Let every part of Merry England be merry in its own way. Death to Hollywood' (quoted in Pick 1991: 108), as well as in the earlier establishment of the British Film Institute (1933), the imposition of an entertainment tax on theatre as well as cinema in 1944, signalling a similar defence against American dominance and quota systems (Curran and Porter 1983). This tax was abolished and replaced in 1960 by the Eady levy on all cinema tickets to help fund British film production. The levy, which would now be worth £25 million a year, was however withdrawn in 1984. Today 90 per cent of the UK cinema box office income is taken by films either originating from or financed by the USA, with British films representing only 5 per cent of this market, in contrast with the 35 per cent represented by French films in France. However despite such intervention, 58 per cent of films shown in France still come from the USA. This fact and the diminishing penetration of the French language has supported cultural policies which saw significant public spending in the media and communications sectors – FF10 million in 1982, doubled the following year plus a further FF50 million for 'new technologies' (remembering that 75–80 per cent of computer communications and databases are in English). After the 1983 general election the French Cultural Ministry was further strengthened with a new Mission, '*economie culturelle et communication*', that took over the cultural industries budget. Culture Minister Lang announced in 1983 that the cultural sector was to be prioritised as part of a FF21 billion package over the next five years, including support and encouragement for micro-enterprises that had difficulties in securing finance for expansion and product development (Looseley 1999: 129). This was of course in addition to the accumulating public investment in the *Grands Projets culturels* initiated by the president (see Chapter 8). French promotion of the cultural economy whilst perceived as a francophone and therefore politico-cultural and heritage move, was also identified with both a wider diaspora and north–south divide, and as Looseley points out: 'objection to inauthentic "multinational cultures", which were rootless and alienating because they were not the natural expression of organic communities but manufactured from a lowest common denominator and then imposed on all' (1999: 79). This 'phoney internationalism' Lang saw in contrast to more genuine exchanges between 'natural cultural allies', therefore distinguishing between the global village and *mondialisation* and the American cultural imperialism and one-way trade associated with *globalisation*.

Culture industry

The coining of the term 'culture industry' has been associated with Adorno and Horkheimer (1943) in wartime Germany (Adorno 1991), as a pejorative view of, again, the Hollywood machine and the associated apparatuses of mass reproduction, and its Trojan horse entry to European culture. In two radio lectures in 1962 Adorno explained that they had first used the term 'mass culture', only to replace it with 'culture industry' to distinguish this concept from 'culture that spontaneously sprang from the masses themselves, that is, the current form of popular art. For by definition cultural industry is distinct from this art' (Adorno and Horkheimer 1964: 12–18). However the support of a thriving trade in cultural goods and 'services' (e.g. live performance) had of course identified those pre-industrial cities of culture and their so-called innovative milieu (Hall 1998), and the early industrial agglomeration seen in Paris, Berlin, Vienna in the late nineteenth century, as the cultural producers of Los Angeles, New York and London (again) were to gather in the twentieth century. Mass production on a truly industrial mechanised scale may not have been available to these early cultural cities – the printing and book trade in mid-fifteenth-century Florence was possibly: 'the first really efficient and innovative pan-European industry' (Johnson 2000: 17). However, a cosmopolitan cultural influence and reach was evident in both courtly culture which was exported between the seats of monarchy, and in the export to satellite towns and cities through colonial, military and also cultural transmission – certain goods attaining exchange value and association with quality and therefore demand, as they are today, through the twin symbolic and economic powers of branding and origin of creation/supply, e.g. Italian designer-goods, French fashion, German machinery, Japanese microelectronics and so on.

How far the city or other concentration of creative activity and production (and as I will discuss later, these are not necessarily interchangeable) develops according to economic phenomena of critical mass, geographic clustering and competitive and comparative advantage depends on the credence given to modern economic analysis (as opposed to its earlier, more communitarian roots – *Oikonoma* versus *Chrematistics*, Daly and Cobb 1989) and also to the cultural city convergence tendency suggested by Hall (1998), Cowen ('wealthy city-state' 1999) and city/globalisation theorists. On the other hand, the conditions which may be historical ('heritage'), even spiritual or sacred – notions that still have lived relevance in non-Western societies/places – and which lead to a build up of creative activity, may also contribute to what Lee refers to in adapting Bourdieu to the spatial sphere as a 'habitus of location'. Here he suggests that cities have enduring cultural orientations that exist and function relatively independent of their current populations or of the numerous social processes at any particular time: 'In this sense we can describe a city as having a certain cultural character . . . which clearly transcends the popular representations of the populations of certain cities, or that manifestly expressed by a city's public and private institutions' (1997: 132). The latter point is important in any consideration of cultural planning, since attempts by municipal and other political agencies to create or manipulate a city's cultural character are likely to fail, produce pastiche or superficial culture, and even drive out any inherent creative spirit that might exist in the first place. In the post-industrial city worldwide, this packaging and pursuit of urban regeneration through cultural activity and buildings has tended to replicate this approach, as I will discuss later, not least in the animation and event-driven programmes that private and public institutions have adopted in order to celebrate and consummate their major project and re-imaging efforts, creating what Handler sees as the 'ushering in of a "postmodern" global society of objectified culture, pseudo-events and spectacles' (1987: 10).

The extent to which imported cultural goods and cultural forms in the economic sense, substituted and crowded-out indigenous production and host-culture (e.g. Hollywood for French film), is also hard to measure in these earlier times, but where ability-to-pay was apparent, as demand was met by increasing leisure-time, spending and concern for the quality of public and private environments, it seems likely that both a real increase in demand for cultural consumption *and* switching to higher quality items and services supported a cultural industry long before Adorno's attack on US cultural imperialism. Indeed the concern for protection of national and regional cultures can also be seen as a response to the early experience of globalisation that predates the post-Fordist boom in the media and consumer goods in the late twentieth century, and the attempt at reinforcement of the nation-state and the identification of a national culture. What the post-War era began to experience however was the acceleration of the scope and speed that new cultural forms and products were disseminated and traded, and the feeling – real and perceived – that any new (especially 'foreign') cultural forms were crowding-out national and local cultural habits and consumption. Aside from the economic rationales – limiting imports, supporting domestic production and export (ironically it was alright to *export* national cultures!) – the underlying commodification of 'culture' itself lay at the root of the concern from the state, rational recreationists and social observers alike. The Arts & Crafts movement was one practical response to mass production (and urbanisation), the support for indigenous cultural production has been another, which France, Denmark, Italy and Germany and emerging nationalist artistic movements such as in Finland particularly embraced.

The expansion of cultural production, commercial entertainment and cultural consumption, both household and out-of-home activity, has also impacted on traditional amenity planning, and together with the new planning response to a changing spatial dimension of leisure activity this has presented a complex set of problems for the planner and for cultural policy and practice. At the same time, an emerging cultural economy that had always existed, but as a benign and unquantified aspect of private and public cultural activity, became not only recognised, quantified and celebrated, but also began to feature as significant elements of national and regional economic plans and of production (Gross Domestic Product, GDP) and employment growth. This occurred in direct relation to the decline in traditional manufacturing, engineering and primary/extraction industries, and to the growing affluence of a consumer class in Western countries and the *nouveau riche* of South America and South East Asia, as working hours gradually declined from the 1950s, paid holidays increased and disposable income and conditions conducive to spend it (despite cyclical economic crises and depressions) prevailed. The cultural economy therefore grew in relative importance as an industry in its own right, associated with activities such as tourism (cultural attractions, venues), export trade (e.g. 'invisibles' such as music, design, the art market, patents and copyright, etc.) and the exponentially expansive broadcast media. Planning for culture therefore no longer just entailed social facilities and amenities, higher-scale arts centres and civic 'flagships' and palaces of culture, but a form of economic planning for both cultural production, consumption and associated infrastructure such as transport, skills/training, workplace and other amenities.

The new cultural economy

The measurement and identification of the arts and cultural industries within city economies, and by extension regional and national macro-economies, had been first carried out in cities that had experienced significant decline and competition in their traditional manufacturing industries and port-based functions (e.g. docks, shipping). The arts had also possessed an economic dimension, not least due to their labour-intensive nature, but whether they were justified by social welfare or community rationales, or whether they were commercial entertainment-based, treating the 'symbolic economy' in financial and economic terms as with other 'industries' had never really arisen. Associating culture and leisure pursuits with the world of work had not been either desirable or a strategy that was likely to gain support, at the extreme a hangover from late eighteenth-/nineteenth-century Romanticism that challenged the encroaching reductive materialism (and the rationalism of the previous era) and viewed art as the antithesis to the prosaic pursuit of trade and reproduction. Adorno of course had something to say on measurement: 'culture might be precisely that condition that excludes a mentality capable of measuring it' (quoted in Jay 1973: 222), but this tension between the aesthetic and bureaucratic administration of culture (Bennett 1998: 196) was not entirely reconciled: 'culture suffers damage when it is planned and administered', *but* 'when left to itself . . . threatens to not only lose its possibility of effect but its very existence as well' (1991: 94). However, as discussed previously, the growing affluent Western society with time and money to spend (and invest) provided the demand, whilst the developing culture and leisure industry saw major opportunity in expanding its range and scale of activities. To cities seeking to revitalise and retain economic activity and life (accelerated by the suburban drift of residents and employers), the culture industry was a timely (re)discovery and which a series of impact studies sought to measure and promote. Whilst the first cities and states to embrace and highlight their cultural economy were in North America, this economic assessment was followed by European cities (Table 6.1), often drawing on the American models which generally adopted a Keynesian multiplier calculation of the direct and indirect employment and wealth creation attributed to selected arts and cultural activities.

National studies of the economic impact of the arts were also undertaken in Germany (1988), in the UK (Myerscough 1988, Casey *et al.* 1996, DCMS 1998), The Netherlands (Kloosterman and Elfring 1991), Wales (Bryan *et al.* 1998) and the USA (Heilbrun and Gray 1993), and today few nations, district or city/regions have *not* brought the arts and cultural industries into their employment and economic development portfolios, often as targeted and priority areas for investment and support (see Chapter 7 on the European Region). These studies focused on subsidised arts facilities or art forms (e.g. theatre), others on the cultural industries and visitor economy, but all stress that this area was both growing and likely to continue growing as other employment sectors faltered and declined. Moreover, the lack of a good range of cultural facilities, it was feared, risked a city or town losing out in the increasingly competitive city-imaging and relocation game that was being played out as firms and the managerial classes – new and old – became footloose.

Rationales and therefore the definitions of the arts and cultural industries that were adopted in the development of arts plans and cultural economy strategies fell into two

Table 6.1 Economic and employment impact studies of the arts and cultural industries

North America		Europe/other	
Location	Year	Location	Year
Vancouver	1976	Basle	1976
Los Angeles	1979	Cologne	1985
Baltimore	1977	Zurich	1985
Nebraska	1978	Amsterdam	1986
Kansas	1980	Bremen	1986
Columbus/St Paul/St Louis/		Liverpool	1987/1999
Salt Lake City/San Antonio/		Dortmund	1988
Springfield	1981	Manchester	1989
Washington	1981	Glasgow	1990
New York/New Jersey	1983/93	Birmingham	1991
Boston	1987	London	1991/2000
Ontario	1987	Hamburg	1998
Toronto	1991	Yorkshire and Humberside	1999
California	1994	London theatres	1998
Montreal	1998	Tokyo	1998

types, with the third, the social welfare and externality argument, losing its place in both resource and planning priorities:

1   Cultural industries – print and broadcast media, recorded music, design, art markets, digital technology/'art' (*sic*) – together rechristened the 'creative industries'.
2   Cultural tourism – arts and cultural venues, heritage sites and monuments, events and festivals as visitor attractions.
3   Arts amenities – arts facilities as public/merit goods, subsidised high/legiti-mated arts, civic and local arts and entertainment facilities.

In planning terms, these require different consideration – the first a more traditional concern for the means of production: workplace/space, distribution, training and invest-ment in R&D; the second, environmental planning that seeks to balance carrying capacity and visitor flows, transport and scale of facilities against cultural policy goals (e.g. artistic content, access, pricing) and finally arts-as-amenity which places civic arts resources, facilities and activities in a local/subregional planning context as considered in Chapter 5. In practice (and see *production chain* assessment below), these types and dif-fering levels of arts facility and cultural activity interrelate, both positively and negatively. Local arts amenities, or those serving a resident population may also combine, particu-larly seasonally, with tourist users, producing conflicts of crowding, price inflation and other environmental problems (e.g. parking, litter), but tourist usage may also provide income that sustains such facilities and employment throughout the year. Cultural tourism arguably also presents an opportunity for human exchange beyond local and national boundaries, as MacCannell claims, tourism is now the cultural component of globalisation (1996, Evans 1995c). The overlapping usage and interrelationships caused by this multilayered city destination is presented spatially (and implied temporally) by Burtenshaw *et al.* (1991) whose European tourist-historic city can now be applied to cultural capitals the world over (Figure 6.1).

Perhaps the key linkage in terms of cultural planning is between the cultural industries, small-scale production and creativity, and local economies – both through arts amenities that take on a cultural production role (e.g. arts and media resource centres), and in dedicated cultural workspaces that support seed-bed and small cultural enterprises (see below). The relationship between commercial arts and entertainment and the subsidised arts is also important if often ignored and unquantified. The hierarchy of arts facility is one example – the creative, participatory and production/performance link between the small- and medium-scale, amateur to professional, and so on. This is also evident in the links between education and training – public investment in 'human capital', in dance, drama, music and film colleges for instance, and between subsidised repertory theatre and commercial theatre and film (feature, television, music videos, etc.). In their study of the condition of theatres in England, the Theatres Trust remarked thus: 'Commercial, self-financing and subsidised theatres form an essential part of a cultural industry, which needs to be seen as indivisible. Their activities should be mutually supportive. They prosper or starve together' (1993: 6).

As town and city planning itself evolved responsibility for economic planning, from central business districts (CBDs), inward investment and employer/industry relocation programmes, cultural activity was therefore seen as a prime economic activity in its own right – where a critical mass of employment/consumption represented a significant

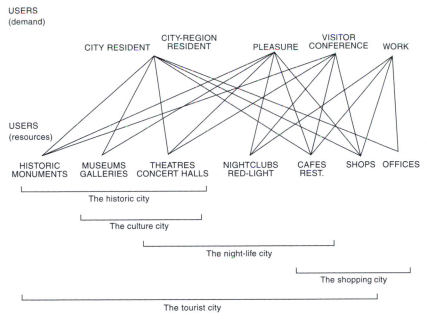

Figure 6.1 Some interrelationships of recreational users and uses in the touristic-historic city
Source: Burtenshaw *et al.* (1991)

proportion of local employment and trade – and/or as a key *quality-of-life* indicator and attraction for employer location and retention. As Rustin observes: 'It is curious that in a commodified world it seems to be *social* and not merely economic factors which determine whether capital investment will take place or not. Attractive locations for individual and collective consumption have become preconditions of production' (1994: 81). This had also been the case in the English Renaissance of the seventeenth century when: 'More fashionable and better housing, better civic facilities and the existence of an appealing new range of recreational services were critical in attracting the wealthy to visit towns and reside in them' (Borsay 1989: 312). The association of cultural and leisure amenities with inward investment and industry location was tested for instance in a survey of middle-managers in the national study undertaken by Myerscough (1988) of *The Economic Importance of the Arts in Britain*. Here, whilst quality of environment, notably proximity to green space, was particularly important, the range of cultural facilities was also felt to influence both the location decision and enjoyment once in place, in contrast to sport and recreation which declined in importance between the location decision and actual residence. Cultural activity was therefore an image and 'draw' and an amenity which was used on taking up residence in a new area.

Home and away

Whilst traditional local and neighbourhood arts and cultural amenities and entertainment venues present few planning problems (aside from the control of popular entertainment; see Chapter 3), the new supply-led developments, particularly larger-scale commercial entertainment such as multiplex cinemas and arenas, late-night clubs, and developments in communication, for example cable/satellite television and affordable digital technology, increasingly 'distort' the basis of traditional public leisure planning. Home-based and digital entertainment also tend to escape land-use controls over the public realm and related infrastructure (with exceptions such as planning control on external satellite dishes, noise pollution), and increasingly, therefore, 'private' recreation and entertainment competes with local arts amenity and skews the cultural economy (Darton 1985) and even other land-uses, industrial and social (Evans 1998d, 1999b). The higher rate of growth of in- versus out-of-home leisure in late twentieth-century Britain confirms this relative rate of change in the location of personal and family leisure activity (Table 6.2). This is part technology driven (enhanced by increased spending on home improvements, DIY) and part a reaction to the decline in accessible (and 'safe') leisure amenities, with factors such as transport, price, quality, environment and opening hours influencing choice and participation in out-of-home and more collective activities (Figure 6.2).

At the same time, temporal changes in demand and leisure provision have also been released by the liberalisation of licensing (e.g. alcohol, dancing/entertainment, admission of children) and shop trading hours, towards a twenty-four-hour and night-time economy (Bianchini *et al.* 1988, Kreitzman 1999). Within a year of the relaxation of allowing children into licensed premises in England, visits to pubs by families with under-age children increased by over 70 per cent (Evans 1993a). Where licensing controls vary across legislative boundaries this can also fuel cross-state movement such as in gambling and alcohol consumption in the USA. Speed of access and reduction in the

Table 6.2 Factors affecting location and enjoying and working in a location

Factors affecting location	%	Reasons for enjoying and working in a location	%
Pleasant environment/architecture	98	Access to pleasant countryside	93
Good transport links	84	*Museums, theatres, concerts and cultural facilities*	69
Outdoor recreation and sports	81	Parks and public gardens	62
Wide choice of housing	80	Fine old buildings	69
Good choice of schools	76	Participation in sports	54
Museums, theatres, concerts and other cultural facilities	74	Pubs, clubs and nightlife	50
		Spectator sports	20

Source: adapted from Myerscough (1988: 140, emphasis added)

real cost of air travel between cities also extends this horizon and the eclectic choice of cultural destination and experience. This has also spread and fragmented the traditional timing and availability of cultural and other consumption facilities in terms of both weekday/evenings and weekends, as working patterns also serve the seven-day work-and-play week. In a survey of *Leisure and Value for Time* (Henley 1998), weekend working was not only highest in the UK and Italy, but also significant in other European countries (Table 6.3). Paid holiday entitlement, which has supported a growing tourism and leisure industry since the 1950s, also overstates the extent of working pressures, with up to 25 per cent of holiday time not actually taken up, even in countries with neg-ligible holiday entitlement such as Japan (five to ten days).

Another 'side-effect' of this is the depressing fact that the workers who service the new leisure class and recreational centres suffer from leisure-time *loss* through shift and

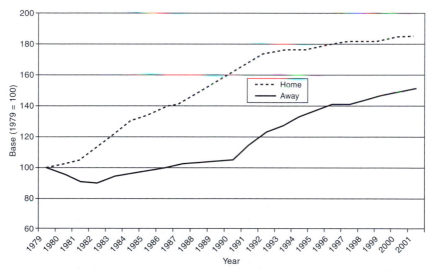

Figure 6.2 Growth of UK leisure spending – in-home and away, 1979–2001

Sources: Henley Centre for Forecasting (1985, 1990), Leisure Consultants (1996, 2000)

Table 6.3 The end of the 'weekend' as we once knew it. Percentage of employees sometimes or
    always working at weekends

%	UK	Italy	France	The Netherlands	Spain	Germany
Saturdays	57	56	36	34	34	31
Sundays	37	18	18	20	14	16

Sources: adapted from Eurostat in Henley (1998) and WTO (1999: 146)

part-time working, antisocial hours, and consequential social atomisation amongst families, reinforcing a widening leisure/consumption divide. As Gorz had already foreseen: 'the economic elite will buy leisure time by getting their own personal tasks done for them at low cost, by other people . . . [which] makes more time available for this elite and improves their quality of life; the leisure time of this economic elite provides jobs, which are in most cases insecure and underpaid, for a section of the masses excluded from the economic sphere' (1989: 5). Barriers to out-of-home cultural activity discussed in Chapter 5 are also of considerable importance if cultural consumption is not to be either ghettoised and dominated by one user-group such as young people on Friday and Saturday nights in town centres (Bianchini *et al.* 1988: 22, Worpole 1992b); mass gatherings (sport events, concerts, 'raves'; Redhead 1999); or made exclusive such as downtown arts and cultural venues and zones, to the well-heeled or tourists from 'out-of-town' (Selwyn 1993).

The evening economy and night-time city is, on the one hand, promoted as a means of catering for flexible leisure-time availability and choice, and as a way of extending the capacity and trade in retail and leisure facilities, and, on the other hand, a possible way of animating towns and cities at night, improving safety and widening access for certain groups, e.g. women, families and older members of society (Comedia 1991b, Bianchini 1994). The spatial dispersion of certain leisure and cultural activities and centres out of city centres, for example to out-of-town/urban fringe areas (e.g. multiplex cinemas, family entertainment centres), has in some respects undermined this strategy and in some cases made city-centre areas less safe and more dominated by younger, predominantly male groups, than previously (Thomas and Bromley 2000). Their study of night-time activity in the city of Swansea, South Wales, found that the majority of residents (62 per cent) rarely visited the city centre in the evening, and from those that did so regularly there was a clear socio-cultural divide between activities undertaken by different participant groups (Table 6.4).

Concern for noise pollution, excessive drinking and rowdiness, and car use and parking has also seen an increase in complaints to environmental health departments and local noise control officers where extended night-time activity conflicts with residential usage and amenity. The location and mix of cultural activities that operate late into the night require careful planning and control if the evening economy is not to create night-time hell for others, and this will also depend on the profile of local residents, their lifestyle and degree of tolerance which obviously varies widely between climates, nationalities, and their work, play and sleeping habits.

Practical access to cultural activities and places of collective consumption also presents

Table 6.4 Reasons cited for visits to Swansea city centre in the evening and at night

Reasons for visit	Percentage of of respondents	Age (years)	Social class/ occupations	Frequency of visit
Theatre	31.2	30–59, > 60	professional and semi-professional	fortnightly and monthly
Restaurants	28.9	30–59	professional and semi-professional	fortnightly and monthly
Late shopping	24.9	16–29, 30–59		fortnightly and monthly
Cinema	22.4	16–29	professional and semi-professional	fortnightly, monthly and weekly
Pubs	21.5	16–29	semi-skilled, unskilled and unemployed	weekly
Night clubs	11.1	16–29		weekly

Source: adapted from Thomas and Bromley (2000: 1417)

major problems of exclusion to those who are outside of the mobile economic and travel social groups. This is starkly seen in the differing rates of ownership/usage of a car between poorer inner-city residents and outer/suburban dwellers – in London for instance this may be between 30 and 90 per cent respectively (Evans 1998d) – whilst the opportunities for foreign, let alone domestic travel and holiday-taking is also highly skewed towards the higher socio-economic groups, which take three to four 'breaks' a year, and a core 30 per cent of the population which never does so (a proportion that has not shifted over the past twenty-five years; Evans 1996b). The importance for cultural activity and amenities at a local level, despite the emphasis on larger-scale but thinly distributed venues and centres (out-of-town, downtown), is therefore critical for those without the ability to 'escape', as is, of course, public/affordable transport to take up opportunities locally and wider afield (Evans 1998d).

The pleasure periphery

As well as temporal change, the traditional land-use distinction and separation between workplace, home and leisure has also begun to blur and overlap. In post-industrial city centres and former industrial zones this has again combined altogether to produce living and work space ('loft living', gentrification, studios, mixed-use, etc.) in proximity to cultural and entertainment facilities (e.g. Le Corbusier 1929), as time itself becomes a precious commodity: 'unlike the Victorian middle classes, today's new (predominantly white) urban professionals no longer wish to escape the urban core. Instead they wish to reclaim the city for themselves as workplace and pleasure zone' (Foord, quoted in Evans and Foord 1999). Becker's theory of the *work–leisure trade-off* (1965) highlighted the economic 'option' (*sic*) raised by this social change – the choice between time-intensive activities (for those whose time was of low economic value, e.g. the unemployed, but who lack disposable income), and goods-intensive activities where the opportunity cost or income foregone was high and therefore consumption was focused on shorter-burst but higher cost and intensive activity (Gorz 1989). The extended core–periphery and widening commuter area beyond cities and major towns – the *100 Mile City* (Sudjic 1993) and what Hogarth and Daniel have coined the *New Industrial Gypsies* (1988), have however become well-established, notwithstanding this belated return of certain groups to the inner-city, often encouraged by public-led investment in housing, transport and cultural amenities. The forms of gentrification and cultural provision that serve and follow these spatial flows have in practice reinforced and even exaggerated the existing cultural and economic divides, as many of the earlier downtown regeneration schemes have found (e.g. the Baltimore waterfront; Levine and Megida 1989). So even as Harvey observes: 'One of the possible benefits from cultural industries in the centres of cities is that, insofar as you can bring back predominantly the suburban upper middle class into the city centre, you will involve them at some level with what's going on in the city', he also admits that 'many people commute into the centre of the city to work and then go off back to the suburbs and are not bothered with what's going on elsewhere in the city' (1993: 8). Audience and visitor profiles for the performing arts and museums repeatedly confirm this social and spatial divide, and the scenario now familiar in the US 'Edge Cities' (Garreau 1991) and in the European and Latin American city periphery and new towns (Evans 1993a, Potter and Lloyd-Evans 1998, Massey *et al.* 1999) where this phenomenon is manifested in the shape of ever-larger shopping malls, urban fringe/green belt leisure 'parks', multiplex, entertainment and arena developments (Evans 1998d). An associated feature of this divide is the 'fortress' development – impenetrable and security-conscious apartment, retail and office buildings overseen by guards and closed-circuit television, designed literally to resist 'common' (*sic*) recreation and culture and limit community access and amenity planning – for instance, large shopping malls that shut-off traditional pedestrian routes for local people when closed. Richard Sennett describes this spatial and experiential shift brought on by the out-of-town entertainment zone:

> We saw a [war] film in a vast shopping mall on the northern periphery of New York City. There is nothing special about the mall, just a string of thirty or so stores built a generation ago near a highway; it includes a movie complex and is surrounded by a jumble of large parking lots . . . one result of the great urban transformation which is shifting population from densely packed urban center to thinner and more amorphous spaces, suburban housing tracts, shopping malls, office campuses and industrial parks.
>
> (1996: 17)

The pleasure periphery is therefore expanding both in terms of leisure consumption and physical usage, and in terms of land-use and traffic generation (Evans 1998d). As Garreau remarks: 'The hallmarks of these new urban centers are not the sidewalks of New York of song and fable. . . . But if an American finds himself tripping the light fantastic today on concrete, social scientists know where to look for him. He will be amid the crabapples blossoming under glassed-in skies where America retails its wares' (1991: 3). One of the original models for the second-generation out-of-town shopping centre, Houston's Galleria, opened in 1970 and 'gallerias' are now re-created in many cities worldwide. These also draw on their association with museums and art galleries, often adorned with obligatory public art installations and water-features whilst the urban fringe and reclaimed industrial and quarry sites chosen for these retail–leisure centres are rechristened in order to evoke a more peaceful and idyllic association, in the UK – Chester Oaks, Lakeside, Meadowhall, Merry Hill, Braehead, Bluewater Park, Cribbs Causeway, White Rose *et al.* One of the largest centres in this *biggest is best* trend is in West Edmonton, Alberta, which blurred shopping with entertainment even further: 'What they offer is a fair which, instead of travelling the world to reach its audience, sits still on one permanent site and waits for its visitors to come to it' (Sudjic 1993: 246). The ultimate one-stop-shop experience and the shopping centre-as-theme park and EXPO combined.

However as Gratz and Mintz claim, malls can only simulate public places:

> The malling of America has malled the culture and homogenized taste. A mall mindset is penetrating the public consciousness in insidious ways, often in the name of good design, improved style, and a perceived need for order. It contributes to the loss of local character. . . . A spreading of sameness, is overwhelming individualism, artistic quirkiness, the marks that distinguish one place from another'.
>
> (1998: 339)

The scale of such developments continues to grow, however, for instance in Minnesota's *Mall of America* with an enclosed entertainment mall based on Snoopy characters, 800 shops, eighteen cinemas, nightclubs, a health club, high-rise hotels and a 70-foot-high artificial mountain. It has 9.5 million square feet of enclosed shopping, entertainment and hotel space – twice that of a city such as Glasgow. Because of its scale, planners expected that people will spend two or three days there – as well as hotels, there is a mobile-home hook-up in the car park which will accommodate 12,750 car spaces. Kahn critically sees these developments as *Anti-Urban* sites:

> In the twentieth century enterprising forces have determined to break the city's robust character and reorganize its abundant small and vulgar structures. Striving to channel and co-opt urban energies while giving the impression of holding the city in place, urban designs become calculated efforts to invest a site with discernible limits, to give it identifiable features. By transforming the city from an unruly constellation into a collection of named places, the wish is to fend off the danger of becoming lost in the flow. . . . Some of these precincts attract global capital investment such as Battery Park City, New York, Canary Wharf, London (Docklands) and Euro-Disney outside of Paris. Others are more localized, such as corporate-sponsored atria, developer trade-off plazas, shopping malls and even cultural (and often publicly funded) museums and libraries.
>
> (1998: 18)

The quality of experience and symbolism presented by this commodification of countryside and out-of-town also represents a converging of home-based entertainment, consumption and *recreation*, where in industrial society they were exclusive, even the antithesis of each other – in time, place and purpose. As Urry (1995), Sudjic (1993) and others have posited, the central question about the out-of-town leisure–consumption phenomenon is whether it is its form that dictates the nature of urban living, or if it is the post-modern city that in fact dictates how the pleasure periphery has developed. The answer appears to be somewhere in between. Vast sheds that serve (some but not all) people from more than one city or conurbation (or country) demonstrate that urbanism has already become an amorphous landscape in which mobility allows anything to happen anywhere (Sudjic 1993).

In contrast, cities, particularly inner-urban and central zones and traditional industrial and crafts quarters, have managed to retain – some barely so – a concentration of spaces for both cultural production and consumption, whether trade- or individual-based. As Montgomery claims: 'Cities have always been the great centres of innovation, both technological and cultural. It is in cities that risks are taken, problems raised, experiments tested, ideas generated; it is historically to cities that creative people gravitate, for employment, stimulus or the comfort of strangers' (Urban Cultures Ltd 1994: 1). This partly sentimental perspective still has some resonance, particularly in some forms of art and exchange, and more seriously so if one considers the cosmopolitan society, ethnic and social mix which cities largely and uniquely engender. Cosmopolitanism itself raises fundamental issues for local governance and for cultural planning if it is not also to be reduced to a mix of ethnic goods markets and city-exotica (e.g. *Bhangratown* in London's East End and relocated/re-created Chinatowns in Birmingham, UK and Toronto – for ethnic 'quarters' read 'ghettos' . . .):

> cities such as London and New York are themselves now being colonized by people whose countries have been physically or economically colonized by the West. . . . This reverse pattern suggests that many global cities will increasingly need to address issues of racial, ethnic and cultural difference. The city, as the contested site of difference . . . must therefore provide spatially democratic frameworks which will support its citizens in order to construct new identities based on difference.
>
> (Mostafavi 1999: 9)

How far a market-based cultural industry policy and planning approach might supersede or even suppress the arts and their educational and social values depends also in part on the definition of the cultural industries themselves – how far 'form follows function'. Not surprisingly a reaction to the cultural industries argument persists, notwithstanding the accepted benefits, such as: 'highlighting things that should be known about the arts [and] a major justification for state involvement' (Wright 1993: 13–14). However, in Wright's view, echoing Adorno, as well as opening up access and experience of the arts to a wider audience and consumer, there has been a tendency for the cultural industries as producers and promoters of popular arts and media, to play down more aesthetic, 'artistic' considerations with 'a retreat from the very idea of artistic value, as a *more or less arbitrary* matter of elite taste and pretension . . . foisted on the public at large' (ibid.). A purely positivist stance would place the cultural industries in a neutral, mechanistic position as regards the creative arts that are 'transmitted'. Given the rationales and ideologies that have chosen urban cultural policy as a saviour or a strategy, the normative approach, including planning standards discussed earlier, places the political economy of the arts as inseparable from modern society and therefore from urban, economic and social policy spheres: 'Urban policy is now inseparable from cultural policy. The one informs the other. Both will depend on creating a working economic base' (Worpole 1991: 143). Von Eckardt, from the American perspective, puts this with more equanimity:

> Cultural planning does not imply any attempt to plan culture, it is the attempt to nurture and cultivate cultural activity so that the arts can grow with vigor and yield abundant fruit. Properly planned [it] will include *all* the arts, which can yield economic benefits, as well as enjoyment and inspiration for everyone.
>
> (1982: 15–16)

Cultural production

A typology of culture *industry* should therefore logically be based upon the *production of culture* and whilst the definitions and philosophical notions of culture shift over time and in relation to political ideology, the means and methods of production and evolution of art forms are on one level a function of technology, place and human cultural exchange, as much as the dynamic nature of cultural form and expression itself, which can be invented, reinvented and re-created. Marx had earlier rejected the object need/desire and consumption process understood by economists, instead constructing *production and consumption* as interrelated, each arbitrating and mediating one another (*Grundrisse* 1973), a relationship never more apparent than in today's consumer culture and cultural consumer. Thus, in his view new forms of production create new forms of response and new possibilities for consumption (Chanan 1980), and which the cultural industries and 'symbolic goods' in particular typify. The utilitarianism and modernist movements, the resurgence of design, crafts and other (artisan) skills have cumulatively shifted emphasis away from the precious and separateness of the arts from society: 'a construct that we make; the transcendence claimed for art in our society gives it status at the expense of influence' (Sinfield 1989: 129). Urban cultural policy and industry development is therefore largely responding to and directed at *influencing* socio-cultural and economic change and markets, rather than in Bourdieu's terms (1993), maintaining culture's status at the expense of, or as the antithesis to, its economic power and value. As Hewison maintains:

> even the most elitist high culture can be a product like any other. . . . This public culture, administered by governments and corporations, has absorbed the traditional values of high culture which it now deploys as a form of niche-marketing sustained by government agencies and private enterprises: museums, publishing houses, recording companies, art dealers, theatre owners and producers, quality newspapers and periodicals and radio and TV.
>
> (1990: 60)

In these terms, Mills's 'cultural apparatus' is both functional and all-inclusive: 'all the organisations and milieux in which artistic, intellectual, and scientific work goes on, and by which entertainment and information are produced and distributed' (1959: 252).

This holistic view of cultural production lends itself to the cultural industry and urban policy developments of the 1980s and 1990s, and also harks back to an urban renaissance epitomised in the eighteenth-century founding of the RSA (Royal Society for the Encouragement of Arts, Manufactures and Commerce), whose founder, William Shipley, proposed in 1754 that its mission should be: 'To embolden enterprise, to enlarge science, to refine art, to improve our manufactures and to extend our commerce' (quoted in RSA 1993). Shipley was a drawing master from Northampton who moved to London and over time assembled a group of eminent artists, philosophers and scientists. The RSA's first award scheme was for innovation in industrial design, notably the thresher machine and power looms (the technological cause of the Luddite's revolt). In 1760 the RSA also staged the first public exhibition of contemporary British artists, which led to the foundation of the Royal Academy in 1768. In 1856 the RSA launched examinations 'for the benefit of the working classes' (later handed on to the newly formed City & Guilds Institute), and established a National Training School for Musicians (to become the Royal College of Music). Contemporary initiatives include an Art for Architecture award, encouraging artists' and crafts' input to urban planning and design. This example is given to emphasise the tradition of the cultural industry approach to urban economic development and the creative economy, long before social welfare planning proper and state arts policy formulation.

Cultural industries – production of culture or the culture of production?

Urry describes the shift to post-Fordist consumption in *Consuming Places* (1995: 150–1), with greater consumer dominance, segmentation (authentic, 'niche', eclectic) and rejection of mass production in favour of more customised products and services and aesthetic tastes (Glennie and Thrift 1992, 1993). However the commercial imperatives of scale economies, branding and replication through franchising and uniformity suggest a more corrupted version. The attraction of the combination of a growing but more discerning market and local production/employment, together with a deep-seated reaction to national high-arts policies and globalisation, has however increasingly underpinned cultural industry strategies and related economic impact studies (see below). These have required a whole-hearted acceptance of the social market, as Worpole perhaps naively maintains: 'The left should stop getting so anxious about the word "market". Markets are mechanisms. They do not produce anything themselves . . . markets *per se* do [not] determine artistic content' (1991: 145; also Hillmand-Chartrand and McCaughey 1989: 45). A pragmatic but conceptual and functional description of what made up the cultural industries was therefore developed in London by Garnham (1983) during the heyday of the Greater London Council's cultural industries strategy (1985), which effectively took a stand against a whole tradition of cultural analysis (Raymond, Williams *et al.*):

> Cultural industries refers to those institutions in our society which employ the characteristic modes of production and organisation of industrial corporations to produce and disseminate symbols in the form of cultural goods and services,

generally, although not exclusively, as commodities – *and more succinctly*: the production and dissemination of symbolic meaning.

(Garnham quoted in GLC 1985: 146)

A distinction, originating with Adorno (see above), is acknowledged between the traditionally pre-industrial creative processes, which then employ mass reproduction and distribution methods (books and records), and those where the cultural form is itself industrial (newspapers, film, television). This definition is largely separate from the traditional performing and visual arts, and therefore from the notion of arts amenity and public/merit goods. These only come into the realm of the cultural industries when they are part of the market economy as tradable goods and when reproduction is achieved or is possible, simplistically the distinction between the 'arts' and the 'media'. Garnham's (1984) 'transmission of meaning' therefore encompassed the following core activities:

- the promotion, distribution and retailing of books, magazines and other printed materials and including the libraries service
- broadcasting
- the music industry, both live and recorded
- the film, video and photographic industry
- advertising
- the performing arts.

Although market based (rather than subsidy-dependent), the cultural industries that emerged in the late 1970s/early 1980s were also linked to community arts development and social action in British cities (Kelly 1984, Davies and Selwood 1999), as they were to be adopted by new urban left city authorities such as Bilbao and Barcelona (Bianchini and Parkinson 1993). Therefore attempts at legitimising all of the 'creative industries' by government runs the risk of conflict with, and control of, their elemental oppositional role, as politicians discover to their cost when they get too close to members of *showbiz*, radical artists or adopt the transient sound-bites as seen for instance in the rise and fall of the concept of *Cool Britannia* (Hitchcock 1998). Although attributed to journalist Mark Leonnard (*The Sunday Times* 26 April 1998: 9), the original context was not quite so flattering. An earlier article in *Newsweek* had described London as the 'World's Coolest City' (4 November 1996: 18), which encouraged visitors to 'go there and enjoy the fun while it lasts . . . this much is certain, it won't last' (ibid.).

Furthermore, the cultural industries are not necessarily benign and reductive production sectors separate from the more effete and precious arts which nonetheless directly or indirectly feed their creative content (human and creative 'capital'). Indeed their mass distribution and populist scope invests them with a power (and threat) which the arts seldom possess today, a fact that fuels the cultural imperialism claims and resistance fifty years on from Adorno and others' reaction to Hollywood. Continuing state controls through censorship, licensing, enforcement of patent and copyright and anti-trust/monopoly laws, bears witness to this power, whilst creative artists who have bypassed traditional art form and communication formats and institutions, and the democratisation offered by new cultural forms, networks and technologies, could not have been foreseen by Adorno at the time (During 1993). The analysis of the Frankfurt School with hindsight and now in another era predated the extent of urban cosmopolitan and multicultural society and its counter-effects to cultural imperialism, but it was also was rooted in a particular German tradition (Taylor 1997) of national cultural development – for instance Max Weber (and his wife) were amongst many other things prime *Grand Tourers*, which included a self-appointed responsibility and role in the promotion and celebration of German high-art (see Chapter 3).

Since the exploitation of cultural products and services is a concern of government/cultural agencies through, for example, creative industries and economic development – which might include policies for education, training/skills, technology and competitiveness – distinguishing between the different types of 'creative products' provides another perspective, which Huet *et al.* in *Capitalisme et Industries Culturelles* (1991) identified as:

1   Reproducible products which do not involve artistic workers (e.g. musical instruments).
2   Reproducible products which presuppose the involvement of artistic workers (e.g. records, books).
3   Poorly reproducible products (e.g. live shows, crafts).

The combination of 'artistic' or 'creative' labour with the 'non-artistic' is a feature of various arts and cultural production types, as the employment analyses reveal in the tables below, and Drucker and others (e.g. Reich 1991) would perhaps allocate these occupational roles between the superior 'knowledge' and supporting 'service' worker (Du Gay 1997). Zallo (1988) also argues for a distinction between cultural industries as such and the production of apparatuses for mediatising cultural consumption. The former consists of a series of branches defined by a profession common to several endeavours (publishing, programming, concerts, etc.), segments and related activities (such as *technoculture*, video production, design). The latter belongs to other areas such as electronic components or other industrial sectors such as non-electronic musical instrument manufacturing. Taking the level of industrialisation as his criterion, i.e. the degree to which labour is subjected to capital and the role it plays in commodity production and the realisation of value, Zallo further distinguishes the following:

1   Pre-industrial activities – mass cultural spectacles.
2   Discontinuous production – book publishing, record production, film and video production.
3   Continuous production – the printed press.
4   Continuous diffusion – radio and television video broadcasting, cable and satellite.
5   Cultural segments consisting of new telematics and informatics production and services for consumption – informatics programs, teletext, videotext, databases, Net/Web.

The last category can be identified with the dissemination as well as the digital creation opportunity offered by the Internet which impacts directly and indirectly on points 1–4. The challenge to live and recorded music for instance and issues of artistic ownership and control are seen in the downloading of music via the Internet, following the successful condensing of music files by a group of Italian engineers known as MP3. Unlike the Luddites, who reacted to the automated loom, the record industry went into denial, then called in the lawyers, then commissioned their own engineers to emulate the technique with copy protection, to be sold at a much higher price than its originator. Governments and industry in this case have little power over cultural dissemination and distribution, however access to the medium of distribution and creation is still an issue and state role in terms of education, and the planning and provision of technology within cultural facilities. As Lewis asserts: 'Neither [market or subsidy] are conducive to a number of cultural values. They suppress, in their very different ways, diversity and innovation in any popular cultural sense and they are only marginally concerned with creating a more harmonious or stimulating environment' (1990: 110).

Creative industries

The identification and later quantification of a series of cultural industries, some discrete, most closely interrelated, has focused on the effect on consumption, cultural values and the risk of homogeneity, but less so on the creative process and production itself. Cultural planning however has an interest in both of these in an environmental as well as a symbolic sense. At a national policy level for example the English Culture Ministry's (DCMS) *Creative Industries Task Force* came up with the following definition which is weighted towards the commodification properties of cultural products, as part of an attempt to coordinate across departments ('joined-up government'), policies to promote the creative industries 'which occupy an increasingly important place in the national economy' (1998: 3).

> Those activities which have their origin in individual creativity, skill and talent and which have a potential for wealth and job creation through the generation and exploitation of intellectual property – these are taken to include the following sectors: advertising, architecture, the art and antiques market, crafts, design, designer fashion, film, interactive leisure software, music, the performing arts, publishing, software and TV and radio.
>
> (ibid.)

The employment-to-turnover relationships of these so-called creative industries varies considerably (Figure 6.3), further undermining the grouping of these activities as either a single 'industry', or as one that is likely to be responsive to a generalised economic and interventionist approach by government, let alone 'planners'.

What these various definitions and approaches are grappling with is the twin notions of 'cultural' and 'creative' in terms of activity/process and economically – production and consumption – and the extent that creativity can be identified as a form of economic and cultural *capital* and therefore national 'comparative advantage'. As British Prime Minister Tony Blair pronounced in a review of creative and cultural education: 'Our aim must be to create a nation where the creative talents of

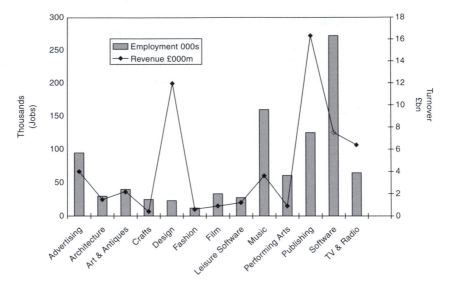

Figure 6.3 Creative industries employment and turnover by UK sector

Sources: DCMS (1988: 8), Evans (1999a)

all the people are used to build a true enterprise economy for the 21st century – where we compete on brains, not brawn' (NACCCE 1999: 5). The basic problem with associating the creative industries with broad concepts such as 'Innovation' and the 'Knowledge Economy', however, is their fluid and subjective status. Furthermore many aspects of economic production and human agency can claim both innovation and a knowledge-base, as Wolff observed: 'artistic creativity is not different in any relevant way from other forms of creative action' (1981: 9), whereas much cultural production lacks both creativity (e.g. originality) and high degrees of knowledge, since it is formulated on reproduction and replication. Becker also maintained that treating the arts as a special case in supply and demand relationships was question-able since: 'attempts to distinguish sharply the market for intellectual and artistic services from the market for "ordinary" goods have been the source of confusion and inconsistency' (1976: 11, also 1996, O'Hagan 1998). Centres of 'creative man-agement' and 'innovation'; industrial designers, engineers, educationalists, sales and marketing people – all could claim the creative tag, so if one is serious about iden-tifying a discrete segment of economic activity or identifiable process, this must be definable, measurable and separable from the 'non-creative' – a merely 'creative' or 'knowledge' economy clearly is not adequate for this purpose (Evans 1999a). The *symbolic economy* is perhaps what is trying to be captured here, the trade in signs, images and symbols (King 1990, Lash and Urry 1994), but operationalising this in specific activity and production terms is also problematic, not least since this version of creativity is increasingly predicated on what Leadbeater termed the 'thin-air busi-ness' and the fickle values assigned to the intangible assets of the knowledge economy (2000).

Cultural 'production chain'

These definitions and distinctions, particularly between the unique, perishable live arts and the mass-produced possibilities of commodification, also raise investment and subsidy dilemmas. Cultural planning of local resources, amenity and enterprise therefore requires a more sophisticated analysis of the arts and cultural industries and their interrelationships, in terms of social, economic, environmental and cultural policy – in fact an urban cultural policy requires an integrated approach to all of these, as attempted in borough 'unitary' planning terms (see Chapter 5 and Appendix I). In order to translate these definitions of cultural industry activity and production, and to provide a conceptual framework for arts and cultural planning and the determination of an arts infrastructure, a 'Production Chain' analysis has been applied to culture (Comedia 1991b: 18–20, Montgomery and Gavron 1991). This attempts to divide cultural economic activities between five interrelated stages and requires an assessment of a city or location's capacity to sustain and distribute cultural activity and products through its *infrastructure*.

1   *Beginnings* – ideas generation, copyright, creativity, training. This examines the capacity of a city or catchment as a site for ideas generation, for the patents, copyrights, trademarks it holds, and for the city's generic creativity. (Infrastructure: education, training, research and development resources.)
2   *Production* – from ideas to products, locations of. This assesses the capacity to turn this 'creativity' into production. Are the people, resources and productive capacities available to aid the transformation of ideas into marketable products? The assessment records the level and quality of impresarios, managers, producers, editors, engineers, as well as suppliers and makers of equipment in film, publishing, design; in-studio capacity; with regard to framemakers, scenery makers and so on. (Infrastructure: entrepreneurs, 'makers', technology, premises.)
3   *Circulation* – distribution, wholesale, marketing, information, circulation. This concerns the quality of agents and agencies, marketing agencies and promoters, distributors and wholesalers (say in film or publishing) or intermediaries/brokers, packagers and assemblers of product. It also includes assessing the quality of support materials such as catalogues, directories, archives, stock inventories, and other mechanisms which aid the sale and circulation of artistic products. (Infrastructure: intermediaries, agents, promoters, publishers, distributors, transport.)
4   *Delivering* – venues, television, cinema, shops. These are mechanisms that allow cultural product and services to be consumed and enjoyed; it is about how the places in which they are seen, experienced and bought. It means assessing the availability of theatres, cinemas, magazines, museums, record shops and outlets of distribution. Increasingly online and e-commerce forms of access and consumption will add and in part replace traditional modes of distributing cultural products, developing its own, more seamless production chain. (Infrastructure: venues, shops, media channels, magazines, museums and galleries.)

5   *Audiences* – watching, listening, viewing. This concerns the public and critical environment within which art works and cultural products are received, and involves the assessment of issues such as markets and audiences, as well as questions of pricing and targets (social market), targeting (including young people, gender and diversity). Tests might includes how far an area's cultural activities reach a wide spectrum of social and demographic groups, overseas markets, and in creating a lively cultural life. (Infrastructure: marketing, pricing, 'access', transport, safety.)

Distinctions can also be made between different types or functions of infrastructure: as a direct factor of production, technology and circulation; as indirect support services, public transport, policing, street cleaning, lighting; and as property – location, space and specialist premises (Montgomery and Gavron 1991).

Making sense of these stages in both human and spatial terms leads us to focus on two particular aspects of artistic and cultural production – *employment* and labour markets and the *places* of creation/production, and subsequent participation and consumption. This former aspect of cultural planning has not surprisingly attracted the interest of governments at all levels, through the economic potential of job creation, retention and enhancement ('quality' and 'skilled' jobs) and for which the arts and cultural industries are felt to hold promise.

Employment in the arts and cultural industries

National economic studies of the arts and cultural industries are generally based on traditional production sectors and occupations as captured in standard industry codes (SIC) classification and production statistics (Pratt 1997). These have developed from an industrial and manufacturing model of production, employment and markets, but one which increasingly does not adequately reflect the cultural production chain, flexible work-practices, or the post-industrial modes of production and distribution, which since the 1980s have been accelerated by new technology (Evans 1999a, Pratt 1998). With the emergence of a large freelance workforce component, multiple job occupancy, small-scale operation and part-time and home-based working, the standard classification of firms and employment therefore seldom reflects either the true size, scope or distribution of cultural employment and production. There is clearly already a gap between official statistics and actual practice which will widen as flexible, distance and contract working develops further. In the UK for instance it was found that 'standard industry classifications were outdated and did not reflect the reality of the new service/information sectors, including much arts and cultural activity. Consequently their value and economic contribution has been understated' (LAB 1992: 6). This is mirrored in the USA where 'the categories of standard classification are rarely fully informative, and this is especially true in the case of the cultural economy' (Scott 2000: 7).

The definitions of the culture industry outlined above also offer a moving target of what to include and exclude, the decision resting generally on the purpose to which such economic data are to be used (or abused; Evans 1999a). Where education and training, investment in R&D and technology is of interest, a wider definition may be more acceptable, but where arts and cultural policy, and related issues of diversity, creativity are of concern, a narrower definition of what constitutes cultural activity may be taken, since several areas of employment within the cultural sector are in fact largely 'non-cultural' or skill based (Tables 6.5–6.9). Applying such definitions in a cultural planning framework might also consider spatial, urban design and distributive issues, as well as socio-cultural and equity evaluation, rather than an activity or production focus.

What these employment and sectoral economic impact studies do not generally emphasise is the generally low pay, poor conditions and job insecurity of cultural workers, notably practising artists (as opposed to intermediaries). This weakness is ironically one of the reasons for: (1) employment growth and (2) high job and income multipliers compared with more capital-intensive industry sectors (Evans 1998e). With professional visual artists earning less than £10,000 a year (Towse 1995, Shaw 1996), 90 per cent of Equity (actors union) members at any one time on the 'dole' or in temporary (non-acting) employment, and the high turnover of creative personnel (GLA 1989), any creative industry strategy that celebrates the size of the sector in employment and value-added terms, but based on a low pay and conditions basis, is questionable and of doubtful sustainability (Evans 1999a). Cultural consumption also attracts high local expenditure multipliers, through the ancillary spending on entertainment, purchases and visits by arts audiences (transport, food and drink, books/programmes, complementary goods, etc.). Analysis of the multiplier effects of the 'arts' are not explored here (Myerscough 1988) not least due to their dubious measurement methods and the problematic assumptions underlying the so-called direct, indirect and induced employment and spending calculations. However a particular spatial issue, crudely analysed in economic geography through input–output modelling and the measurement of 'leakage' of economic activity out of an 'impact' area (whether a venue, cultural quarter, borough, region or even country), is the displacement and transference effects that multiplier calculations tend to ignore, for example substitution and switching from one locality to another and between cultural activity, e.g. theatre to cinema, CD for a concert ticket, and the inherent difficulty in establishing true *additionality* in public investment programmes (Connolly 1997, Evans 1998c). On the other hand the flexible work-practices, interrelationships, informalities and critical mass of cultural activity, as well as labour mobility, renders most analyses of the firm (e.g. sales–employment ratios) and standard industry and occupation assessments of the impact and profile of cultural production and consumption – in both economic and distribution terms – of limited use (Evans 1999a).

The spatial dimension to the cultural industries is however one long recognised, historically in the ascendance (and decline) of cultural cities (Hall 1998), and consequently most of the cultural industry policy and strategic developments have arisen in and been implemented by the individual city/district, such as in mid-1980s Spain, France and Germany (Bianchini and Parkinson 1993), and in London, Birmingham, Sheffield and in Liverpool for example: 'realising and developing the political, cultural and economic significance and benefits of the arts, as part of the cultural industries, in relation to economic development and planning' (*Resolution on Leisure Services Policy, District Labour Party Conference*, Abercromby Ward; LCC 1987). It is at the city and 'cultural quarter' level therefore that most in-depth research has been undertaken (see below) and any attempt at macro-economic modelling and measurement therefore needs look to these in order the better to understand both the structure and relationship between small-scale cultural production, and the public and commercial sectors.

Cultural industry quarters offer key entry points to understanding how the cultural economy works (e.g. the production chain effect), and even where new technology might reduce the importance of location – cost, proximity to markets and suppliers – it is significant that there is a clear preference for close location between notionally competing cultural industry firms (a return to the clusters of crafts firms in medieval towns). As Scott maintains: 'locational concentration enhances both its [cultural economy] competitive performance and its creative potentials' (2000: ix). This is also a phenomenon evident in concentrations of larger media and entertainment firms in Soho, London, the 8th *arrondissement*, Paris and Times Square, New York and the equivalent in other regional cities: 'The economic and spatial structures of the entertainment industry increasingly calls for the specific functions provided by cities. Global cities in particular, are emerging as strategic centres for both consumption and production' (Sassen and Roost 1999: 153).

Country and regional estimates of employment do, however, provide a comparison between sectors, between national and regional levels, and over time, the relative decline and growth in particular sectors. These may indicate technological, socio-cultural or demographic change, shifting tastes and competition, but may also signal constraints to cultural activity which policy and planning measures may address (such as investment in soft and hard infrastructure). Table 6.5 offers a picture of the scale and sectoral distribution of employment in a variety of cultural activities, in this first case, of the UK and the capital, London. The extent of cultural-specific occupations within these sectoral employment figures indicates the considerable difference in the degree of dependence on artistic versus support staff, notably between the performing and self-employed artist professions and the more functional publishing and museums and gallery employers, which tends to validate the distinction provided by Huet and Zallo (see above). (In 1992 the Museum Training Institute estimated that 88 per cent of staff in museum and heritage employment were non-curatorial – generic rather than museum-specific posts.)

As indicated, London dominates in several employment sectors – media, live arts and art markets (and this is swelled by the import of commuting cultural workers into the city and whose commissioning clients are generally based there), with 42 per cent of all national cultural sector employment and an estimated 75 per cent of all practising visual and craft artists. As King has noted: 'Culture . . . is a major export from the UK, with over one-third of British books exported and one-quarter of the world's records emanating from the UK. Some sectors of the printing industry such as newspapers have a much higher representation in London, with 40% of national employment in news-paper and periodical production' (1990: 150). Other UK regions however have a more even distribution of albeit a minority proportion of national employment in the cultural industries, with the adjoining (*commuter belt*) Southern, South Eastern and Eastern regions having the next highest proportion with between 8 and 10 per cent. The remainder however only host between 4 and 7 per cent with the West Midlands and Scotland mirroring the major city/regional concentration, with Birmingham and com-petitive Glasgow and Edinburgh respectively hosting the lion's share of regional cultural production and employment (O'Brien and Feist 1995).

Table 6.5 UK employment in the arts and cultural industries

Sector	Total for the UK (000s)	Percentage with cultural occupation	Percentage of UK workforce in London
Film	163	47	50
Publishing	83	17–25	26
Radio and television	28	44	58
Performing arts	26	86	47
Museums, heritage and galleries	16	19	20
Visual arts	14	75–100	30
Architecture	10	50	28
Crafts	9	100	6
Art and antique market	8	n/a	66
Fashion design	7	n/a	18
Music	3	5	44
Total employment	367	42	23
Self-employed with cultural occupation	255	'100'	
Total	622	66	31

Sources: Comedia (1991b), Evans (1998e), DPA (2000)

Note: by far the largest group of the self-employed with a cultural occupation is composed of performing artists, composers, writers, etc. (about 75,000), followed by architects and crafts and visual artists

In another cultural capital and *world city*, New York, employment data on the cultural industries and practising artists are also drawn from census, targeted employment studies and also from membership of trade unions and guilds. The estimates in Table 6.6 to some extent exaggerate the extent of cultural labour activity due to bodies where inactive members are included, such as musicians and actors which have both declined since the 1950s – musicians by over 50 per cent, whereas the numbers of writers have in fact increased. The city dominates the region's (state) employment in these cultural fields,

Table 6.6 Employment in the cultural sector of New York City

Sector	Total workforce (000s)
Film	41
Actors	15 (40% of USA)
Musicians	14 (about 4,000 'active')
Book publishing	12
Cinemas	3
Visual artists	7
Writers	4
Dancers	2
Graphic artists	2
Total	100

Sources: Port Authority of New York (1983, 1993), US Census of Service (1987), Comedia (1991b)

with over 75 per cent of performing, visual and media industry regional employment located in the city. Unlike London, however, New York has established competition from the West Coast in Hollywood as the centre of the film production industry where over 100,000 people are employed, but although New York has witnessed increased employment in actors and directors and exploits its comparative advantage as cosmopolitan, entertainment capital (Sassen and Roost 1999), Los Angeles and Chicago gained more economically active creative artists and workers during the 1980s. Where technical and creative skills are either mobile and/or where they exist in sufficient critical mass elsewhere with an infrastructure to support them (e.g. studio facilities), Hollywood's film production dominance can be broken, as the cities of Toronto, London, Dublin as well as New York have proven, particularly where cost advantages (e.g. tax breaks) are also made available. (Los Angeles itself first developed as a partially illicit or at least less-restricted location for filming, including its proximity to the Mexican border; Hall 1998) However, with decentralisation of cultural activity through regional cultural development, university and design colleges and contracting-out to lower cost production areas: 'New York City's mission may now be to sell and display rather than make art' (Zukin 1995: 149–50). The retailing and display of cultural goods and services no longer implies proximity to its creation and production, however this does sustain a considerable intermediary and advanced producer service economy, and a key role in the mediation and valuation of culture in all its forms and in its dissemination.

The urban concentration is also evident throughout the USA in terms of employment distribution in specific cultural production sectors (Table 6.7).

Table 6.7 Employment in selected cultural-products industries in US metropolitan areas, 1992

Industry	Employment in metropolitan areas (000s)	Metropolitan areas as percentage of USA
Book publishing	53.2	67
Jewellery, silverware	29.2	64
Motion picture production/distribution	241.2	97
Producers, orchestras, entertainers	58.5	85
Architectural services	93.7	77
Total – all cultural products	1,543.2	50

Sources: US Department of Commerce, Bureau of the Census (adapted from Scott 2000: 9)

Note: of the US population, 53.2 per cent lives in the forty designated metropolitan areas

A third cultural economy example is provided by francophone Canada, in this case the metropolitan region of Greater Montreal made up of five regions, and therein the city represented by *L'ile de Montréal*. Here again the city's dominance of employment in the cultural sector is further accentuated by the presence of the majority of public arts administration staff in this cultural (if not administrative) capital. The relationship between Montreal and the provincial capital, Quebec city, is also a competitive as well as cultural one, between the cosmopolitan, international festival city, and the historic seat of the francophone parliament and emotive capital of New France, but one struggling with its parochial heritage and self-conscious Québecois separatist status

(Laperièrre and Latouche 1999, Evans 2000a). Montreal wins this particular battle with the concentration of contemporary performing and visual arts centres, and as hub to cultural production, trade and tourism – both leisure and convention based (Table 6.8).

The spatial concentration that arises through the historic clustering of cultural activities, markets and production is therefore evident in both smaller countries, where cities dominate in certain cultural services, and in major cities such as New York, Paris and London, where geographic as well as sectoral changes can be observed since the 1980s and where technology and corporate strategies together create a core–periphery relationship in terms of the packaging of work and out-sourcing generally. This is particularly the case in the so-called 'knowledge economy' where the growth of independent creative production has been the corollary of contraction in head office and institutional employment in creative professions such as broadcasting and publishing. As Sassen points out, the clustering of information and cultural industries is counter-intuitive, given their freedom from physical proximity to markets and distribution channels, and therefore their ability to bypass the costs and limitations of a city location (1994). This does ignore however the hub-and-spoke relationship that producer services do maintain with the surrounding city–region markets and the strength of personal and cultural networks, and as she concedes: 'economies occur when they locate close to others that produce key inputs or whose proximity makes possible joint production of certain service offerings' (ibid.: 66). Specialist service and creative processes within several types of production chains – crafts, design, performing arts production – all exhibit clustering, as well as lifestyle effects (Evans 1990), and as Myerscough (1988) and others

Table 6.8 Employment in the cultural industries in metropolitan Montreal, 1992–3

Sector	Metropolitan Montreal (five regions)	L'ile de Montréal	Ile/five regions (%)
Design and fashion	20,242	12,418	61
Arts interpretation	6,238	5,556	89
Publishing	5,700	4,415	77
Film	4,651	4,337	93
Heritage	1,638	1,386	85
'Cultural'	1,198	1,070	89
Recording	1,084	1,038	96
Visual arts	475	434	91
Total 'culture'	41,432	30,670	74
Media	4,161	4,103	99
Publicity	4,624	2,838	61
Television and radio	2,099	1,948	99
Television distribution	995	865	87
Total media	11,879	9,754	82
Public administration	8,502	7,967	94
Total	61,813	48,391	78

Source: Juneau (1998)

Note: the five regions of the metropolitan region include Montegrie, Montreal, Laval, Lanadiere and Laurentides, with a total population of 4.01 million (1995)

have discovered: 'concentration arises out of the needs and expectations of the people likely to be employed in these high-skill jobs who tend to be attracted to the amenities and life-styles that large urban centres can offer' (Sassen 1994: 66). Like earlier centres of crafts production, locational preference and proximity can be seen visually in the case of Paris where both film production companies and the location of creative talent and agencies is closely aligned (Figures 6.4 and 6.5), with concentration in inner-city districts. Dominance in the location of the US music industry is another example, both in terms of the majors and also independent labels which also cluster in the five cities of Los Angeles, San Francisco, Chicago, Nashville and New York, and growing centres in Miami and Atlanta, as well as in Austin (new Country/Blue Grass) and Miami (e.g. Cuban). Synergy with related cultural industries such as broadcast, film, media and advertising combine with historical association (see Hall 1998 on Memphis) to reinforce their competitive advantage and act as magnets to others (Scott 2000).

As Scott therefore observes: 'In locational terms, firms subject to this sort of productive-cum-competitive regime typically converge together into transactions-intensive agglomeration . . . most importantly those large metropolitan areas that are rapidly becoming the principal hubs of cultural production in a post-Fordist global economic order' (ibid.: 7).

As noted, however, concentration of cultural activity and employment is not limited to the global or cultural city, since this scenario can be seen in Wales where the capital Cardiff ('south' region), ensures above-average employment in the performing arts and media, but not in traditional cultural amenities such as libraries and museums (despite the location of national museums in the city), or in the higher employment sectors of literature, publishing and visual arts and crafts which are more evenly spread amongst the regions and more rural areas of Wales. These cultural activities are maintained by stronger local and subregional markets than the visitor-based and larger broadcasting institutions (e.g. BBC Wales; Fuller-Love *et al.* 1996, Bryan *et al.* 1997) and multimedia companies based in the Welsh capital, which claims 4 per cent of all employment attributable to the 'arts', double the national rate (City of Cardiff 1994, Thomas and Roberts 1997). The ratio of the actual number of people working to full-time equivalent (FTE) employment also shows the extent of part-time working in the cultural industries, in all but the largely public library and museum sector (Table 6.9).

Employment and sectoral change in the cultural industries

The spatial and temporal shift in cultural production activity is particularly evident in London (Table 6.5), where following major structural change in cultural production and employment during the 1980s, 'jobless growth' is forecast to be the norm in the next phase of post-industrial employment in cultural and advanced producer services (Urban Cultures Ltd 1994: 12). Table 6.10 shows the major increase in the numbers of self-employed and freelance cultural and creative workers over this decade, at the cost of direct employment in music (as in New York), print/design and publishing sectors. What these changes in absolute employment belie is their distribution and the concentration of employment within the city. A drift eastwards of cultural activity is apparent in printing and publishing, including the transfer from traditional print to IT-based multimedia/desk-top publishing and design, advertising and visual arts/crafts –

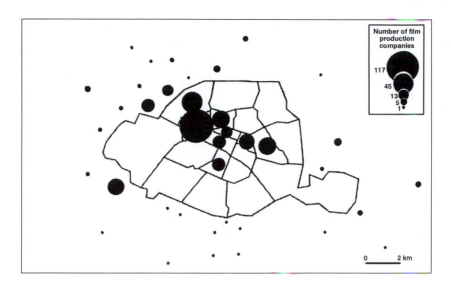

Figure 6.4 Location of film production companies in Paris
Source: Scott (2000: 100–2)

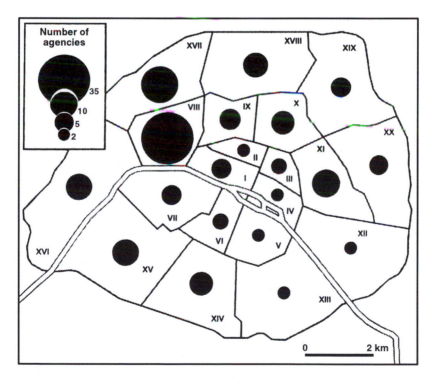

Figure 6.5 Location of creative talent and casting agencies in Paris
Source: Scott (2000: 100–2)

Table 6.9 Subregional estimates of arts and cultural industry employment in Wales

Wales region	Performing arts	Visual arts, crafts, design	Literature publishing	Media	Libraries, museums	General cultural	Total	Subregion share of all employment
North	18.1	23.1	20.4	15.2	21.7	16.5	19.8	23.3
West	13.2	24.2	21.7	10.0	19.9	16.6	18.3	26.2
South	61.5	45.9	38.4	71.0	42.0	54.5	52.7	43.5
Mid	7.1	6.9	19.5	3.8	16.4	12.4	9.3	7.0
Total (%)	100	100	100	100	100	100	100	100
Total jobs	4,985	11,038	2,413	3,691	2,885	3,574	26,585	
Total FTEs	2,623	5,421	1,243	2,906	2,331	1,609	16,134	

Source: Bryan et al. (1998)

Note: subregion 'south' includes the cities of Cardiff and Newport; 'west' includes Swansea

Table 6.10 Employment in the cultural industries in London by sector and subregion, 1981–91

Sector	1981	1987	1991	Percentage change, 1981–91	Inner West (%)	Inner East (%)	Outer West (%)	Outer East (%)	Outer South (%)
Audiovisual	52,594	59,035	58,530	11	20	7	-1	-29	-19
Printing and publishing	109,318	93,861	88,290	-19	-45	31	-23	13	-13
Music	5,190	5,605	2,376	-54	-58	-67	-35	-85	-47
Advertising	19,722	23,544	22,306	13	-7	91	229	70	46
Self-employed artists	3,049	4,740	4,810	58	54	59	81	7	90
Visual arts and crafts	13,686	15,363	14,042	3	1	13	No change	-18	12
Total	203,559	201,968	190,354	-6					

Source: adapted from Urban Cultures Ltd (1994)

influenced by cheaper rents and public regeneration programmes (see Chapter 7 on London Docklands), whilst the reduction in music production and instrument-making reflects the absolute decline in this activity, caused by technological advance, overseas competition (e.g. Japan) and restructuring (e.g. out-sourcing). Notwithstanding these structural and locational changes, the Inner West region of London (West End, Soho) *still* retains over 50 per cent of employment in the cultural industries, with 14 per cent in Outer West, followed by 25 per cent in Inner East, 3 per cent in Outer East and 8 per cent in Outer South London regions. This concentration is due to the high retention of audiovisual and advertising firms and support services in the Soho cultural industries 'quarter' which has developed, unlike its emulators in regenerating cities and other London areas, largely through the market and private capital (Sassen 1994). It also reflects, however, the historical location and preference of creative and cultural production, predating recent technology and planning legislation, and the presence of museums and galleries both public and private – in short this distribution reinforces the inheritance of the east–west socio-economic and land-use divide which London, as other early industrial cities, typifies.

The importance of cities in their concentration of cultural activity can also be measured in comparison with national economic estimates of the cultural economy, expressed in terms of total employment (Table 6.11). Here one can see the higher rate of overall employment represented in cities than nationally, and within cities the higher rates in local areas/districts, than the city as a whole. This is a concentration common to developing country cities, which also dominate their national production and services output, such as Sao Paulo with 36 per cent of Brazil's GNP and Lima with 43 per cent of Peru's GDP (Sassen 1994: 30–1).

Table 6.11 Cultural sector as a percentage of total employment in Europe

Country	Percentage of total employment	City/region, borough	Percentage of total city/ regional employment
Belgium	1.5	Catalonia, Spain	1
Finland	1.6	Helsinki, Finland	3.7
France	3.4	New York, USA	6–9
Germany	2.8	Boston, USA	4
Italy	2.5		
The Netherlands	2.2		
UK:	ACE 2.5	North West, UK	2.4
	CITF 5	Manchester, UK	3.5
		London, UK:	5–6
		London Borough of Haringey	11
		London Borough of Islington	10
Ireland	1.9		
Wales	2.6	Cardiff, UK	4
EU 15	2.3–3		

Sources: Evans (1989, 1993d, 2000b), Comedia (1991b), O'Brien and Feist (1995), DCMS (1998), EC (1998)

Elsewhere in Europe, the cultural sector is also a growing feature of city economies, along with financial services and (cultural) tourism activity. For instance in the Baltic States of Lithuania, Latvia and Estonia, the cultural industries account for 5 per cent of employment and GDP in Tallinn, Vilnius and Riga (Cooke *et al.* 2000).

As the profile of occupations within cultural production sectors varies, distinguishing the creative component of cultural industry activity is also important both in terms of education and training and in developing a greater understanding of the creative process itself and the type of support and infrastructure that might be usefully directed at each cultural sector. For example, from surveys of employment in the two London boroughs noted in Table 6.11, whilst not in the central 'core' of the city, they still host an above-average proportion of cultural employment than the city as a whole, whilst the division between arts and non-arts staff, and between full-time and fractional employment, reveals significantly different profiles between each production/activity sector (Table 6.12).

Table 6.12 Employment in the arts and cultural industries, London Boroughs of Haringey and Islington

Haringey – Outer East Sector	Average number of staff		
	Full-time	*Part-time*	*Freelance*
Performing arts	4	4.5	6
Visual arts	6		
Crafts	4		
Literary/publishing	6	4	7
Audiovisual/photography	5	4.5	
Graphic design/printing	3	2	

Note: in Haringey only thirty per cent of employees worked full-time, twenty-two per cent part-time, the remainder were freelance/on contract

Islington – Inner East Sector	Average number of staff	Full-time: part-time	Ratio of 'arts' to support staff
Performing arts	11	70:30	1.7:1
Visual arts	5	59:41	1.1:1
Crafts	10	87:13	2.1:1
Literary/publishing	21	83:17	1.1:1
Audiovisual	10	46:54	1.1:1
Print and design	11	91:9	4.2:1
Total (average)	12	79:21	1.7:1

Sources: Evans (1989, 1993d)

What these local cultural employment studies also confirm is the substantial micro-enterprise economy that operates outside of the major cultural institutions that tend to be located in the 'islands of culture' and CBD zones, and the extent of flexible (part-time, freelance) working. The intensification of cultural production and consumption between macro- and micro-levels is therefore a feature of the post-modern city-as-cultural workshop. Higher concentration of cultural activity is evident at lower, more localised zones than wider areas, both as centres of production and exhibition/performance. This is counter-intuitive in terms of the de-urbanisation drift, but also reflects the traditional economies of scale and process specialisation familiar in Fordist production, and the central core of arts and entertainment zones and museum quarters which have evolved historically and survive today. Industrial and economic geographers who have focused on the spatial dimension of labour and production, notably Massey (1984/95), Sassen (financial services, and with Roost 1999 on global media entertainment location) and on the post-industrial city/region (King 1991, Soja 2000), have surprisingly perhaps not seriously considered the arts and cultural industries which bridge both production *and* service sectors in new and old modes, although the cultural industries in terms of commercial entertainment (Hannigan 1999) and cultural production (Wynne 1992, O'Connor and Wynne 1996, Pratt 1997, 1998, Scott 2000) have started to receive attention. The economics of the arts has on the other hand been a minority study, focusing on the subsidised sector and state involvement (O'Hagan 1998, Frey and Pommerehne 1989), and on features such as the *cost-disease*, pricing, taxation and inflationary effects (Baumol and Bowen 1966, Peacock *et al.* 1984), as well as specific types of organisation such as museums and theatres. These have in consequence been primarily concerned with micro- and public-sector economic policy and organisational analysis, rather than macro-level planning and the arts industry *per se*.

The attention on city–regional cultural economies as summarised in Table 6.1 has therefore arisen due both to economic development and regeneration imperatives, notably cultural tourism, and in response to industrial–commercial decline in other sectors of city economies. The tendency today is to conflate the arts with cultural industry employment and multipliers (Myerscough 1988), however traditional spatial and industry models which academics and governments continue to employ are still rooted in both the delineated regional and industrial/sectoral models, despite the historical, 'pre-industrial' formation of cultural production and its complex behaviour in terms of labour, dissemination/distribution and capital flows. What distinguishes cultural production now is the high proportion of small (SME) and micro-enterprise activity (including self-employment) and their symbiotic if unequal relationship with a small number of powerful cultural organisations who commission, distribute and disseminate mass cultural products and performance. (This also has a historical parallel to the power of court, Church and city-state over the employment and commissions to the small artisan, crafts-guildsman and architect.) For instance, the contribution to the turnover of SMEs in the UK represented by *one* main client or customer was found to be over 50 per cent, and the *two largest* clients over 80 per cent (Stanworth *et al.* 1992), a case of the Pareto-effect in action where a small concentration in monetary or volume terms, represents a disproportionately high element of the turnover or level of activity, e.g. an 'independent' television production company working 75 per cent for one commissioning channel.

Performing and visual art galleries and museums also serve the growing cultural tourist market, which has ensured the viability of venues against declining domestic and local arts attendance generally. Spatially these commercial and public institutions (including governmental and educational) operate an effective hub-and-spoke system, supporting a range of specialist suppliers and creative companies and artists in a flexible labour/contract market. In many cases proximity between supplier/producer is important, such as performers (actors, musicians), technicians (stage, seamstress, designer) and venues (theatres, studios), as well as art teachers and colleges/students, but less so say between writers/composers and venues; visual artists and galleries; and film production, post-production and exhibition. The impact of the Internet and information communications technology (ICT) on both cultural production and distribution has yet to be fully experienced or measured, however in the sectors most susceptible to this technology, employment fragmentation and work-practices have been pronounced, such as in print and publishing (including newspapers and magazines), and this is expected also in the visual arts, education and training, music recording, broadcast media and in archival activities such as libraries and museum collections. In terms of consumption, over 60 per cent of online sales – which have risen exponentially ($1 billion in 1997, $15 billion in 1998, about $30 billion in 2001), driven by the US market-user – are from retailers who had businesses that predated the Internet rather than Web-based e-commerce operators such as the loss-making Amazon books. Two types of retailers have therefore emerged:

1   Pure 'play' whose only outlet is on the Web, e.g. books, CD/videos, music download.
2   Multichannel retailers combining traditional outlets (e.g. shops, catalogues).

The late twentieth century has of course witnessed a dose of technological determinism which was last experienced over a century ago. However to put this in some perspective, in a 1996 national survey of the impact of new technology on households in the USA the results were not the Net, satellite television or Sega Gameboy, but the humble microwave and VCR, as Fernandez-Armesto claims: 'generally societies only get the technology they want or need' (1996: 707). The spatial impact of e-commerce and online living is less apparent however: 'In the age of mass media identities are de-territorialized, hybridized and constantly shifting. The electronic society engenders new "wish-landscapes" through tourism and migration. But the fluidity, flexibility and decentralization which are often cited as the hallmarks of the new virtual global economy do not always find their equivalence in physical space' (Mostafavi 1999: 8–9). The reality of late twentieth-century living in the advanced West and emerging mega-cities is perhaps a long way from the futuristic millennia scenario which sees holidays in space and labour-saving robotics. The 'end of cities' armageddon theorists also see a residual role for cities as purely ceremonial centres (Fernandez-Armesto 1996), network or technopolis cities (Castells 1996) and nodes for international tourists and business people, but this seems just as deterministic. Predictions of de-urbanisation and the basis of the last thirty years of urban policy, with the benefit of hindsight, have been exaggerated – the *new millennium* city is perhaps as robust and just as paradoxical as it was at the turn of the eighteenth and nineteenth centuries, and the demand for and provision of the means of both collective and individual cultural consumption persist and continue to gain force. It would be hard to reject the association between this and the privatised, atomised forms of individual cultural consumption and recreation that have evolved in the era of post-industrial or late capitalism.

Whilst mass produced cultural forms are by definition not constrained by physical proximity between artist/creator, producer and end-consumer, even the pre-industrial live and visual arts/collections are potentially able to be toured, provided venues and networks of receiving cultural houses exist at the destination. Cultural planning therefore encompasses not only production, consumption and participation in the arts in a bounded geographic sense, but also the wider dissemination, exchange and therefore the notion of an arts infrastructure discussed in Chapter 5. These conceptual and functional distinctions and production phases require testing in practice since they present opportunities for intervention through planning and arts policy mechanisms. They have also underpinned the cities of culture and urban regeneration strategies which have looked to the arts and cultural flagships, as discussed in Chapters 7 and 8, as well as locations for specialist cultural production and distribution. A particular type of cultural industry production and premises-use, however, that draws on both traditional artisan and resurgent artist-led activity is the small-scale workshop or studio.

Cultural work space planning

The planning for cultural production in one sense has not surprisingly looked to the planning for industrial activity generally. However as already noted, a feature of post-industrial cultural activity has been its small-scale nature, its linkages and proximity to other forms of cultural activity (e.g. live performance/venues, design, print and publishing, IT, financing) and markets – trade and consumer – and therefore traditional industrial planning models that separated production from work force and consumer no longer hold good. Furthermore, the role of artists and cultural activity as part of wider regeneration processes such as the reuse of redundant buildings increasingly requires the integration of living, working and lifestyles not witnessed since the practice of medieval crafts workshops and quarters: 'It is after all the artist and not the bureaucrat who provides the catalyst for much change in our city by colonising redundant buildings, informing and challenging the design of the urban environment, and animating the street or square with performance' (LAB 1993: 26). In London, as in other cities, not-for-profit organisations were first established in the early 1970s (during an earlier property market boom), for example the ACME studios which manages over 460 studios in 230,000 square feet of converted industrial property. These include former meat pie, cosmetics and cigarette factories. As Worpole observed: 'In addition to the performance-based arts, small-scale workshop production is back on the agenda again both in handicrafts and hi-tech cultural forms such as video animation, computer graphics, electronic music, desk-top publishing' (1991: 143). One almost iconic type of space for cultural production is therefore the artists/crafts person's workspace or studio. This long-established mode of production has been a growing feature of post-industrial urban development, but one that has attempted to mediate within a largely inhospitable property and entrenched land-use separation and use-value-system (Jencks 1996). Some cities have however retained stronger provision and protection for artist workspaces, whilst others have developed planning policies that support and recognise the integration of uses, every day living and the cross-trading/production possibilities and attraction for consumers and visitors.

The development of a high concentration of cultural workers and facilities for public consumption has also been a familiar aspect of arts and entertainment zones such as theatre-land and cinema-land, and less structured entertainment zones such as

Amsterdam's 'red light' district (Burtenshaw *et al.* 1991), but this can also be seen in 'non-public' cultural activity that focuses on production separate from distribution/dissemination, such as in London's Soho (film/media production) and Clerkenwell (print, design, publishing). At a more local level, versions of agglomeration and cultural industry quarters can be seen (or not, i.e. they are hidden but none the less active), bringing together a range of compatible elements in the particular production chain, whether audiovisual, design, crafts, visual arts or producer services based. For example, the Arts and Crafts settlements or artisans villages in the Modena region of northern Italy have played an important part in the area's renaissance since the late 1970s and 1980s through the flexible production of individual settlements made up of a wide variety of small manufacturers (Lane 1998: 158). These form a network in which companies are competitive with and complementary to one another, in common with small crafts producers in managed workspaces (Evans 1990). These producer zones in turn form a 'polycentric grid' throughout the region, which has ensured their competitiveness over manufacturers (e.g. furniture) in traditional unplanned areas, such as in East London. As Worpole maintains: 'a new dynamism . . . is strongest where inventiveness and industry combine, as they do in Milan, Frankfurt and Paris' (1991: 144). The support and promotion, including planning policies, of local and regional cultural industry production and work space has also sought to redress the imbalance between capital and central production areas which nationally (and internationally, e.g. London, Paris and New York) dominate and act as an 'unfair magnet' for skilled workers and creative artists. Examples in the UK include Sheffield's Cultural Industries Quarter, Birmingham's Jewellery Quarter and Custard Factory, and similar clusters and regional city networks in Cardiff, Manchester, Liverpool and Yorkshire (Fleming 1999), often supported by designated cultural industries development agencies. As Fisher asserted: 'For years our cultural life, like almost every other aspect of British life, has been hugely weighted towards the south-east. Despite cities like Manchester developing a strong cultural voice, the capital has kept most things to itself, theatres, galleries, television companies, publishing houses, agents, work, investment. Now other cities are fighting back' (1991: 6).

   A particular feature of workspace and site-based regeneration, both new-build and conversion/renovation, has been the mixed or multi-use designed development, where a variety of public and private functions take place within a complex, such as arts and entertainment, retail, office and workspace, as well as residential – in what Coupland optimistically terms *Reclaiming the City* (1997). This contrasts the Utopian models of modern planning philosophy (e.g. the Athens Charter[1]) which ignored the interdependence of everyday activities, recognising instead that 'Our most enjoyable cities are those which quietly weave together a rich and complex pattern of different uses and activities' (Zeidler 1983: 9). Artist and small-scale cultural workspace increasingly coexist with other services and light industrial production, which have similarly expanded as a result of contracting-out to 'independent' producer and ancillary services. Indeed Lash and Urry argue (1994) that 'all industrial production, being design-intensive, is increasingly similar to cultural production' (McGuigan 1996: 88). The issues raised by planning for the seemingly fragmented small and micro-enterprise and work-practices embracing home-working and *tele-cottaging* might suggest that cultural planning is largely redundant outside of the surviving cultural industry quarters and zones, as Montgomery notes: 'The role that land-use planning policies can play to help foster economic development is arguably marginal' (Urban Cultures Ltd 1994: 2). However he goes on to suggest that:

> in this case the link between creative businesses and the city environment which sustains them is of paramount importance . . . the land use planning system in seeking to support economic development of the creative industries [should] foster an environment for SME growth, risk-taking and innovation. This means a more flexible attitude to studio and managed workspace developments and above all encouraging the mixture and diversity on which creative industries thrive.
>
> (ibid.)

Experience of the consideration, planning and protection of artists workspaces and studios in cities in North America and Europe reveals a variety of planning approaches which reflect both the historic planning regime and degree of control exercised, and the attitude towards cultural production and the role and status afforded the artist (and the historic conservation of the urban industrial landscape). The attraction and availability, albeit transitory, of former industrial buildings also coincided with the shift to large-scale work by contemporary artists. In SoHo, Manhattan lofts averaged 2,500 square feet: 'The large windows of cast-iron construction flooded each floor with natural light. Freight elevators provided useful access. Rents were affordable. A perfect prescription for artists. The transformation of SoHo had begun' (Gratz and Mintz 1998: 297). The effects of global capital and in particular, the property-led regeneration cycle are however evident in all of these cities to a greater or lesser extent, as the following review of artists work space development and planning indicates in these selected *cultural cities*.

Toronto, Canada

During the 1980s, the city of Toronto commissioned studies into several aspects of arts and cultural provision (*Cultural Facilities*, de Ville and Kinsley 1989; *Cultural Capital: The Care and Feeding of Toronto's Artistic Assets*, Hendry 1985), prior to a planning consultation exercise towards its *City Plan '91* (see Chapter 5) which took a planning approach employing a combination of spatial, catchment, gross demand and hierarchy concepts to assess its cultural facility requirements. Metro Toronto (the metropolitan city authority) effectively opened up their city arts plan to consultation ('planning as debate'; Healey *et al.* 1988), with contributions from special interest groups (*Aboriginal and Ethno-Racial*, Lee 1991) and local areas (*Area Municipalities*, Nagata 1991). Specific studies were also commissioned by the city arts council, such as for artists' workspaces (Social Data Research 1991, Stephen-Wells 1991) and public art (City of Toronto 1991), but the key departure from previous plans was the integration of *arts* planning with the *city plan* itself: 'The very choice of the name *Arts & Culture Plan* rather than *Policy* or *Strategy* reflects this willingness to accept the vocabulary and principles of other Metro initiatives. This is the Trojan horse theory of cultural policy-making' (Bradley 1993: 3). In Toronto, members of the city Arts Council were drawn from the artist community itself (150,000 people were estimated to work full- and part-time in the arts in this secondary *world city*; TAC 1992a), rather than solely civil servants or arts administrators.

In Toronto, the exploitation of planning and zoning laws and procedures has facilitated the development of artists' studios in former industrial buildings, and the creation of live-in studio-housing (e.g. Arcadia, Beaver Hall developments) using specific planning zone categories for this purpose (Stephen-Wells 1991). Like most post-industrial cities, planning was still largely rooted in the industrial past. In Toronto a battle had been fought in the early 1980s over bye-laws permitting multi-use buildings (see above), against criticism that such developments were uneconomic and single-use buildings were the most efficient (and easy to finance). However as Zeidler claims (1983: 98):

> If we view this within the context of the city, we realize that such efficiency really does not exist. Single-use structures and their districts are occupied for only part of each day or week and stand empty and unused the rest of time. Multi-use structures bring people together at different times – a much more efficient use of urban space.

Like in London, such as the not-for-profit ACME (1990), Clerkenwell Green (Evans 1990) and Space studio organisations, the intervention and brokerage of a social property development organisation – *Artscape* – facilitated the conversion of buildings for artists' workspace, housing and other mixed-use development (BAAA 1993). Whilst the fluctuating cycle of commercial property development has been just as apparent in this city (Hendry 1985, Social Data Research 1990) – Toronto was the 'home' of the property developer Olympia and York/Reichmann Brothers, whose questionable Midas touch was also seen in London Docklands and New York's Battery Park – planning protection and targeted capital investment by the city and metropolitan government has ensured greater security for practising artists and given them more control over their own destiny (TAC 1988). However as the graffiti-artist quotation reproduced on the cover of Toronto Arts Council's annual report warned: 'Artists are the storm-troopers of gentrification' (ibid.) – the capitalisation of artist-occupied property which turns the negative value of unused and undesirable buildings/locations to 'hope value', is a universal threat, as London, New York, Berlin and even Paris amongst others have found to their cost. The regeneration of Toronto's harbourfront in the 1980s for instance has followed the now familiar pattern of creating waterfront-based visitor attractions, shopping malls and recreational activities. However in this case gallery and arts centre developments have been provided as a form of 'planning-gain' creating further opportunities for craft and visual artists to establish work–residence accommodation, with the benefit of sales outlets for their work. Some such studios are partially 'open' in design allowing visitors to observe production in progress and to commission work directly from the artist or craftsperson personally. The vagaries and opportunities thrown up by the international property market and footloose capital also affected a range of amenity developments and provision in the central areas of this city, such as parks/open spaces, but which arose largely outside of any plan, locational preference or assessment of 'need'. Opportunistic negotiation with the development process may therefore present a pragmatic response where public resources and planning measures are inadequate, however as Harvey observes: 'The historical geography of place construction is full of examples of struggles fought for socially just investment (to meet community need); for the development of "community"; expressive of values other than those of money and exchange; or against deindustrialization' (1993: 8). Amenity planning in this city as in other liberal systems is largely reduced to negotiation, and reconciling and adjusting for variations arising from the 'plan', including obtaining compensation (e.g. planning-gain, reduction in impact, design, etc.) for such breaches of *a priori* city plans.

USA

The USA and to a lesser extent Canada have witnessed a de-urbanisation, suburbanisation and 'edge city' phenomenon, which as Sennett (1994) describes, itself created the iconic shopping mall and out-of-town settlement now familiar in urban sprawl and conurbations in other countries. Whilst this drift outwards continues – for example the capital Washington, D.C., which has lost over 115,000 residents (18 per cent of its population) over the past twenty years (see Chapter 8) – even American cities are seeing an increase in their inner-city populations, such as Denver, which forecasts a quadrupling of its inner-city residents by 2010, as well as Houston, Chicago, Seattle, Boston, Philadelphia and Cleveland. The focus of this repopulation has often been 'downtown' or 'midtown' areas, former industrial districts and buildings (e.g. brick-built warehousing) which lend themselves to residential 'loft-style' conversion (in Denver 'LoDo' living is promoted using the slogan *Kiss the 'burbs g'bye*), and which have stimulated new housing development in previously moribund residential markets. It is no coincidence that these areas have been active in developing mixed-use leisure–retail schemes, renovating arts venues and building new venues and museums for their new/returnee residents and visitors. Resident artists and groups, in addition to those incoming and art college graduate practitioners, have also played a direct part in the reuse and occupation of industrial buildings and near-derelict areas, through studio developments and squatting. The story of artist loft gentrification and commercialisation over recent years in New York is described for instance in Sharon Zukin's seminal study of *Loft Living* (1988). This artist colonisation of SoHo[2] began in the 1960s, but as Worpole (1991: 148) comments:

> Unknowingly the artists were used by developers and real-estate agents to create an ambience and a buzz in SoHo and other downtown industrial areas, which was then capitalized over their heads through the rise in property values. In short, the artists whose activities had created a desirable place to live in, displaced themselves in doing so. They could no longer afford to live in the neighbourhoods they had revitalized.

Artist Martha Rosler also writes on this situation: 'artists were a pivotal group in easing the return of the middle class to the area, although artists themselves were displaced by the wealthy clients who followed them into the newly chic neighbourhood (1991: 31 quoted in Miles 1997: 107) and Miles sees this as a 'contemporary form of the purification of parts of the city for bourgeois life which took place in the eighteenth and nineteenth centuries' (ibid.). Displaced New York artists who could not afford these upmarket rents moved on to West Broadway and TriBeCa, but this enforced outward drift has continued, with only Brooklyn still providing affordable studio space. The New

York experience therefore paralleled similar property regeneration movements in London (City, Covent Garden and Docklands areas), although greater use of 'fair rent' which is traditionally used in social housing, and also in countries such as France, Germany and Spain for work space, gave some protection to resident artists. In other American cities the cycle has similarly been played out, with artists unwittingly acting out the role of footloose 'storm troopers' of the property developers, not by choice nor benefiting from the longer term regeneration of the rundown areas or redundant premises into which they initially moved. The desire to gain security and put down roots has been the motivation behind the work space developments noted here. However, mechanisms such as planning measures supporting and protecting residential artist communities are still the exception and in practice interventions which do occur are often unplanned. For example, in Philadelphia artists had already been displaced from recently gentrified industrial workspaces, similar to the SoHo, New York and Clerkenwell, London experience (Evans 1990). Security ultimately depends on owner-ship. In this city an artists group, the Greene Street Artists Corporation and the Philadelphia Historic Preservation Trust secured vacant industrial premises with a grant from the Pew Charitable Trust in order to create new accommodation with long-term security. In this case the interest of the Trust and charitable funds was key to this studio-housing development. The value of combined living–working premises for this group of artists should not be understated. As one tenant commented:

> Personally, I see that we will no longer be paying three rents each month and trav-eling back and forth between work, studio and home every day. We will be building equity, and will have the freedom to invest substantial improvements to our studios. We will be consolidating our art careers into one home address and phone number . . . and enjoy contact with the other artists in the group.
>
> (Fisher, quoted in PHPC 1992: 3)

In Philadelphia the intervention of local architects also led to the creation of a Foundation that aims to encourage public input to the design and planning of the city and to inform local citizens about planning and design issues. The Foundation sponsors symposia and panel discussions about issues affecting the quality of Philadelphia's phys-ical planning and urban design, ranging from the height of city centre buildings, density standards and pedestrian activity, through to cultural development and facilities and his-toric resources: 'By encouraging the dreams and ideas of design professionals, artists and the general public, a provocative dialogue about the physical form of Philadelphia was begun' (Cowan and Gallery 1990: 43). In Massachusetts a similar initiative, *Boston Visions*, was developed, again by the city's architects, while in California a similar *City Visions* programme took place in San Francisco, supported by the National Endowment for the Arts (NEA). The USA is associated with lower direct state funding of the arts than in Europe (choosing higher individual and corporate support and taxation incen-tive regimes). However, the NEA, which was established in 1965 by the US Congress as an independent federal agency to encourage and assist the nation's cultural resources, supported 450 public art projects and an *Art-in-Architecture* programme which com-missioned over 300 artists across seventy-five cities/states in its first twenty years (Harris 1984).

Owing to its liberal taxation and sponsorship rules, the US has also been more successful than other countries in promoting 'percent for art'[3] and public art input to
property development schemes (Shaw 1990b), including artists/crafts involvement in
urban design, rather than as an add-on or after-thought to completed buildings (e.g. the
ubiquitous concrete sculpture or water-feature fronting new offices and squares; Shaw
1989, 1990a, also Garreau 1991). However as Harrod (1991: 16) maintains, this is not
necessarily a salvation in the building design process and conception:

> Architects arrogantly continue to ignore the contribution which artists can make
> and architecture continues to fail in its traditional role as the mother of the arts.
> The modern movement has long been berated for having expelled the artists – but
> the architecture of the late C20th shows no sign of welcoming them back.

It could be argued that much contemporary public art is little more than inadequate
compensation for the poor quality and variety in urban design and modern architecture,
including the mixed-use and cultural quarter developments that lack both 'soul' and
physical engagement with their users. Public art in its own right, or based on 'percent for
art' schemes, like most betterment mechanisms, also arises where development takes
place, and therefore tends to be absent in areas of decline and inactivity. The nature of such
'public art' (*sic*) has also generally been limited in scope: 69 per cent of local authority
commissioned public art work in the UK between 1984 and 1988 was of only two
types – either sculpture (47 per cent) or murals (22 per cent) (Shaw 1990a). More creative
schemes for community benefit (e.g. programme-based versus static 'art') arising from
development, private and public, have developed in cities such as Seattle and Los Angeles,
and also in France where endowment funds and a degree of ownership and self-sustain-
ing cultural provision is provided, as opposed to the one-off capital scheme financing
public artworks (Percival 1991, City of Toronto, 1991, case-studies in BAAA 1993).

Berlin, Germany

Another city undergoing major and rapid development is the newly unified Berlin,
whose historic and traditional cultural centre and facilities lay in the former German
Democratic Republic – East Berlin (see Chapter 8). New public and commercial offices
and hotels were developed very rapidly, in most cases inevitably placing a strain on existing usage and capital values, and the situation has been complicated by reversionary land
settlements relating to pre-Communist (and in some cases pre-Nazi) land ownership
claims (Evans 1995b). In *No Art, No City*, Kotowski and Frohling (1993: 1) describe
this situation thus:

> Land prices and commercial rents have skyrocketed in the congested areas of the
> new Federal Lands – most noticeably, but not only in Berlin. The consequences for
> the fine arts are catastrophe there: workrooms for artists are becoming prohibitively
> expensive. Art is threatened: no studio, no art.

Typically in Berlin, artists themselves have developed a coordinated response, as a
defence against the loss of the infrastructure they see as necessary to support creative

practice, rather than the city or federal government, or cultural agencies themselves. The 'creative city' argument has been to the fore in this: 'As regards its reputation as a cultural metropolis, Berlin largely relies on its visual artists. They are important for the urban quality of Berlin' (ibid.: 4). As in other major cities, the estimated 4,000 to 5,000 visual artists in Berlin were threatened by rent increases and a chronic shortage of space (which predated unification): 'one thousand studios were lacking in the western part of the city . . . several hundred cases of eviction from studios housed in commercial buildings must be assumed every year' (ibid.). From investigations undertaken by the Berlin Senate, commercial premises renting from DM12 to 15 per square metre (unheated, about £0.50 to £0.75 per square foot) were no longer available, while studies by the Studio Commissioner showed that a rent of DM7 per square metre (about £0.30 per square foot) was the maximum most artists can afford to pay (ibid.). Recommended responses to this problem include investment in new studio developments and accommodation (*live-work*): 'the cultural infrastructure must likewise be included in the planning of major investment projects in Berlin on the same level as the social infrastructure' (ibid.: 4.1). Special residential forms of studio flat are to be included in the City's First Promotion programme for new housing, with a target of two hundred artists' workplaces over the first five-year period.

Strategically, the *Kulturwerk des BBK* (Cultural Institute of the Professional Association of Visual Artists in Berlin) sought protective clauses in the structural plans of the *land* of Berlin and direct artist representation on the committees considering major investment projects (as in Toronto and Los Angeles), and the transfer of artist-occupied premises as special assets to a *Development Company for Cultural Areas*. The BBK rents, leases and, where possible, purchases property and studios for letting to artists at controlled rents (cf. ACME and Space Studios in London): over sixty studios and apartments in four large complexes are managed in Berlin. Planning law was also looked to protect the change of use of studios, as sought without success in Clerkenwell or other UK cities: 'Cultural infrastructure, in particular for visual arts, must be a self-evident part of urban planning, publicly subsidized housing construction and urban renewal supported by public funds' (ibid.: 6). The rationale for public intervention is stated again by Kotowski and Frohling (ibid.: 4:4):

> In view of the structural magnitude of the problem it will be necessary for the public promotion of studios to ensure without restriction, to the benefit of all professional artists whose financial situation does not allow them to survive in the free commercial-rent market. The public promotion of studios and the public allocation of studios are of basic importance for the cultural infrastructure.

Elsewhere in Germany, such as in Munich, artist's studios are retained as seed-bed and sabbatical retreats for artists, and let rent-free for a fixed number of years. Given the hot-house Berlin scenario, however, it is clear that unless such workspaces are held in public or independent ownership, and protected from the pressures of the property and land-use market wherever they may be located (i.e. in high value zones), their initial revaluation and rent rises will swiftly be followed by change of use to more lucrative occupation or redevelopment, as London and New York have experienced, with a permanent net loss of light industrial and accessible workspace in the central areas.

Paris – an artist's haven?

Finally, given the strength of France's commitment to cultural-led and urban regeneration, it comes as no surprise that the response of Paris to the infrastructure needs of artists is both comprehensive and interventionist. The city uses public land for the erection of housing units, including some *ateliers-logements* or artists-residence studios, with building regulations specifying minimum ceiling heights, storage areas and separation between living and workshop areas. Over 1,000 such units combining studios with living accommodation were built in Paris between 1977 and 1992 by which date the annual budget allocation to this programme was FF22 million. More basic studio accommodation is also planned, aimed at offering cheap premises for first-time artists. In addition to this building programme, two major cultural complexes provide studios for artists in residence; the *Cité Internationale des Arts* (265 studios – minimum two months, maximum one year residency), with shared central facilities such as an engraving workshop and rehearsal rooms (annual subsidy of FF2.65 million in 1993). The *Cité* is jointly supported by the city and state cultural departments, offering artists the opportunity to work, exhibit and establish contacts with Parisian artistic communities, on subsidised terms. The second, at *La Ruche-Seydoux Foundation* (seventy-two studios) was created in the late nineteenth century by the sculptor Boucher, again with city government revenue funding (Berger-Vachon 1992). Short-life properties turned over to temporary studio use include the *Hôpital Ephemère*, which took over the former Bretonneau Hospital and converted it into studios, pending conversion into a geriatrics centre (cf. in Gothenburg, Sweden – a hospital to studios conversion *The Epidemy of Arts*; Konstepidemin 1993), and various 'open-door' studios across the city (Berger-Vachon 1992) (Table 6.13).

Table 6.13 Artists' studios in Paris

City of Paris's total number of studio-flat combination units (30 per cent built since 1977)	1,071
Government-owned studio-flat combination units within the City of Paris	309
Number of studios built annually:	
Until 1985	20
Since 1985	25
Total	45
Pending applications for studio-flats	500
Total	1,925

The rationale for Paris's involvement in studio premises for artists is pragmatic and well-established: 'As an international artistic capital, Paris has and attracts thousands of plastic artists, who may stay there permanently, temporarily, or for long periods' (Mairie de Paris 1993: 1). Support for the arts through systematic state intervention and patronage dates back to the *Ancien Regime*, with subsidies available for individual artists in the form of the *Prix de Rome*, and with the foundation of institutions such as the *Comédie Française* and the Paris Opera. The city's *ateliers-logements* are intended for painters, sculptors and engravers living in France. Eligibility follows a housing allocation system: applicants must prove that they have applied for housing to the town hall or *arrondissement*, or the central housing department for those coming from outside of Paris. Two consultative artistic committees led by contemporary artists, curators and administrators meet twice each year to consider applications for studio premises, which give their opinion on the applications. A similar artistic jury selects applicants for the *Cité Internationale des Arts*. Grant-aid is also available for the costs of converting other premises into workshop studios. The city's policy also targets specific groups for assistance, for example artists over the age of sixty-five have a number of housing units created for their use.

From a position of an absence of basic planning documents in the late 1970s, Paris had begun to revise its planning strategies and procedures, including establishing Regional Development and Ground Use Plans. The latter articulated detailed land-use and densities and the preference for developing new *quartiers* which would more modestly respect the dimensions of surrounding buildings and, wherever possible, parks, gardens and public amenities would be positioned at the heart of new housing blocks. Land-use Plans include Ground Use Ratios which together with Mixed Development Zones are recognised as being more costly and more space-consuming than previous practice, but these are well accepted by inhabitants who are at last beginning to find a quality of life to which they have always aspired. All of these measures are therefore underpinned by a plan-led approach: 'Paris is still the city which places the greatest faith in the planning system to create and enable the city and its region to progress harmoniously towards a new millennium' (Burtenshaw *et al.* 1991: 267).

Despite this tradition and support of artist studios, Paris like London (e.g. Clerkenwell), New York and Berlin is experiencing gentrification of its airy nineteenth-century lofts and studio buildings, with demand for space exceeding supply and artists unable to meet regular rental payments. In the areas of Montmartre and Montparnasse artists communities have taken to squatting, since as one artist states: 'We have no place to work and the city has thousands of empty buildings. . . . Historically Paris has been a home to artists, but to work you need a workshop, to get a workshop you need a state certificate, and to get that you need to know the right people' (quoted in Henley 2000: 15). Like the newly fashionable city districts elsewhere, it is the new creative brands of fashion houses, design and media firms that are taking over the former Bohemian haunts and cultural workshop areas, while 'rich young professionals in search of out-of-the-ordinary accommodation have also begun infiltrating the *cités des artistes*' (ibid.)

* * *

Cultural planning that engages with public and creative artists and designers (including crafts in building finishing and features) is best practised where design briefs and guidelines are explicit and where local area and site plans incorporate policies towards public and per cent for art. Left to their own devices, the training of and pressures on the planning and design professions preclude this is in practice, not least where 'build and design' (*sic*), space-efficient planning and computer-aided design becomes the norm and where 'The artist is not treated as an intellectual contributor to development proposals . . . [but] brought in late to "decorate" rather than being integral to the process' (LAB 1992: 1). This is in contrast to the *arkhitetron* as masterbuilder, and the stone masons who built the first gothic cathedrals – craftsmen as well as designers, whereas the modern secular cathedrals of art self-consciously bear the signatures of their architects rather than the art they are supposed to celebrate.

The Artist in the Changing City (BAAA 1993) whilst retaining a symbolic and even sentimental place therefore also seeks engagement with space and resource allocation to avoid the nomadic and insecure existence which crude property and planning uses otherwise create:

> Planning at all levels of government can greatly assist the development of flexible working and living spaces for artists. The crucial thing is that artists should be visible, that they should be consulted directly, and that the solutions to their needs should be designed to be long-term and integral to all urban cultural planning.
>
> (ibid.: 47)

Conclusion

Size matters in economic development and in order to gain serious consideration within national/global industrial and related policy intervention. Cultural production, notably crafts trades and live entertainment, have in the past formed significant aspects of the economic as well as social life in cities, but it has been the mass production and commodification possibility that has raised certain cultural goods to the status of global industries, or even just to industries in their own right. Defining and delineating cultural industries is however problematic and emotive ('draining the arts of their meaning'; Hughes 1989), but this has been an irresistible consequence of promoting cultural strategies and in arguing for cultural resources both within the state-supported sectors and as elements in urban regeneration and industrial investment. The proportion that culture, even where generously defined, represents in national employment and other macro-economic indicators as detailed above, is fairly small – less than 10 or even 5 per cent of employment and disproportionately less in GDP (suggesting that 'cultural work' is below-average in terms of productivity; Heartfield 2000: 53). However it is where spatial agglomeration occurs that cultural production comes to represent more significant and visible proportions of city economies and therein cultural industry clusters in local areas, where cultural activity is most pronounced. In planning terms this is therefore important since it is at the city and local area level that cultural activity in all of its physical and material functions, whether theatre, graphic design or film production, is manifest. The cultural economy has also gained attention in inverse relation to the declining importance of other sectors, notably manufacturing but also traditional services sectors. There are and will be more in the future, major shifts between cultural industry sectors (e.g. employment, dissemination, consumption), as technology, mobility and fashions and taste work through. The 'shape of things to come' that deterministic predictions prefer (e.g. holidays in space, robotics, virtual reality) are even less reliable, however, since they have failed to materialise in the past, not only due in part to self-interest, but also by underestimating the human needs and preferences for both diversity and subcultures, as well as for collective activity which the culture-houses of the late twentieth century have again tapped, from Bilbao to Bankside. The sizes of the cultural industries are also growing, if unevenly and unpredictably. Short-term variations occur where technology meets rapidly growing demand – how long will Web page designers be in such demand, once software becomes user friendly and sites established? More importantly, how far will the creative skill and content be required as opposed to the maintenance of images and information? This is behind the selective but crudely drawn map (DCMS 1998) of the scope of employment and turnover in the 'creative industries' that on the one hand ignores the arts in their educative role (e.g. a dance teacher), but values a stallholder selling bric-à-brac in a touristic street market and a 'leisure software programmer' (Evans 1999a). In order to be effective in planning, including the provision of arts education and training, the cultural industries need to be deconstructed from their consolidated form used in state and industry advocacy, and distinguished in terms of their cultural and creative content and processes and analysed according to both production chain linkages and cultural planning assessment which visualises them within a city or area locale. Grouping these subsectors of both the arts and cultural industries in these terms and then considering them spatially, I would argue, moves closer to a robust understanding and model of what culture is in a creative and productive sense, and how far the devices of planning and democratic resource distribution might be applied in best protecting and supporting it. Ambitiously this might also help to diffuse the entrenched high-arts versus popular culture dialectic by perhaps concentrating the scope of both of these rather than imbuing virtually all forms of production and recreation as 'creative' or 'cultural' (Scott 2000).

In terms of urban policy, the shift from the arts-as-amenity to the cultural economy, and from high-arts to cultural industries policies, whilst not 'seamless', has been justified as a pragmatic *and* ideological response to the decline in public cultural services and amenities, as well as in traditional commercial and industrial employment at a regional level, and at the same time to the disempowering effects of globalisation and commodification of much cultural production and consumption. As well as at the city–region (and translated to the national–regional tier), the advantages of critical mass and agglomeration at the local level through the phenomenon of the cultural industry quarter/district is emulated almost as a panacea for the economic and environmental survival of cities and urban areas – of 'holding down the local' in the global (Amin and Thrift 1994). Politically, the rediscovery and talking-up of the cultural economy and formulation of targeted economic development and creative industries policies has offered national, regional and local governments a growth sector on which to focus its planning – physical/land-use, economic and social. In particular the state it is felt can 'unify small-scale cultural producers to give them power in the market to combat multi-nationals on distribution and allowing consumers wider choice by combatting oligopolies' (Henry 1993: 51).

The potential that the cultural industries promise has also coincided with, on the one hand, the recognition of the small enterprise economy, and, on the other, the role that the arts and urban regeneration arguably offer in reinforcing cultural identity and diversity, notably through cultural tourism and flagship cultural projects. The intervention and planning framework within a harmonising but widening European Community is therefore explored in more depth in Chapter 7, including cultural policy and planning rationales within regional development generally, through which supra-national, geopolitical and nation-state policy goals have been pursued. Since the urban renaissance has both European roots and is and was manifested in the city-state, the model of the arts and city regeneration is explored in this and in Chapter 8 in an international perspective on cultural planning and its incidence in Europe and other regions of the world.

Notes

1 The Congress International Architecture Moderne (CIAM) was formed in 1928 in Sarraz Switzerland. In their fourth meeting in 1933 after an examination of thirty-three cities the 'Athens Charter' was established which implied a complete overhaul of the city advocating the dispersion and segregation of a city's parts – dwelling/habitat, work, leisure (mainly sport and recreation) and circulation – identified with architects such as Le Corbusier, Frank Lloyd Wright and Sigfried Giedion: 'If in an industrial age the various functions of daily life cannot be clearly separated, that fact alone spells the death sentence of the great city' (1963, quoted in Zeidler 1983: 15).
2 SoHo (South of Houston Industrial Area, New York) became a symbolic model for other downtowns, which emulated the acronym: LoDo (Denver), SoDo (Seattle), SoMa (San Francisco), SuHu (Chicago) (Gratz and Mintz 1998: 303).
3 'Percent for art' is a voluntary scheme whereby a percentage of the development or building costs is dedicated to works of art, public realm or design aspects of a building (e.g. 1 per cent of capital cost). Percent for art is also supplemented in some US states by hotel/motel room taxes, local lotteries and bond issues. Trust funds have been created from 40 to 80 per cent of the art contribution, and between 0.5 and 2.0 per cent of the capital construction cost for ongoing arts programming and maintenance, e.g. festivals.

7   European common culture and planning for regional development

Introduction

Given the roots of the first urban renaissance in Western Europe, which subsequently spread through international trade and colonisation, the adoption of its past glories and possibilities by the European nation-state and collectively through geopolitical formations provides a useful basis for an analysis of how far cultural planning, urban regeneration, and the processes of regionalisation and globalisation have manifested themselves. This is not of course restricted to Europe – east or west – but is a universal/ist phenomenon in part driven by global economic as well as cultural movements (Hobsbawm 2000).

The European Project itself can be identified with the various efforts of the expanding European Union (EU), other Europe-wide supra-national bodies and agencies, and regional alliances (e.g. Franco-German) to reaffirm and reinforce the notion of European culture – social, political and economic. However, the 'imagined community' associated with the projection of the nation-state (Anderson 1991) stretches the imagination even further when applied to Europe, but whilst the arts have been a peripheral aspect of EU policy and programmes (Evans and Foord 2000b), the bias of the European *Economic* Community (EEC) was not necessarily that envisioned by one of its founding fathers, Jean Monnet, who had allegedly foreseen culture as having a more central role in European harmonisation (Gowland *et al.* 1995, Shore 1993). 'Culture' had never been, from the Treaty of Rome onwards, a technical competence of the European Community and therefore no definition of, or discrete policy for culture, has been created. In terms of its designated arts, cultural and related media and cultural industries programmes, these have also in practice been a very minor aspect of European policy and represent a tiny proportion of the total expenditure by EU directorates. On the other hand, the promotion of European tourism – domestic, intra-regional and from overseas – has focused on *cultural tourism*, notably heritage and the visitor-based arts (e.g. festivals), and whilst the EU is the prime policy and funding executive, the Council of Europe and others have also sought to focus on cultural and heritage tourism as a means of celebration and exchange. These regional interventions have however been implemented largely outside of both a cultural policy framework and a cultural plan – national *or* Europe-wide (Evans and Foord 2000b).

Common culture and identity

As Chapter 6 concluded, it has been at the city–region level of planning and particularly economic development that cultural planning has been most apparent and where European regional development and regeneration policy has been most effectively adopted (Bianchini and Parkinson 1993, Evans 1993c). Since the late 1970s, European Regional Development ('Structural') funds in particular have provided leverage and direct investment in cultural and heritage facilities, particularly those linked to city regeneration, in place re-creation (Ward 1998) and in visitor-led strategies, and indirectly in encouraging the promotion of European and regional identity through substantial levels of grant-aid. These funding schemes have been driven by regional economic and employment policy rationales in areas of industrial and rural decline and high unemployment, but not necessarily in the areas of highest need or deprivation. In contrast, direct support for designated artistic and cultural activities by the EU represented less than 8 per cent of all support to arts and culture during the 1990s (Wates and Backer 1993). The major impact on arts facility provision in Member States has emanated from these regional development and social funds, thus neatly bypassing the controversial and contentious imposition of a universal European cultural policy on the nation-state, where the notion of a national 'common culture' itself has increasingly been under question or been seen as a perceived threat. In many respects this can be seen as an attempt to (re)define the supra-national state as the nation-state diminishes in power and resonance. Mulhern for instance writes on *The Logic of European (Dis)Integration*: 'Out of control, yet not chaotic. . . . The theme of a new "European" identity is increasingly current. The real probabilities of such an identity are either weak or dangerous' (1993: 200–2). Marquand (1994) has also argued that the European project now finds itself confronted by four paradoxes, the first being *identity*:

> Born in the shadow of the Cold War, the identity of the EU was implicitly accepted, originally, as essentially western European, developed, mainly Roman Catholic. Expansion towards the Protestant north and the non-western and underdeveloped East obviously forces reconsideration of a European identity . . . what it is to be European as a citizen of the EU has now become more complex and problematic, culturally, socially and historically.
>
> (Gowland *et al.* 1995: 284)

Fontana (1994) has documented how the European identity was always constructed against the 'other', the barbarians of different kinds and different origins. Europe's self-image has consistently been defined in opposition to a less civilised non-European 'other' (Jordan and Weedon 1995), but as Said maintains: 'Most histories of European aesthetic modernism leave out the massive infusion of non-European cultures into the metropolitan heartland during the early years of this century' (1994: 292). The re-presentation of the European Renaissance has historically reinforced this spectre of emergence from the Dark Ages and cultural wasteland left over from Byzantium, and its (secular) rediscovery of classicism. However as recent reworkings suggest, countering Giorgio Vasari's seminal text *The Lives of Artists* (1550): 'it makes more sense to think of the Renaissance as a culmination rather than a rebuttal of certain medieval tendencies. . . . If no attempt is made to understand the mixed origins [Christian, Moorish, pagan] . . . then the richness and much of the beauty of its art will remain unappreciated and misunderstood' (Graham-Dixon 1999: 13). The cyclical model of art history and visual styles therefore reflects the modernist position which rejects what went before, in Vasari's case: 'in order to exalt the art of his own time, [he] found it essential to derogate the Gothic that preceded it as the art of the barbarians who destroyed the classic Roman art he admired and his *rinascita* revived' (Smith 2000: 81). The Italians never used the term 'Renaissance' (*rinascita*) at the time. It was first coined by French historian Michelet in 1858 and was later used by Burckhardt in *The Civilisation of the Renaissance in Italy*: 'Thus a nineteenth-century term was used to mark the end of a period baptized in the sixteenth century' (Johnson 2000: 3).

Today, the 'ethnic quarters' of most European cities now have African-Caribbean and Asian cultural centres, Jewish museums and multicultural arts centres – spaces more or less independent of the dominant society (and its funding regimes) and alternative to white-European cultural institutions (Jordan and Weedon 1995). Few of these are flagships in the equivalent sense to the established temples of High Art (an exception might be the Arab Monde in Paris) and as Owusu observes: 'For many black artists working in the city, the city itself is a terrain of contested spaces, and that changes the whole equation for many of them, because one does not assume one's own space within the city in the way that a white or European artist might' (quoted in BAAA 1993: 22). Ethnic communities that have been well-established in European and colo-nial cities still exist therefore within urban systems which ignore their own personality and aspirations, as British-Asian architect Rajan Gujral comments from Southall, West London: 'Ethnic communities are a permanent part of the society in the major cities of the country. There is no mistaking the areas favoured by the various ethnic groups; the writing on the shops, the rhythm in the streets, the faces, the dress. But somehow the communities live in spite of their environment rather than shaping it' (1994: 7). As the architectural landscape changes incrementally in old cities, Methodist halls, converted to synagogues, now function as mosques. The visitor is therefore a two-dimensional voyeur in a temporally and spatially three-dimensional cityscape. The historicist cultural ('Grand') tour still however represents an impression of unchanged and unchallenged heritage of a supremacist civilisation, and the acquisitions of the colonial conquests from the Orient, which together with legitimised Western classical and mobile modern art make up the prime museum and gallery collections (Evans 1998a). As the Dutch archi-tect von Eyck earlier observed: 'Western civilization habitually identifies itself with civilization as such on the pontifical assumption that what is not like it is a deviation, less advanced, primitive, or, at best, exotically interesting at a safe distance' (1962, quoted in Frampton 1985: 22). This is a far cry from, or at least at odds with, the essential dia-logic nature of European unity that Morin presented in *Penser l'Europe* (1987): namely the value of the combination of differences without homogenisation as not only the basis for cooperation, but as a cultural feature in itself (Sassatelli 1999: 598–9, see also Derrida 1991, Habermas 1992).

Behind this embedded Eurocentric sentiment lies the tension today between aspira-tions and expectations of assimilation and integration by successive generations of immigrant communities. This is of course kept alive by new immigration, whether political (e.g. refugees) or economic, and by the shifting position and response of second and third generations in terms of assertion of cultural rights, forms of expression and therefore an equitable place in resource distribution, including places and control over their designated cultural facilities. Where diaspora are well connected they can draw on community wealth, e.g. in some Arab communities private support for cultural projects is more evident (Islamic arts centres, mosques), although this is not equally so amongst differing national groups (e.g. between say Algerian and Bangladeshi, and Saudi and Kuwaiti). Generally, however, the development of cultural facilities by ethnic minorities relies more on individual subscription and community support than that of the state and is therefore relatively marginalised (and often not treated as part of the arts funding regime but as a 'community', or 'religious' activity). Exceptions are the growth of arts festivals building on religious celebrations (e.g. Diwali, Chinese New Year, Carnival Mas) promoted in touristic itineraries as well as opportunities for community celebration and display. The growth and popularity of some of these festivals has created tension and planning problems, which have resulted in their rescheduling and resiting away from core inner-city areas (e.g. Toronto's Caribana; Evans 1996c), whilst the demand for ethnic festivals has stretched city authorities such as in New York where weekend road closures have proliferated (Plate 7.1). Studies of youth culture and ethnic group attendance in mainstream arts and entertainment venues also suggest a form of social exclusion (e.g. the suppression of black/music nights in clubs under pressure from the police; Boese 2000: 16), which has in turn created an alternative circuit of cultural activity. These tend to operate outside or on the fringe of the city centre and take place not in arts and entertainment venues but in function rooms, community centres or multipurpose halls attracting large audiences from a wide catchment – an indication of latent demand and exclusion from other events and venues. As Trienekens concluded from a study of Rotterdam: 'The absence of cultural diversity in the programming of the established venues and how migrant groups perceive of these venues seem to be part of the explanation for the relative absence of migrant groups in established venues' (2000: 62).

The largest traditional Hindu temple outside of India is in Neasden, North West London (Plate 7.2) in a nondescript outer London borough that also hosts Wembley Stadium, and which was one of the first local authorities in the country whose profile became majority black and Asian in the early 1990s. This same borough was also the proposed site for a 1 million square foot Sun City, a giant £210 million Asian arts and media centre that was to include a Bollywood eighteen-screen multiplex cinema, a studio to broadcast Asian Sky television and a nightclub for booming Asian dance music. At its heart a 3,000-seat arena would cater for weddings of all religions: 'It was no surprise Asians wanted their own cultural centre when their tastes had been largely ignored by the mainstream. In TV, theatre and museums, there is little sense that Asians play a strong role and have made great achievements' was one comment. Another commentator said: 'This project sounds like the grand gesture of a Latin American style dictator.' How far integration, the celebration of plural and diverse cultures and the natural fusion of cultural activity can coexist and, more importantly, receive recognition within cultural planning and arts resources is a particular issue and concern within multicultural Europe, which still trades on its Renaissance, high-art and heritage past.

Plate 7.1 Italian street festival, Manhattan, New York

Common culture?

Notwithstanding the contested ideological and historical base and operational difficulties of pursuing common cultural objectives, 'the idea of "Europe" as the foundation of an identity has been stimulated by the EU's search for instruments of legitimation' (Sassatelli 1999: 593; also Smith 1992, Garcia 1993). Reinforcing the notion of a 'common European heritage' was for the first time linked to the continuing move towards harmonisation between European Member States, with the following cultural goals being enshrined in the Maastricht Treaty on European Union:

1   The Community shall contribute to the flowering of the cultures of the Member States, while respecting their national and regional diversity and at the same time bring the common cultural heritage to the fore.
2   Action by the Community shall be aimed at encouraging co-operation between Member States and, if necessary, supporting and supplementing their action in the following areas:
    • improving the knowledge and dissemination of the culture and history of the European peoples;
    • conservation and safeguarding of cultural heritage of European significance;
    • non-commercial cultural exchanges;
    • artistic and literary creation including the audio-visual sector.
3   The Community and the Member States shall foster co-operation with third countries and the competent international organisations in the sphere of culture, in particular the Council of Europe.

Plate 7.2 Hindu Temple, Neasden, north west London (1998)

4   The Community shall take cultural aspects into account in its action under other provisions of this Treaty.
   (*Source:* HMSO 1993 – entered into force 1 November 1993, now incorporated under an Article of the Treaty of Amsterdam 1997)

The promotion of cultural tourism in Europe has been a particular political as well as an economic tool in pursuit of this common cultural heritage and exchange mission. This growing element in European tourism, in contrast to sun, sea and sand holidays (Evans 1993b, 1998a, b, Richards 1996), is therefore an attractive mechanism through which these somewhat contradictory objectives might be fulfilled. Castell's (1996) evocation of the twin trends of globalisation in economy, technology and communication, and the parallel affirmation of identity as the source of meaning are arguably both manifested in the international *cultural* tourism process (Evans 1998a). As well as the Commission, other European-wide initiatives have also promoted European identity through tourism exchange, such as cultural routes, e.g. Santiago de Compostela, Galicia in Spain (one of nine European Cities of Culture 2000) and the pilgrim trail, and trans-border Routes des Vignobles – itineraries and architectural heritage schemes that have been supported by the Council of Europe under twenty themes and across the forty-seven countries which have signed the Council's Cultural Convention. Economically these policies also support a 'marketing Europe' strategy that draws on the European identity manifested through its heritage – real and re-created – and the primacy of the European Renaissance and its late twentieth-century revival (Arts Council 1986) with the objective of reversing the decline in Europe's global market share in international tourist flows (WTO 1998). As Ashworth maintains: 'History marketed as tradition is a predominant element in the national tourism promotional images of most European countries' (1993: 15). *Whose* history and heritage is of course the question where notions of common culture are to be represented to visitors, since as Ashworth also says: 'You cannot sell *your* heritage to tourists: you can only sell *their* heritage back to them in your locality. The unfamiliar is sellable only through the familiar' (1994: 2). The urban renaissance is therefore central to the cultural component of European integration and the reaffirmation of European culture and heritage through city and regional economic development programmes and tourism promotion. This association is not coincidental since the Grand Tour (Evans 1998a) and the Enlightenment itself both created the conditions and desire for cultural products and services (e.g. travel and tourism), and as Jardine suggests in her analysis of the Renaissance era with its prosaic and powerful promotion of *Worldly Goods*, she makes no bones about the roots of European cultural tourism:

> In London as in every other European capital, springtime heralds the arrival of its cultural pilgrims, who throng Trafalgar Square and its surroundings, following trails well laid by the international tourist industry to lead us to the supposed roots of our Western intellectual and artistic heritage. Cameras at the ready, we trawl the museums and galleries ready to record all relevant items from our guidebook inventory of important vestiges of Europe's collective history. High on our list are the treasures of the period which formed what is broadly known as the Western tradition in art and learning – the European Renaissance.
>
> (1996: 3)

Furthermore, the European Project is not limited to a periodically enlarging EU membership (including those 'short-listed' and on probation, such as the Czech Republic, Hungary, Lithuania, Cyprus and Turkey), since European Commission programmes and initiatives also encompass eligible countries within the wider European Economic Area (EEA); Central and Eastern Europe; peripheral groupings such as Med-Cities (e.g. Malta, Cyprus, Israel, Jordan and Egypt); and overseas development programmes supported by the EU's Lomé Convention, which funds development projects in former European colonies in Africa, the Pacific and the Caribbean. The re-adoption of Prague and Budapest as Renaissance cultural capitals on the circuit of cultural tourism and the speed at which international hotel operators and Western banks moved into Central Europe after 1989 are also signs of the powerful public and private hegemony which has been instrumental in the cultural commodification of Eastern Europe (Evans 1995b, Evans and Foord 2000b). The European Project can therefore also be presented as a venture whose horizons spread far wider than the participant and contributory EU members and in a similar manner to national cultural promotion bodies such as the British Council, the German DAAD and Goethe Institute, the Institute Français, and international arts and heritage organisations such as both the Paris-based UNESCO and ICOMOS. The World Bank's recent foray into cultural development (1998), which focuses on World Heritage Sites and cultural cities (Evans 1999c), also replays this role in Eastern Europe (e.g. St Petersburg, Russia and Butrint, Albania) as well as in lesser developed countries, applying the dualistic strategy of 'universal patrimony' and heritage tourism in creating an economic development opportunity as part of its financing regime. For example, Rojas cites the case of the city of Salvador, Bahia in Brazil, a UNESCO world heritage site where despite the developers introducing free music and theatre performances to attract 'customers' to this historic quarter: 'recreation and tourism activities expelled residents and craftspeople that used to live in the historic center. It is doubtful whether [they] can survive only on the basis of these activities whose demand is volatile and may, in recession, induce abandonment of the center by merchants and entrepreneurs' (1998: 7). This is a familiar scenario in European and North American historic zones (Evans 2000a) – the historic and cultural quarter as 'theseum' (Batten 1993), and where conservation, property gentrification, including tourist hotels, and corporate investment in architectural heritage has ensured that there are very few living communities in the touristic centres of Venice, Florence or in the fashionable museum quarters of London, Paris and Madrid. One effect of this sterilisation of cultural heritage areas is the lack of regular (or 'authentic') exchange between host and guest, the absence of community amenities, and a largely faceless and privatised built environment (Evans 1998a: 2).

Moreover, this pursuit of a 'common identity' and identification with a 'common cultural heritage' potentially conflicts with, or at least raises the issue of the protection and celebration of cultural diversity within Europe, particularly in areas hosting recent migrant, and 'non-European' (*sic*) communities. This concern for the expansion of the Europeanisation project (Evans 1995b) into the cultural dimension had been expressed early on: 'It is unthinkable that the Community should attempt recommending a European cultural policy. . . . It is equally out of the question for the Community to propagate the idea of a "European culture"' (Dumont 1979: 9; also Loman *et al.* 1989, Mulder 1991). Politically and in planning terms this can be interpreted as the divide between centralism and regionalism, or the notion of *subsidiarity* – that is, the consensus that fiscal, administrative and other legislative responsibilities should be vested as closely to the level of impact as possible, and that action be assumed by the Commission only if it cannot be taken more appropriately at national, regional or local levels. The primacy of the European supra-national state had however already been forecast, perhaps optimistically by the then European Commission President, Jack Delors: 'By the mid-1990s, 80% of all economic and social legislation in the EC will be determined by the Community and not nationally' (quoted in Lintner and Mazey 1991: 28). This intervention extends to the cultural sphere (Evans and Foord 2000) and although the 'identity' and 'Union' is simplistic and reductive, 'it cannot be denied that the European Union increasingly monopolises discourses of "Europe"' (Sassatelli 1999: 593). Given the EU's legislative superiority, the European Central Bank and partial monetary union (EMU), this measurement may be technically the case. However culture exists and persists outside of the political economy as well as within it, and the local and cultural responses to Europeanisation resemble those arising as a result and part of globalisation rather than with an overtly European cultural policy or identity. What peripheral regions and socio-cultural groups have pragmatically embraced is the notion of pluralism within a democratic Union which has empowered marginalised groups within their host countries, attracting a degree of autonomy over cultural affairs and identity. Examples in Spain include the Basque and Catalan 'autonomous' regions; in France the Provencale and Breton communities; the Celtic 'fringes' in the UK, and the Barents Sea coastal regions of Norway, Sweden, Finland (e.g. Lapland) and Russia. The coming together of 'second cities' in peripheral regions, for instance, seeks to share their common experience and strategic solutions (including making the case for regional aid), such as the regions of the Atlantic seaboard of western Portugal, Spain, France and England, Wales, Scotland and Ireland. As Bianchini notes, these latter cities 'are, far more than London, active exploiters of EC resources and active members of international urban networks, and are also keen to initiate projects which would improve their ambience internationally' (1991b: 3). What this overlooks however is the European regional development policy that since the late 1980s specifically targeted such provincial city–regions through the redrawing of the map of Europe instead of the national assessment of regional development areas (Jones and Keating 1995, Evans and Foord 2000b) and which effectively passes over pockets of deprivation in otherwise 'wealthy' cities. Conversely, the cities of culture benefiting from regional aid have maintained and attracted middle class and gentrified enclaves, including of course the cultural intermediaries who have stood to benefit most from the injection of cultural capital.

Lingua Franca

Perhaps the most powerful element in cultural autonomy is language and this is therefore one test of the success of regional cultural determination which any cultural planning process obviously needs to reflect and respect. In Catalonia, for example, as a result of the 'normalisation' policy enacted in 1983, the percentage of the region's population speaking Catalan rose, it is claimed, from 64 to 75 per cent between 1986 and 1996; and from 61 to 72 per cent of those who actually read the language (Cubeles and Fina 1998), although 40 per cent still claims Castillian ('standard' Spanish) as its first language, it being one of four officially recognised in Spain. A new regional government law in 1998 proposed minimum quotas for film and radio (50 per cent of new films and total radio output to be in Catalan, including 25 per cent of songs played on music stations). Provence and Wales have moved towards bilingualism (including school curricula), whilst Breton and Gaelic (in Scotland) are not so established and effectively are minority- if not virtually dead-spoken languages. The support of Gaelic art programmes in Scotland for instance has focused on supporting performing arts and literature projects in the Highlands and Islands areas where cultural facilities are sparse in comparison with the city-conurbations of the Strathclyde and Lothian regions. In the Basque region, Euskera, only a robust written language since the late nineteenth century, is being actively developed and promoted, albeit from a low spoken base. In the reassertion of Basque identity, a network of *casas de cultura* and civic centres in *barrios* are funded by the municipality (Gonzalez 1993: 78), and which are, as in other Spanish and Latin American cities (e.g. Sao Paulo), often community/voluntary sector operated with funding from private foundations and philanthropic institutions. Respect and provision for dual and minority languages can be seen in countries such as Finland where over 5 per cent of the population still speaks Swedish (7 per cent in Helsinki). Swedish-language theatres are maintained in the old west coast capital of Turku (Plate 7.3) and in Helsinki where the Swedish Theatre is one of the largest of twelve, along with the National and Helsinki City venues (HCP 1997).

Language is also the most sensitive and symbolic manifestation of European common culture – the LINGUA programme for instance was established in 1990 to encourage

Plate 7.3 Swedish Theatre, Turku, west Finland (1999)

bilingualism (or multilingualism) in EU countries (although controversially it was wound-up in 2000). This was in the context of an earlier EC directive pledging Member States to take appropriate measures to provide free and adequate tuition for the children of migrants to learn the 'official language' of the host country. The Maastricht Treaty (Article 126) also stressed the Community aim of developing 'the European dimension in education' (Gowland *et al.* 1995: 236), especially through the learning and development of the languages of Member States. The current French quota system for film and music could be argued as being almost as imperialist as the Anglo-American domination of mass culture, but one whose impact has in fact been deleterious to the French cultural product and consumption. (The oft-quoted figures of 80 per cent of e-mail and contracts and 75 per cent of the world's letters being in 'English' are of course heavily weighted by North America, whilst in Europe the balance between English, French and German is much more equal.) French films are being relegated to daytime and unattractive slots with consequential attendance decline in contrast to the rising demand for Hollywood and other non-French movies. Music quotas require 40 per cent of radio playlists to be in the French 'language' ('lyrics'), with 20 per cent being for 'new bands' and 20 per cent for 'other Francophone' music. This may be good for French rap (i.e. a US import/hybrid), which might have been played anyway, but less so for Algerian or Moroccan bands not wishing to rap in French or patois! The balance between cultural protection and prescription is notoriously hard to get right, not least as forms of dissemination and distribution increasingly take place outside of terrestrial or national control.

The development of cultural policy and funding initiatives in Europe during the 1980s had not surprisingly tended to promote and favour cultural unity, but perhaps as a reflection of the post-Maastricht mood against further European centralisation and the realities of a widening Europe – culturally, geographically and economically – the recognition of diversity is now pragmatically accepted and respected. Tensions are apparent however between region and city-state (or cultural cities) in terms of cultural policy and regional development. Examples include Barcelona, perhaps the leading exemplar of culture-led regeneration, and the Catalan region. Here a cosmopolitan city, attracting overseas as well as Spanish visitors – particularly following the promotion of very successful Olympic Games in 1992 – pursues an international cultural agenda, whilst the Catalan regional government, a conduit for European and other regional assistance, pursues a Catalonian identity and cultural policy – it promotes the Catalan language (see above) in drama, its promotion being a condition of regional grant-aid. Most Barcelona theatres on the other hand prefer Spanish and/or English language work – resulting in much dance, mime and physical/non-literal drama and musical theatre in the city. Seventy-five per cent of theatrical productions (including musicals) are required to be in Catalan in order to qualify for regional arts funding – from over twenty theatres in 1950, by 1993 Barcelona supported only four. These cultural 'laws' are primarily politico-cultural rather than artistic since they are also resisted by artists themselves, both because of the institutionalised censorship they imply and because prosaically the artist wishes to 'speak' to as wide an audience as possible, as a Catalonian novelist claimed: 'I write in Castillian because that way I can reach 400 million people around the world, rather than 6 million in Catalonia' (quoted in Gooch 1998). Meanwhile, 'indigenous' sports facilities built by the Catalonian government for the Olympics also lie under-used. A similar tension exists between Quebec's francophone cultural policies (De la Durantaye 1999) and the cosmopolitan all-year festival city of Montreal, which pursues a more pluralist approach to arts provision. Even here the tension is apparent. For example, the annual international Festival of Comedy attracts over a million people (20 per cent of whom are tourists) at a cost of C$16 million, of which C$1 million is from public funds. Suggestions that the festival is now the city's biggest *English*-speaking cultural attraction and has overshadowed the quality of French-language programming are played down by the director: 'We're not an English event. We're not a French event. It's an international event that happens to take place in Montreal' (quoted in Hustak 1998). This illustrates the conflict arising where art venues supported by European and national regeneration funds are justified through a universalist artistic policy and programme – appealing to tourist/visitor markets and a cultural elite alike – as opposed to one that is rooted in regional or local cultural identity and production.

European planning systems

The freedoms enshrined in the founding Treaty of Rome: 'free movement of goods, services, capital and people' and articulated further in the Single European Act and the Maastricht and Amsterdam Treaties, suggest that sooner rather than later, physical planning will be seen as an activity beyond national boundaries (Antoniou 1992: 12). Indeed town planning is explicitly mentioned in the Maastricht Treaty which established the concept of a trans-European infrastructure network, whilst town and country planning is directly involved in European environmental policies such as European standards for the Environmental Impact Assessment (EIA) of major development projects, as required since 1985. Tendering for contracts over a certain size, e.g. building and design projects, requires Europe-wide advertising, thus opening up competition to architects from other EU member countries. This has enabled a small group of 'star' design firms to feature in several major arts and other public building and regeneration projects, and the adoption of their individual 'styles' by cities and venues seeking to replicate their design signature, e.g. Foster, Rogers, Coates from the UK, Gehry and Meier (USA), Calatrava and the late Enric Miralles (Spain).

Historically, land-use planning has evolved under several, albeit hybrid, systems with the French civic code providing the model for the Napoleonic Empire, the basic elements of which after independence in 1815 'were retained in much of Europe, notably Belgium, the Netherlands, Luxembourg, Spain, Portugal and Italy' (Newman and Thornley 1994: 51). The Germanic system reflected the already semi-autonomous states operating from the fifteenth century, which Austria, Greece and Switzerland largely inherited, whilst the Nordic countries combined this with aspects of Napoleonic centralism (regional agencies of central government) and strong local, municipal planning powers. Planning systems today not surprisingly vary widely across Europe, with some important differences also within nation-states themselves. In the UK, for example, the systems of Scotland on the one side and of England and Wales on the other; the semi-autonomous German *länder*, as well as the social, economic and cultural disparities and diversity across and between European regions. In Scotland, for instance, greater regional and 'structural' land-use planning and adherence to local area plans is evident than is the case in England, alongside regional-level economic development and a more integrated land-use and amenity planning system, including arts and cultural facility standards (Feist 1995). However as Burtenshaw *et al.* have observed, population, catchment and hierarchy of facility models (as discussed in Chapter 5) have been widely adopted:

> Although large differences can be detected between cities within Western Europe, attributable to differences in economic priorities, political traditions and social preferences, urban planners have responded in recognisably similar ways. . . . The monitoring of the adequacy of provision led to the study of the effective range of demand, the estimation of catchment areas, and ultimately the creation of scale hierarchies of provision.
>
> (1991: 194)

Despite these historic and legislative differences, further developments in supranational planning policies are expected as the European Commission promotes greater harmonisation and coordination. States such as Germany and The Netherlands have appointed ministers for land-use and physical planning. Others such as Portugal and France combine land-use planning with regional policy. The chief exception to this continuity in Western Europe, from the early days of planning, has been the UK. Development Plans since 1947 have not been a form of legally binding zoning plan, and national and regional planning guidance is not administratively binding on local government (Davies 1994a, b). Of fundamental difference are Britain's constitutional position and land rights in contrast to the French *Code Napoléon* – the English lack a written constitution: 'we are subjects not citizens and as such we have virtually no rights. We are allowed to develop land at the discretion of authority' (Antoniou 1992: 12). The uniqueness of the British planning system includes an absence of legally binding plans; the separation of development control from building control and the discretionary approach to development between planning policies and actual control decisions, i.e. the flexibility of the planning system to allow development contrary to approved land-use plans – the difference between policy and practice. In contrast the Continental model is essentially plan led. In Denmark, France, Germany and The Netherlands a proposal conforming to the plan ensures a right to develop land, and planning and building controls (including design) are combined in a single permit. The consideration and special treatment of artists workspaces and studios is also evident in some North American states and on the European Continent, as already discussed in Chapter 6. In the USA, some states and cities are highly *dirigiste* – this has entailed specific land-use zoning and the protection of artist studio facilities within town planning and property rental markets – while others largely reject 'planning' on ideological/libertarian grounds ('the new frontier'). Canada and Continental Europe also benefit from stronger city and regional plans – a weakness of the British approach which lacks integrated action on regional planning, despite its importance in structural adjustment: 'employment and economic change lie at the heart of regional planning' (Cullingworth 1979: 234). From the late 1970s town planning in Britain has shifted markedly from the 1940s (cartographic) plan-led model to social and economic ones, based far more on descriptive objectives, for example a focus on the impact of the sustained growth of unemployment and post-industrial economic change. Local authorities have exercised employment generation and economic development functions, which have commonly been managed as part of environmental planning, in recognition of the relationship between land-use development, economic regeneration and employment creation, a corporate strategy approach to urban socio-economic problems. The days of Abercrombie's grand designs for London or Hull are long gone. As Waters notes, however, British town planning is now largely reactive rather than proactive in approach: 'As with other aspects of [British] town and country planning policy, planning in the sense of vision and opportunity is noticeable by its absence. Planning policy is reacting to the market, not anticipating or controlling it' (1987: 59). The development of model planning policies for unitary development (land-use) plans in London in the early 1990s (LPAC 1990b) has however signalled a greater appreciation of both the plan-led approach generally and the integration of cultural planning within the mainstream environmental planning process (see Appendix I).

Simon Jenkins writing on the centralisation of British government from the 1980s also suggests that the French central-state example is: 'no longer relevant. . . . Communes and mayors enjoy wide discretion in planning and local budgets. . . . The same is true in Italy, Spain and Portugal' (1995: 257), and he also compares the Scandinavian 'free commune' system and German *länder* with power of veto and opt out from national legislation. Britain's urban concentration with 92 per cent of the population (from 86 per cent in 1960) living in cities, towns, suburban zones and large 'villages' – the entire population of Britain lives on 10 per cent of the land mass – suggests that in terms of structure planning, and *hierarchies of need*, amenity and other distributive planning strategies are not as comparable with other Western countries and that universal models are not wholly transferable. In Greece (63 per cent), Italy (68 per cent), France (74 per cent) and the USA (74 per cent) a much lower proportion of the population is urbanised and even the higher urban densities of cities in smaller countries such as Belgium (97 per cent 'urban') and The Netherlands (89 per cent) do not compare with the urban concentration of mainland Britain (Population Reference Bureau 1995).[1] For example, only 12.7 per cent of the national population in Belgium lives in large cities, compared with 39 per cent in the UK, which is concentrated in eight city–regions.

In further contrast to Britain, regional tiers in France have considerable status with directly elected regional assemblies having major responsibility for infrastructure and development (the European Commission's approach to the distribution of EU funds is largely based on the French integrated system of regional economic development). In several European countries, from Spain to the former Czechoslovakia for instance, planners and architects are a single, combined profession, whilst in Britain planning was separated from architecture and engineering from the earliest days of its professional recognition in 1914. British planners (who therefore numerically constitute 90 per cent of Europe's specialist 'professional planning' workforce) are seen to be, in the words of Robin Thompson – a former President of the Royal Town Planning Institute – 'the aliens of European planning. . . . We practice discretionary planning. . . . We also engage in a range of activities notably in economic and environmental action which our European neighbours generally assign to other professions' (1994: 18). Conversely, whilst regional planning in most of Europe is the domain of the professional economist, engineer or geographer-planner, other city planning practice combines architectural and urban design with 'planning', a factor perhaps in Thompson's observation that 'the best European practice outdoes our own. It is strategic, imaginative, fluid and cultured' (ibid.). Concern has also been expressed about 'the lack of urban design training for British town planners, whereas it is a central element in European professional training' (LAB 1992a: 1; also Landry 2000). Britain has therefore not produced the master-planning architects, as opposed to design and build developers (e.g. Sir Christopher Wren), such as Le Corbusier (Raeburn and Wilson 1987), although more recent but exceptional cases include the late planner and architect Francis Tibbalds and architect Richard Rogers (Rogers and Fisher 1992, Tibbalds 1992). Significantly, until recently much of Rogers's built work has been outside of Britain, such as the Pompidou Centre in Paris (with Renzo Piano) and the British-based architect Zaha Hadid has also had to make do with her temporary internal structure for the maligned Millennium Dome (designed by Rogers) – the 'Mind Zone' – whilst abroad she is 'fêted, allowed to stretch herself on arts centres in Cincinnati and Rome' (SPACE 2000: 3). Until the opportunities thrown up by the reintroduction of a National Lottery in Britain, which has co-funded capital arts, heritage and other public projects since 1995 (Evans 1995a, 1998e), modern(ist) architects had little acceptance in urban design and new build schemes. Witness the red brick mass of the long-awaited British Library in London in contrast with, say, the radical design for the *Grand Travaux* Bibliothèque Nationale in Paris. Even with the injection of 'free' lottery funds (i.e. not accountable through either public borrowing, central or local taxation regimes), architects such as Hadid (Cardiff Bay Opera House; Crickhowell 1997), Daniel Libeskind (V&A extension, London) and even Rogers himself (South Bank Arts Centre) had ambitious and costly schemes rejected and/or pilloried due to an inherent suspicion and conservatism in design and in cultural 'risk-taking' amongst politicians, planners and key decision-makers. As Bird *et al.* therefore suggest:

City life in Britain has never conveyed the alluring resonances of the great centres of European modernism – Paris, Barcelona, Madrid, Milan, Hamburg, or the glittering but brittle spectacles of American urbanization. Neither the left nor the right has laid claim to the city as a site for the construction of subjectivity and political identity other than as the backdrop for the enactment of ritual and tradition: the ceremonial commemoration of privilege, national identity or loss.

(1993: 121)

European regional development

The prime objective of European intervention in regional policy has been the reduction in the disparity in socio-economic development between the various regions of the EU. This policy has therefore sought to contribute towards stability within the EU as well as to promote high employment against a region's uneven capacity for generating sustainable development and in adapting to new labour market conditions and global competition. This development also reflects the growing impact of regionalism within and across European Member States, as well as a political opportunity for the European Commission itself to assert a more direct influence, to an extent bypassing national governments, several of whom had a political antipathy to the European project and its expansion, e.g. the UK, Denmark and The Netherlands. Regionalism is also an ambiguous term, and as Harvie observes: 'It is difficult to separate the cultural, economic and propagandist elements of "regionalism" and to subject it to the same sort of critique which has come the way of the nation-state' (1994: 5). Dialectical regionalism therefore also contains a paradox: 'On the one hand it has been associated with movements of reform and liberation . . . on the other, it has proved a powerful tool of repression and chauvinism' (Tzonis and Lefaivre 1981: 178). All of these elements have been adopted, however, in presenting regional cultural and economic agendas within autonomous and other regions, often in direct proportion to resistance of the centre (e.g. Catalonia and Basque, Spain; Scotland and Wales, UK; Lombard League, Italy; and regional capital cities). European structural assistance has been a key tool in both legitimating and financing major infrastructure and investment programmes in eligible regions, with culture a secondary but symbolically important rationale in the re-imaging and promotion of regional and, in particular, city identity. Cultural projects and facilities have therefore been supported, Trojan horse-style, on the back of regional economic development programmes, to the benefit of both the European centre and regional political movements (e.g. 'new urban left'; Henry 1993) during the 1980s and 1990s.

However, since no explicit cultural policy objectives are referred to in the regional policy agenda, regional development has been mainly executed through two major programmes: the European Regional Development Fund (ERDF) and the Cohesion Fund, and these have been used to co-finance programmes and projects that target structural assistance at the more disadvantaged EU regions in partnership with national and/or regional authorities in the Member States. The ERDF is the largest of four programmes representing over 50 per cent of all Structural Funds, with assistance aimed at four priority objectives:

1   supporting small and medium-sized enterprises
2   promoting productive investment

3   improving infrastructure
4   furthering local development.

The micro-enterprise cultural industry economy highlighted in Chapter 6 also reflects the structurally weak, European small-firm economy generally (Table 7.1), which is also increasingly reliant on a small number of major institutional, central-core and trans-national organisations which in most cases were the prime inward investors and beneficiaries of public investment programmes.

Table 7.1 Indicators of enterprises in the European Union

	Micro	*Small*	*Medium*	*Large*	*Total*
Number of enterprises (000s)	17,285	1,105	165	235	18,590
	93%	5.9%	0.9%	0.2%	100%
Employment (000s)	37,000	21,110	15,070	38,220	111,410
	33%	19%	14%	34%	100%
Average size (number of employees)	2	20	90	1,035	6
Turnover (ECU millions)	0.2	3	16	175	0.8
Value added per person occupied	30	40	50	55	40
Share of labour cost in value added (%)	38	63	60	53	53

Source: EC (1997)

For example, crafts/designers and visual artists are traditionally self-employed and sole-trader-based (Knott 1994, Towse 1995, Pratt 1998, Evans 2000b) with a growing freelance and contract work relationship with larger cultural employers in the perform-ing and media arts (e.g. venues, broadcast and print media) and specialist retailers (including exhibitors and galleries). Self-employment generally grew within the labour markets of countries such as the UK and Spain during the 1990s, in line with growth in cultural sector employment, as flexible, piece and project-work, and contracting-out became the norm in certain service sectors. In the 1990s the phenomenon of zero-base growth has seen new technology and related creative activities substitute for low-tech and pre-industrial forms of cultural production (see Chapter 6) in contrast to the ear-lier real-terms growth in employment within the arts and cultural industries. Other sectors such as audiovisual production also conform to this profile with the concentra-tion, diversification and globalisation tendency of the medium or large conglomerate. In the case of Hamburg, for instance, they are 'fed by an increasingly dense network of small(est) enterprises and structures of self-employment' (Henriques and Thiel 1998: 19), and also in Lisbon, Portugal, where over 63 per cent of firms in the audiovisual sector employed less than ten people in contrast to television/radio and news agencies which employed on average over one hundred workers (ibid.: 22).

The lion's share of European Structural Funds, whilst supposedly targeting small enterprises, has in fact provided substantial investment in major cultural and heritage facilities, particularly those linked to city and regional regeneration and urban tourism strategies. This includes support for the major investment in the high-profile cultural cities of Barcelona, Seville and Madrid; Dublin, Glasgow, and northern English and Italian cities (Bianchini and Parkinson 1993, Evans 1993a, b; see Chapter 8) and latterly East Germany, notably Berlin. National and regional governments have also applied the arts and urban and regeneration formula linked to cultural tourism activity in the regional cities of Frankfurt, Hamburg and Cologne, Germany; Bilbao, Spain and Lisbon, Portugal, as well as regional cities of France (e.g. Grenoble, Rennes, Lyons, Montpellier), which have emulated the *Grands Projets* of Paris. The support of arts, heritage, cultural tourism, and related training and regeneration projects through European funding pro-grammes therefore supplements funding from national and regional sources (and vice versa). In practice, national and local government funding acts as partnership and lever-age to European funds under matching funding criteria, but the absence of a real control or counterfactual base for comparison to establish true additionality (or an in-depth understanding of the resource allocation and decision-making processes at each level) effectively limits the evaluation of both national and European funding policy regimes (Evans 1998c). Elite and pluralist approaches to an understanding of urban politics (Stone 1993) and the distribution of power do not adequately capture the complexities where in this case the supra-national, national, regional and local tiers interact with business (large and small), voluntary and a host of other community sectors including the arts. A theory of urban regimes (Judge *et al.* 1995) and governance needs to take into account these multidimensional relationships and interests and the mechanisms by which resources are first bid for, allocated and then distributed. In these different stages and devolving levels of decision-making – European, national, regional, province/county, city, local, project/organisation – culture has symbolic if not economic power in adding value to regeneration programmes and projects. In practice, however, the suspicion between the EU and some Member States (and between national and local government) is that real additionality has not always been transparent, i.e. all or most of the investment would have taken place anyway (without regional aid). In the case of culture, this is fur-ther hampered by the bypassing of relevant agencies and local communities at national, regional and city levels, thus undermining a planning or needs-led approach to cultural provision and risking such facilities being in the 'wrong place' and/or of the 'wrong type' (see below). As the Commission itself also admitted:

> It is not possible to provide precise information on assistance given to culture under the mainstream operations of the structural funds. This is because the Commission's role is to adopt and co-finance programmes. The individual projects making up the programmes are selected and managed within the Member states. In addition, the facts that the cultural sector is not homogenous and that there are sig-nificant variations in definition and statistical classification of culture make precise and systematic data collection impossible.
> (Wulf-Mathies, in *Official Journal of the European Communities* 1999: 55)

This situation conveniently obviates the Commission, beneficiary members and regions, from justifying the choice and location of cultural projects that received European and national support. The availability and criteria for European funding has not however been benign in terms of either regional autonomy or the cultural heritage that has been put forward as part of regional development programmes. This has not been a simple case of national versus regional political freedoms, but one which has divided regional, provincial and local districts in terms of the support of cultural development and, more importantly, how it is planned and resourced. For example, in the Castilla-Leon autonomous region of Spain, an eligible ERDF area (with a Gross Domestic Product of 70 per cent of the EU average and population of about 2.5 million), cultural spending increased by 63 per cent, or 5 per cent a year, from the mid-1980s to 1997. This however masks a major redirection of the type of culture receiving public funding, which was directly influenced by the European regional funding regime and the promotion of a type of 'common heritage' and identity (see below) by the higher level regional government, in contrast to local areas (i.e. the nine provinces and 2,200 municipalities). Table 7.2 shows the extent of the switch from cultural investment in local arts ('cultural diffusion' – performance, festivals, small scale) to museums and heritage sites, including the funding of two new museums in Leon and Zamora.

Cultural planning in this scenario suffers where the diversity and aspirations of local areas, including cities and larger towns which host major new cultural projects as elements of cultural tourism and 'regional identity' strategies, are deprioritised and therefore where local area planning is not reconciled with strategic and structure plans.

Community support of 'culture'

Although the European Commission has long-resisted adopting a specific cultural 'competence', the Council of Europe itself had initiated a number of projects in the 1970s, for example around socio-cultural animation and studies of the cultural sector such as taxation, the protection of cultural workers, art trade and copyright (Mennell 1976, Dumont 1979, Goodey 1983). Resolutions in 1974 and 1977 and adopted by the European Parliament in 1979 laid down the first Community action in the cultural sector (Dumont 1979), although these Parliamentary resolutions had no executive or legal power of implementation. More recently, in recognition of both citizenship and duties, the European Declaration of Urban Rights also included 'culture' alongside nineteen other urban environmental rights, which also identified the importance of integrated urban planning and functions. This declaration arose from the European Urban Charter adopted by the Council of Europe's Standing Conference of Local and Regional Authorities of Europe (CLRAE) on 18 March 1992 in Strasbourg:

8. CULTURE – to access to and participation in a wide range of cultural and creative activities and pursuits.

11. HARMONISATION OF FUNCTIONS – where living, working, travelling and the pursuit of social activities are as closely interrelated as possible.

17. PERSONAL FULFILMENT – to urban conditions conducive to the achievement of personal well-being and individual social, cultural, moral and spiritual development.

Table 7.2 Allocation of the cultural budget of Castilla-Leon, 1985–97 (figures are percentages of total budget)

Programme	1985	1986	1987	1988	1989	1990	1991	1992	1993	1994	1995	1996	1997
Historical and artistic heritage	50	50	31	52	56	57	57	54	51	52	56	61	57
Museums, archives and libraries	9	7	16	24	26	27	25	27	27	27	19	21	25
Cultural diffusion	31	33	35	16	7	4	7	8	9	16	11	12	12
General cultural services	10	10	18	7	12	11	11	12	13	6	14	7	6

Source: Devesa (1999: 9)

The thrust of the Council's earlier studies rested on the expectation that local author-ities increasingly needed to shoulder the burden of public patronage of culture and leisure, both high-art and popular and traditional culture, leading to the question: 'how is a town to distribute its limited resources to the best advantage?' (Mennell 1976). This met the familiar problems and complexities that resist a standardised, uni-versal approach to arts and amenity planning. As Burtenshaw *et al.* note, there is divergence between European cities, for instance: 'large variations in the popularity of entertainment media and the responsibility of urban authorities' (1991: 180), whilst participation rates, e.g. for cinema, vary widely between European countries (see Chapter 2). Cross-national and 'cultural' comparisons have been questioned in other chapters of this book, not least when resource allocation is contrasted between national systems that treat and define aspects of the arts, heritage and culture quite differently (Evans 1993a). The balance between central and local/regional levels of funding pro-vide one indicator of subsidiarity and the relative power retained in planning and resource terms, whilst the proportion of arts to all public spending and to national GDP indicates the element that culture has in public provision *financially*, and, by the same token, the relative importance of the private sector in cultural activity (e.g. USA, Spain). Per capita arts spending perhaps provides the 'acid test' comparative, with high local/regional support in Germany (£56 per head) and Finland (£59) accounting for the highest levels of spending per person, with a middle level of subsidy in France (£38), Canada (£30), Sweden (£38) and The Netherlands (£30), less than half of this in the UK (£17) and Australia (£16), and minimal amounts per head in Ireland (£6) and the USA (£4) (Feist *et al.* 1998). The extreme differential between say Germany and Ireland, which funds only about 10 per cent of the former country's level, is obviously one indication of a wide gap in the relative importance afforded to public culture, notwithstanding social, historic and artistic variations between countries and where it is 'credited'. These include, for example, cultural preference and the 'non-traded' and pri-vate cultural spheres, as well as largely hidden forms of participation: 'Being a member of a choir, taking part in a community play, making pottery, performing in a carnival or religious festival, or being on the planning committee of an arts centre would be . . . invisible' (Brinson 1992: 73). Spending on culture, whilst one quantifiable indicator, is primarily an 'input', i.e. resources, not an 'output' measure such as audience/participant numbers and profiles, let alone an indication of the 'outcome' in terms of cultural cap-ital and other impacts on a nation's cultural development and relative health (Evans 2000b). These comparisons give rise to the periodic but ultimately subjective ques-tions – are the Irish more or less 'cultured' than the Germans, or does Ireland produce more/better culture than Germany?

Identifying and quantifying the cultural component of regional economic develop-ment investment is also complicated, as conceded above – both in data analysis and political terms. Where culture is neither defined nor planned, support of cultural schemes and projects is seldom promoted or consolidated at the macro-level – to do so might invite difficult questions as to the rationale for the type of culture being funded (e.g. heritage, city-based arts venues), the lack of a planning or needs-led framework (e.g. locations benefiting and those not) and therefore the absence of a European cultural policy. However, in a rare study of cultural funding from the EU as a whole, it was estimated that Community funding benefiting the cultural sector totalled ECU494 million a year or ECU2.47 billion from 1989 to 1993 (Bates and Wacker 1993), and by the mid-1990s still only 8 per cent of the £350 million a year spent on arts and cultural activities in the EU originated from the designated 'cultural' office. (The EU's Culture Unit formerly DGX *Information and Culture* was merged in a 1999 reorganisation under an *Education and Culture* Directorate.) This cultural spend corresponds to less than 0.8 per cent of the total Community budget for the period. As with tourism, by far the greatest amount of funding for culture has been provided by the European Structural and Regional Development Funds, as discussed above (Evans and Foord 2000b). Moreover, ERDF funding in the cultural sphere is mostly related to the conservation of cultural heritage (notably the built heritage), as well as to the development of cultural trails and itineraries. Given the EU policy of promoting cultural tourism – both to reinforce European common culture and heritage and to celebrate regional cultural diversity – the funding of tourism development through regional and structural programmes has also been a prime source of capital investment in cultural projects and infrastructure. This aim was articulated by the European Parliament and implemented by the Commission in its successive Structural programmes:

> The [European] Parliament notes that the tourist activity least subject to seasonal fluctuation is cultural tourism, which has very considerable development potential in Europe since it continues to attract citizens of non-member countries as well as strengthening the feelings of Europeans of belonging to the same community; the Parliament urges the commission to give preference to applications for ERDF assistance from Member States involving projects *which develop sites of cultural interest and which include cultural programmes.*
>   (EP VI Resolution on a Community Tourism Policy; CoE 1991, emphasis added)

The support of cultural tourism development, in the eyes of policy-makers at least, was therefore seen to have the potential to provide a more even spread of economic activity; to widen the European inbound and intra-regional tourism markets, as well as to reinforce European identity and pride. Following this policy statement, the EU Tourism Unit supported a major trans-national survey of cultural tourism at arts and heritage sites across nine EU member countries between 1992 and 1994 (Evans 1993b, 1998b, Richards 1996). This confirmed, in Bourdieu's terms, not only the high 'cultural capital' of cultural tourists (education, employment, prior visitation, etc.; Bourdieu and Darbel 1991), but also the divergence between the cultural habits and motivations of visitors – European and 'others' – and a clear preference for visiting museums and heritage sites over the live and visual arts. In the case of England, for example, the preference for live arts productions was also evident from a survey of visitor intentions (BTA 1995) amongst English-language visitors when compared with visitors from other origins who eschewed theatres for museums and galleries (Table 7.3).

However, arts venues have also been the prime recipients through ERDF funding of regional arts centres, theatres and galleries (Table 7.4), but whose viability rests on significant visitor (domestic and overseas tourist) rather than local audiences. It is no surprise therefore that foreign audiences make up a significant proportion of theatre audiences in London, or that the need to appeal to as wide a linguistic and cultural visitor has ensured that commercial as well as subsidised venues (e.g. Royal National

Table 7.3 Importance of the arts as a factor when visiting Britain

	Total by area of residence (%)				
	All	*Europe*	*America*	*Other English speaking*	*Other non-English speaking*
Museums	59	57	66	53	67
Galleries	37	33	40	34	50
Theatre	37	31	53	48	38
Concerts	21	21	14	21	28
Ballet/opera	12	12	10	13	16

Source: BTA (1995)

Note: 59 per cent of all tourists rate museums as 'important or very important' in their decision to visit Britain

Theatre) rely on the ubiquitous musical – original and revival. Between 1987 and 1997, attendance at West End theatres in London rose by 5 per cent; however, when the 'modern musical' is excluded, attendances actually declined by 26 per cent (Gardiner 1998). Cultural tourism can therefore have a direct effect on the nature of programming itself and the relative demand for activities, but which may not reflect cultural preferences, aspirations or, importantly, the creation of new work (Evans 1999e, 2000b).

Bourdieu's survey of museum visitors in the 1960s also confirmed in his mind that the possession of cultural capital was even more important and therefore that cultural development was an unlikely outcome from so-called cultural tourism (Bourdieu and Darbel 199): 'As one opportunity among others of expressing a cultivated inclination, cultural tourism, that is tourism in which museum visiting plays a part, depends on level of education even more than ordinary tourism' (ibid.: 23), and: 'if it were simply a question of giving the initial impetus, tourism cannot compensate for the lack of an artistic or intellectual education' (ibid.: 24). Another notable observation from the EU study carried out in 1993–4 was that nearly 15 per cent of all cultural tourists actually worked within the cultural and heritage sector in their home country – the 'culture vultures', or what McGuigan identifies amongst the 'Professional-Managerial Classes' as cultural intermediaries: 'those particular sections that are directly employed in practices of cultural mediation and consumer management' (1996: 39; also Bourdieu 1984, Featherstone 1991). In the development and management of urban cultural strategies, as well as the largely benign influence of town planners, the power and influence exercised by arts and cultural mediators is one that reinforces the existing hegemonies and legitimisation of both high-art and contemporary versions of the type of cultural activity which the arts and regeneration process best 'needs'. This is seen in the transfer of key ('footloose') staff from major arts institutions to regional projects, mirroring perhaps the international transfer market in artistic and executive directors amongst theatres, opera houses, orchestras, museums and galleries (Evans 1999g). Whilst regional cultural development has stressed decentralisation and devolution from the 'centre' therefore, in France for instance this was 'accompanied by an inverse tendency to recruit those with specialist talents and abilities from national level, often [the capital] Paris' (Negrier 1993: 142).

The spatial concentration of visitors to a small number of museum and heritage sites in Europe is also demonstrated in Table 7.4, and as Frangialli complains, the creation and promotion of such *honey-pots* also results in a drop in the quality of visitor services, higher prices, congestion, long queues and a marked degradation of the sites and monuments themselves, 'and the imposition of an imported cultural model which distorts the original' (1998: 8). With this over-concentration and unsustainable promotion of a few heritage locations, many located in cultural capitals (e.g. London, Madrid, Paris, Rome/Florence/Venice), the majority of sites and buildings languish in neglect and find it impossible to attract either public or private investment, or significant visitation (cf. Southern Italy, Mariani 1998; and Northern Spain, Evans 1998b, Devesa 1999).

Table 7.4 Visitors to heritage sites in selected European countries

Country	Numbers of heritage sites	Annual number of visitors
France	38,000 historical monuments, 5,000 museums/sites	1,500 receive over 20,000 visitors 15 receive over 100,000 visitors
Italy	1,700 public museums, 700 private museums	15 receive over 300,000 visitors 8 receive over 800,000 visitors
Spain	1,250 museums, 7,500 heritage sites	26 receive over 100,000 visitors 25 receive over 100,000 visitors
England	1,900 national monuments (50% private)	17 receive over 150,000 visitors

Source: Frangialli (1998)

In terms of the distribution of European structural funding of arts and cultural projects, the preference for city-centre arts venues can also be gauged from a survey of schemes in the UK allocated ERDF grants from 1990 to 1996 (Table 7.5). Here a small number of schemes, often a single flagship project in regional cities, dominated European funding to the eligible region as a whole, with projects forming part – sometimes a central part – of wider regeneration and image-development strategies. Given the leverage system whereby European grants are used to match targeted national and regional regeneration programme funds, this also follows the national pattern of urban regeneration support for the arts, and can also be seen as *anti-planning* where selective regional assistance areas are not congruent with arts and cultural 'need'. This distribution also tended to reflect existing arts activity and legitimate centres for the performing and visual arts, even where demand and audiences for these were in decline (Evans *et al.* 1997, 2000). A counter argument would claim that such investment is required to address this decline and improve both the quality and quantity of cultural activity. However experience at one new facility, the Centre for Popular Music in Sheffield (Plate 7.4) (see below), jointly ERDF and Lottery funded and designed by radical architect Nigel Coates, has seen visitor numbers and therefore income fall well short of forecast, thus resulting in staff redundancies and a serious financial crisis within weeks of its opening in 1999. A rescue package only a year on sought to develop strategic partnerships with the city authority, the established Cultural Industries Quarter and local university, and to rebrand and shrug off any elitist image and for the centre to become part of the region's 'cultural fabric' by involving local people in its events programme. One might ask why such an approach was not part of the original development and therefore why no cultural planning consideration was evident in this high-profile post-industrial cultural city and its original public funding conditions.

Table 7.5 European Regional Development Funding of major arts projects in the UK, 1990–6

Major projects	£ (000s)	Major projects	£ (000s)
Liverpool Institute for the Performing Arts	5,900	University of Nottingham Arts Centre	993
Philharmonic Hall, Liverpool	3,800	Notts Foundation for Music & Media	507
Tate Gallery, Liverpool	1,500		
Percentage of North West Region (*n* = 9)	93	Percentage of East Midlands Region (*n* = 5)	74
Sunderland Empire Theatre	1,300	Old Malt Cross Music Hall	570
Northern Sites	682	Percentage of Eastern Region (*n* = 2)	99
Dovecote Arts Centre	500		
Laing Art Gallery, Newcastle	469	Glasgow Gallery of Modern Art	7,515
Percentage of Northern Region		Harland and Wolff Theatre	
(*n* = 15)	58	conversion	487
		Glasgow Celebration of Visual	
National Centre for Popular Music,		Arts 1996	3,148
Sheffield	1,880	Dundee City Arts Centre	8,969
Percentage of Yorkshire Region		Pitlochry Festival Theatre Arts	
(*n* = 3)	93	Complex	4,874
		Percentage of Scottish projects	
National Centre for Literature,		(*n* = 28)	77 (total cost)
Swansea	1,360		
Percentage of Welsh projects (*n* = 7)	58		
New Art Gallery, Walsall	4,500		
Grand Theatre, Wolverhampton	2,000		
Centre for Cultural Enterprise, Coventry	1,700		
Percentage of West Midlands Region (*n* = 5)	91		

Sources: Evans (1997, 1999f), Evans and Foord (2000b)

European expansion

Since EU enlargement from twelve to fifteen Members in 1995 to include Austria, Finland and Sweden, a sixth ERDF programme 'Objective' has been created that incorporates the lesser-populated regions in Sweden and Finland, which have both embraced the potential of the arts and urban regeneration through regional development aid, e.g. Helsinki's turn as one of nine European Cities of Culture in 2000. Cultural industry quarter projects in this case attracting EU urban funding include the 1930s' Glass Palace Media Centre in Helsinki (Plate 7.5), an inheritance from the ill-fated Olympic Games, which was converted to contain an art house cinema, art book shops, cafés and

Plate 7.4 National Centre for Popular Music, Sheffield, South Yorkshire, 'temporarily closed' (2000)

Plate 7.5 Glass Media Palace, Helsinki (1999)

media production facilities to form part of a cultural triangle with *Kiasma*, the new museum of contemporary art, a multiplex cinema and a planned tennis palace museum (Verwijnen and Lehtovuori 1999: 219). In the north of the city, the famous Arabia ceramics factory, still producing versions of the Aaltos' now-classic designs, is the location for a cultural workshop development, a University of Art and Design and the Sibelius Music Academy. Manufacturing industries are decreasing in number and proportion here as elsewhere – 80 per cent of the city's employment is in the service sector and premises are often taken over by information industries and cultural institutions. The 1992 Helsinki Masterplan was drawn up during the country's worst ever recession and long-term strategic planning was adopted which presented scenarios for the city to the year 2020. This included the goal of improving the city's international image as a science, art and congress city, as well as creating lively and multidimensional arts and cultural pursuits for its citizens (HCP 1997). In 1998 *Blueprint* referred to Helsinki as the new Bilbao (giving a frighteningly short shelf-life to the Guggenheim satellite-city), with 'Kiasma as a poetic interpretation of site, a building which could result only from its precise location', and as Ryan goes on to say: 'What is being branded in these cities is not just the immediate institution, or anything so arcane as a collection, but the city itself. The museum becomes an icon and magnet for post-industrial urbanity' (2000: 91).

Prospects for future EU funding to the incumbent eligible Members are inevitably diluting as newer and 'poorer' members call on limited EU central funds and EU geopolicies shift. This will inevitably present difficulties for countries such as Greece, Spain, Portugal and Ireland (as well as Northern Ireland and Merseyside in North West England), which since the 1980s have relied on ERDF and other grant assistance (not least the Common Agricultural Programme, which also supported regional language and rural crafts schemes in peripheral regions) to support culture, heritage and related infrastructure investment in both rural areas and cities. An indication of the distribution of structural assistance in the late 1990s is given in Table 7.6, which summarises the gradual shift in country ranking from the previous five years (although Spain is still the highest recipient), with over 50 per cent going to only three countries and over 80 per cent targeted at regions lagging behind in economic development and employment terms.

Table 7.6 EU Structural Assistance (1994–9) at 1994 prices (rank in 1989–93 allocation)

Country	ECU (billions)
Spain (1)	34.44
Germany (−)	21.72
Italy (2)	21.66
Greece (4)	15.13
Portugal (3)	15.04
France (5)	14.94
UK (6)	13.16
Ireland (7)	6.10
The Netherlands (−)	2.62
Belgium (8)	2.10
Finland (n/e)	1.65
Austria (n/e)	1.57
Sweden (n/e)	1.38
Denmark (9)	0.84
Luxembourg (−)	0.10
Total	152.45

Sources: CEC (1996), Evans and Foord (2000b)

n/e: not an eligible member

The unification of Germany has had the most dramatic effect in fund distribution, as Table 7.6 illustrates. Germany is a country of two halves, in 1998 containing both the richest regions (figures are percentages, 100% = average: Hamburg, 195; Bremen,

153) and one of the poorest (Thuringen, 60) – measured by their GDP per capita as a percentage of the EU average ($n = 100$). Here visitor-led tourism is also adopted in the regeneration of post-Communist Eastern Germany, for example in Chemnitz (Karl-Marx Stadt during the DDR period), ECU62 million has been spent on an opera house refurbishment in this city where employment reached over 20 per cent and more than 30,000 residents have left to seek work elsewhere (Evans 1995b). In the former industrial city (including a centre for film production) of Dessau, post-unification unemployment stands at nearly 25 per cent, whilst out-of-town retail parks have accelerated the decline of the city centre. East Germans could be forgiven for feeling they have been colonised by the West (and see Berlin, Chapter 8).

Ireland presents another paradoxical 'success story', a booming economy (albeit from a low population and economic base) built on inward investment and tax incentives, post-Fordist production – new technology and tourism, and a well-qualified workforce. Over £1 billion of European Structural Funds has been awarded and matched with private investment in cultural and tourism programmes which have been reflected in traditional heritage developments and museums in rural and historic towns and sites (37 per cent of all 'product investment'), as well as the typical flagship and cultural quarter projects in the capital, Dublin (Deegan and Dineen 1998). Increases in visitor activity outside of Dublin have not however materialised and few of the heritage schemes are likely to be viable (or were ever desired locally, aside from the Euro-funds they attracted), whilst it is in the one 'cultural city' that urban tourism and cultural consumption combine to make Dublin now the most frequented destination out of London airports (overtaking New York and Paris). As Worpole declaims: 'Dublin's new glitzy, chic, international image is closely tied to the Temple Bar regeneration project, in which a run down part of the inner city has been transformed into a hive of small businesses, record companies, design offices, coffee houses, hotels and restaurants' (quoted in Levine *et al.* 1997: 115). Here, as in other post-industrial cities, it is a highly concentrated, Marshallian district that represents both the symbolic and economic power (Zukin 1996) and centre of the new-found cultural city (e.g. Bankside in London; CELTS 2000). This has been facilitated, whether benignly or deliberately, through the concentration of resources and a narrowing in the notion of European cultural identity and activity – retail–entertainment and heritage dominated – which has typified ERDF and urban programme distribution to date, where, to quote Worpole again, 'form has followed funding' (quoted in Levine *et al.* 1997: 114).

Unlike the five-year plans operating under the Structural Fund programmes, the EU cultural budget has been based on annual allocations and there has therefore been little opportunity for long-term planning. Table 7.7 outlines the Cultural Budget (excluding various media programmes) for the five years before the departmental reorganisation and shows the comparatively small amounts allocated to specific cultural programmes in contrast to the Structural Fund programmes outlined above.

The EU's five-year 'Culture 2000 Programme' also lacks a financial commitment or integration with other planning and regional development programmes. Preparatory actions before this programme's adoption in 1999 supported fifty-five projects (out of over 400 proposals) totalling €6.07 million. These were predominantly trans-national cultural cooperation projects and events, as in the past, e.g. celebrating the 250th anniversary of J. S. Bach's death: 'bringing the works of this great figure of German and

Table 7.7 EU cultural budget 1994–9 (ECU millions)

Year	Cultural cooperation with third countries	Architectural Heritage (Raphael from 1996)	Kaleidoscope (replaced by Culture 2000; see below)	Literature/translation (Ariane from 1996)	Total ECU (millions)
1994	1.4	8.0	4.4	1.0	14.8
1995	1.95	8.85	7.25	1.35	19.4
1996	6.0	10.0	7.5	2.5	26.0
1997–8	–	30.0	26.5	7	63.5
1999	–	nil	10.2	4.1	14.3
Total	9.35	56.85	55.85	15.95	138.0

Sources: Ellmeier and Rasky (1998), EU Education and Culture Directorate-General (www.europa.eu.int, 1999)

1994 expenditure; 1995–9 committed (not expended) funds

Note: Kaleidoscope was a scheme introduced by the European Commission (DGX) in 1990 to promote cultural events with a European profile with awards of ECU50,000 – examples in the UK include the Women's Playhouse Trust for 'Crossing Boundaries' and the Bradford Festival celebrating immigrant communities in Europe. Financial support is provided to the European Youth and Baroque orchestras, and the high-profile 'European City of Culture' and 'European Month of Culture' awards

European music to a wider audience'. The Programme also focuses on the free move-ment and mobility of cultural workers in keeping with free trade, cultural exchange and heritage Treaty goals, whilst the cultural industries are recognised both by their sectoral importance – an estimated 3 million or over 2 per cent of all jobs in the EU by the mid-1990s and growing (see Chapter 6) – and employment prospects essentially based on the development of the information society, technological progress and overall growth of the service sector. The European cultural economy is therefore predicated on a view of technologically determined transmission of 'culture' (far less cultural content and expression), and therefore regional development that looks to the cultural industries for job creation does so in the hope that such activities enhance their knowledge and know-how, promote social interaction and make the region more attractive for new enterprises and residents alike.

Conclusion

One conclusion from this critique is that the main European Structural and Regional funds have dwarfed direct intervention through European arts, media and literary/lan-guage schemes. European regional assistance has therefore largely been directed at the arts infrastructure and in particular at major (and politically high profile) regeneration and visitor-led *grand* projects and to a lesser extent at employment, training and tech-nology schemes. The nature of European regional support requiring a scale of development (e.g. the creation of at least ten jobs) obviously dictates this, although this appears to undermine the objective of supporting small enterprises (see above). The nature of the cultural schemes and their location within eligible regions is however a national/regional decision rather than a centralised European one, since no *cultural* assessment of projects is undertaken at this level. Since such programmes operate as part of national regional assistance and other funding regimes (most notably the National Lottery in the UK since 1995), they can be seen as part of urban and regional economic policy rather than as arts or even wider cultural policy. Lacking a planning or needs-led framework, such investment rests on the success of economic regeneration in the cities benefiting from public funding, but as Cheshire and Hay's (1989) analysis maintained in the earlier experience of Structural Funding, there has been a lack of 'spatial con-gruence' between areas qualifying for ERDF assistance and the worst areas of urban deprivation and need (e.g. Southern Italy), concluding that such funds do not neces-sarily benefit *urban areas* as opposed to city-centre/downtown zones and historic quarters. This tendency towards regional fragmentation or devolution in practice reverts to a zero-sum scenario of inter-city/regional competition, an unsustainable situation that regions such as the Caribbean discovered with islands out-doing one another in incentives for inward investment, a game being played out within Europe for inward investment (e.g. Japanese, US) and between competing subregional development agen-cies (e.g. Scotland and Wales).

Experience in Europe also suggests that greater autonomy, devolution and regional development is likely to fragment analysis and interpretation, as resistance to centralised reporting and standards (amenity, planning) within the sphere of culture intensifies, weakening attempts at more sophisticated analysis and evaluation of policy and effec-tiveness of development plans and schemes. As the cultural funding analysis also concluded, surely an understatement and mirroring European policy separation: 'the Ministries of Culture are not always informed about activities benefiting the culture sector administered by other Ministries' (Bates and Wacker 1993: iii). On the other hand, the strengthening of regional policy and development, which ERDF infrastructure funding has underwritten in several EU member countries, has also seen the growing importance of regional planning and governance (Lowyck and Wanhill 1992), and as Akehurst *et al.* (1993) observed, there seems to be a shift to a regional level with a national structure becoming less important and the EC and individual regions gaining in importance at the expense of national organisations. At a European-wide level, some moves towards a more thematic policy and executive structure (European 'joined-up government') might offer some hope that culture and related areas of tourism, transport and land-use planning, which cut across several policy and programme areas and EU Directorate responsibilities, but which are central to Treaty and supra-national objectives, may be dealt with in a more integrated way. At a national level, maintaining a policy overview and realistic measure of additionality in public investment programmes in this field is likely to require a more substantial and structured cooperative effort between central, regional and local levels of policy and resource allocation. This may be essential if national and European policies in the area of culture and tourism, such as access and cultural diversity, are to be implemented and their impact measured, and if regional policy is to meet the highest areas of need rather than the concentration of resources and a narrowing in the notion of European cultural identity which has typified ERDF and urban programme distribution.

Notes

1 Since 1944, the South East of England increased its urban area by 44 per cent (470,000 acres), an area more than the size of Greater London itself (CPRE 1993). By the same token, government attempts to bring back into use derelict land in existing urban areas – through the Derelict Land Grant scheme – have so far failed: since its inception in 1974 only 6 per cent of designated derelict land has been brought into use.

8 Cities of culture and urban regeneration

As we have seen historically, the celebrated cultural city and capital is neither a new phenomenon nor one that necessarily outlives particular empires (Hall 1998) and the effects of social, political and other forces of change. The post-industrial era is however witnessing a more self-conscious and self-styled re-creation of the renaissance city, however superficial or questionable this may seem to residents and outsiders. As discussed in Chapter 7, since the late 1980s the encouragement and assistance given by the European centre to 'regions' and regionalism – both economic and cultural – has benefited urban and particularly city–regions, and within cities, major central and regeneration area cultural flagship and quarter projects. Politically this has also empowered city authorities over central and even regional ('meso') tiers of government (Balchin *et al.* 1999), echoing the power of merchant and early industrial cities in the late urban renaissance period, and as representative sites for the European 'common heritage and inheritance' to be displayed idealistically for internal (resident) as well as external (i.e. tourist – business, leisure/cultural) consumption. According to Le Gales and Lequesne: 'This is not surprising . . . modern Europe was in part invented in the cities of the Middle Ages' (1998: 250), and as Newman and Thornley maintain: 'cultural displays also serve to reinforce the assertiveness of city governments and highlight the relative weakness of national planning' (1994: 16).

An inescapable focus of the process and practice of cultural planning has therefore been on the role and exemplar that the city presents over time. Where supra-national regions such as the European Union (EU) have developed and grouped around trading blocs and geopolitical allegiances – e.g. NAFTA (USA, Canada and Mexico), Inter-American, Caricom (Caribbean), ASEAN (South East Asia), MERCOSUR and other regional associations and leagues (Arab, African, Latin American) and their development institutions – within these areas, spatial concentration and determinism is also evident. This core–periphery divide has fuelled and supported policy intervention to compensate marginalised subregions, improving transport links to the centres and the funding of their cultural aspirations (e.g. language, crafts) – largely tokenistically and marginally, however, relative to national and city cultural resources. However, at the same time this divide has also strengthened the scale hierarchy of cultural activity and facilities in cultural capitals and former industrial cities that have managed to retain cultural production, consumption and visitation levels. At the highest (global) level, these cities have arguably turned their cosmopolitan society and associated movement of people (and cultural influence) to comparative advantage. As King comments on this contradiction presented by the global city:

At once the centre for the production and diffusion of a 'Western' mass culture, it is also through the diversity of its peoples, its ethnicities, its sub-cultures, its alternative cosmopolitanisms, its representations of both core and periphery, also an instrument for changing that 'Western' culture. . . . It is not only the economy which is being restructured, but, also, the nature of the national culture and identity.

(1990: 150)

Cities that have used culture, whether architecture, design (including public art/realm schemes), event/animation or cultural production-based, are celebrated and looked to as successful proponents not only of culture-led regeneration, but also of urban regeneration generally. Regional capitals such as Barcelona and latterly Bilbao in Spain, Glasgow in Scotland, Frankfurt in Germany, and several English (Huddersfield, Manchester, Sheffield) and French (Lyons, Grenoble, Rennes, Montpellier) secondary cities have used aspects of the urban cultural planning formula, which smaller towns and cities have sought to adopt in these and other European countries, as industrial cities of the USA have done through their version of boosterism in waterfront and downtown city areas (e.g. Baltimore, Boston, St Paul, Lowell, New Jersey) (Boyle and Meyer 1990). Indeed, the celebration of American urban arts and regeneration was promulgated in Europe in the late 1980s through the British American Arts Association (BAAA) conference and publication series (1988, 1989, 1990, 1993). Prescient writers in the USA such as Kevin Lynch (1960) and Janet Jacobs (1961) had also influenced these initiatives. Jacobs maintained that American cities had been rendered incoherent by the inappropriate application of the international style: 'the practice of breaking with the past . . . had robbed American cities of their natural sense of order and space' (Vickers 1999: 166). McNulty, founder of *Partners for Liveable Spaces*, cites an early example of a response to economic decline in the State of Kentucky:

Traditionally American cities have prospered economically and then with flourishes of boosterism and unabashed civic pride, have created the amenities that define a city as great – parks, museums, sporting arenas, public plazas, tree-lined boulevards. In fact that is what Louisville Kentucky did in the age of prosperity. But when economic prosperity flagged the city made the unorthodox decision to see if the tail can wag the dog – to see if by concentrating on amenities, quality of life and the tourism, it hopes those things will engender, it can assure prosperity back in a lively revitalized Louisville.

(McNulty *et al.* 1986: 95)

Louisville's city population had declined by 17 per cent since 1970, but in 1984 proposals for a Kentucky Centre for the Arts were to include a 2,400 main- and 610-seat theatre and three rehearsal spaces at a cost of $26 million, that would host the Louisville Symphony Orchestra, civic ballet, children's theatre and the Kentucky opera companies. Some argued for the dispersal of arts facilities, others for the renovation of the existing theatre, but a new-build mega-project was chosen, thus signalling the preference and strategy for arts facilities to be tourism- and leisure–retail-oriented, as has been the case in major upgrades such as the Louvre in Paris, and waterfronts in Barcelona, Albert Docks in Liverpool, the Lowry Centre/Salford Quays and Gateshead Quays in Northern England, as well as *harbour fronts* in Baltimore, Toronto and Montreal (Bruttomesso 1993).

In Newark, New Jersey, for example, a twelve-acre mixed-use redevelopment stretches from the Passaic River to Military Park as part of a masterplan which links the city's twenty-block downtown arts district with the city museum and library. Ten years on from McNulty's comments (see above), the combination of cultural facilities within a retail, hotel and office complex continues to be a strategy adopted where others, such as office development, have failed. Newark had also seen a major economic and employment decline – half of its private sector, three-quarters of manufacturing and half of retail firms being lost since 1960 (Newark experienced street riots in 1967). The New Jersey Performing Arts Center costing $180 million is therefore the flagship project that, it is hoped, will attract not only local residents, but also visitors from New York and the surrounding suburbs. A 6,000-seat baseball stadium will complement this centre, both linked to the main train station by a 2-mile-long $75 million riverfront esplanade. This continuing embrace of the arts and urban regeneration has not been entirely reactive and pragmatic, but it has also been a reflection of the failure of urban renewal programmes since the 1960s and even earlier, where federal funds flowed into 'problem' cities, but where arguably this pattern has contributed directly to the decay of urban core areas and the rise of suburbia (Norquist 1998).

From this more recent US perspective, in a major retrospective exhibition – 'Building Culture Downtown: New Ways of Revitalizing the American City' held in Washington in 1998 – numerous projects were modelled and celebrated, largely by their architects and city mayors ('champions') who proclaimed, as if this had not been occurring previously, that 'now cities are capitalizing on their traditional assets – art and culture – to revive their downtowns. They are turning to museums, performing arts centers, theaters, opera houses, and concert halls to spur economic growth' (National Building Museum 1998). Presenting major development projects in San José (Silicon Valley), Fort Worth, Kansas City, Cincinnati and Minneapolis: 'unlike the arts centers of the 1960s [see Table 4.3] such as the Kennedy Center and New York's Lincoln Center, the architecture of the new cultural buildings is designed to reinforce connection to the city . . . not idealized monuments to culture, but street-savvy accessible buildings that often reflect the idiosyncrasies of their urban settings' (ibid.). A look at the visitors' book in this national exhibition suggested that these claims were not only exaggerated, but also that the human dimension was lacking: 'what about the local community'; 'culture-led regeneration projects have not benefited residents'; 'projects have created negative physical impacts in these "po-mo" [post-modern] arts and entertainment zones'. It is also doubly ironic that the exhibition was held in the capital city designed in the grand manner in the 1790s by Pierre Charles L'Enfant, where the suburban drift has turned into an exodus and residual ghettoised communities exist in a double-life, crime-ridden but with Capitol Hill and the national arts, library, museums and monuments (e.g. Kennedy Center, Smithsonian museums, National Theater and galleries) and other symbols of American constitutional and military history nearby. Washington had undertaken an 'animation' plan of its arts and cultural provision in the mid-1980s (Cuff and Kaiser 1986) that saw an expansion of its institutional and 'representative' (i.e. multicultural) facilities. However this was integrated neither with a city or cultural plan, local arts provision and need, nor with broader solutions to the urban decline and sterility caused by an over-concentration of government and national activity and land-use, in what is now a divided city 'surrounded' (*sic*) by an ever-widening Edge City and suburban drift.

Urban regeneration that looked to the arts, heritage and inward investment programmes has had fifteen to twenty years of life in the USA and warrants both long-itudinal and objective study to assess not just the cultural strategies adopted and the respective regime models, but the changing landscape, community and economy which has emerged from their late-industrial and post-industrial states. In Lowell, Massachusetts, for instance, this birthplace of the American Industrial Revolution and model New England mill town sought refuge from its industrial (textiles) decline from the 1920s and high unemployment in the 1970s (15 per cent). This was found through its newly designated Urban National Historical Park in 1984, and the relocation of Wang Laboratories' new world headquarters there, which also attracted smaller tech-nology and support firms. Lowell's Cultural Plan was also a model of public–private partnership, which ambitiously asked whether culture should be broadly defined: 'a plan to import culture to a place, or to value and reinforce the culture of that place?' (Halabi 1987, and in BAAA 1988: 13). The plan was deemed successful, maintaining that: 'Culture is important; cultural expression is the marker of our time . . . In Lowell eco-nomic development and quality of life were important . . . Spiritual and human development in cities are now being recognized as vital, and cultural planning offers a way to bring this about' (Kreiger 1989: 182). Even ten years on from the incorporation of the Lowell Plan, it was deemed too early to evaluate the results of the planning effort, although the process was judged to be inclusive and raised the importance of the arts amongst both business and the community: 'Old New England City Heals Itself . . .' was part of the *Wall Street Journal* lead article on 1 February 1985 (which contrasted Lowell's success with the failure of Akron, Ohio; Zukin 1995). Wang has since closed, leaving Lowell's heritage industry as its main asset, but with heritage and other forms of cultural tourism fast multiplying in city, historic towns and natural heritage sites, and the ubiquitous high-tech industries being universally pursued, Scott sounds a note of warning: 'As the experience of many actual local economic development efforts over the 1980s demonstrates, it is in general not advisable to attempt to become a Silicon Valley when Silicon Valley exists elsewhere' (2000: 27). The same may be said of cultural industries and the arts that are not sufficiently rooted in community life.

The extent to which differing models and approaches to urban regeneration through cultural development are apparent in North America, Continental Europe and more recently in Australia and South East Asia (see below) depends in part on the obvious social and political differences that pertain, and the influence of historicity and symbolism which individual cities retain. Global capital, international tourism, acculturation and other forces of cultural convergence suggest that the factors leading to culture-led or at least culture-influenced city regeneration are largely common, whilst endogenous cultural preferences, levels of participation, planning and amenity systems will also dictate the nature and scope of cultural city formation. In Asian Pacific cities, given their pace of growth and rapid urbanisation, it is the urban condition itself that is forcing architects and planners to think holistically. Here the Japanese model where high land costs and population density have created a system of connected nodes is contrasted with, say, Bangkok as an example of poor planning and unrestrained development which has brought this city to a virtual economic standstill (see below). In response to second-world cities aspiring to first-world status and membership of both cultural and economic networks, they have looked to the experience in the West in their own versions of urban regeneration and city renewal.

Here, international design practices now specialise in 'entertainment-based retail placemaking' as a formula that 'many city planners believe has since become a standard way of revitalizing urban centres' (Levine *et al.* 1997: 124). In the forerunner of this phenomenon of the latter half of the twentieth century, in the USA – and this reflects the now familiar hardening core–periphery and technopole city predictions – a few highly successful, high-profile regeneration projects in well-known downtowns and waterfronts are surrounded by mile after mile of continued decay and despair. Just as the expansive out-of-town shopping malls created an over-supply and ghost town scenario in the USA (a risk that the multiplex/leisure–retail park may emulate; see Chapters 3 and 4), city regeneration which relies on an external formula for sustained social and economic revitalisation will inevitably presage a game of winners and losers which no amount of culture-intervention alone will prevent, as Bianchini and Parkinson concluded: 'Experience from cities both in the USA and in western Europe suggests that cultural policy led regeneration strategies – particularly when they are focused upon city centre-based 'prestige' projects – may bring few benefits to disadvantaged social groups' (1993: 168). The range of city case-studies contained in this edited collection provide perhaps the best comparative of how urban cultural policy emerged and was rationalised politically in 1980s' Western Europe. In terms of economic development and in particular employment generation, there was little evidence of any sustained improvement in *local* employment arising from the investment, itself generally not sustained, in these examples. Bianchini notes one response to this divide in the spatial distribution of cultural provision in the creation of neighbourhood-based arts facilities, citing Hamburg and Bologna (1993: 201). However where local cultural amenities already formed the basic infrastructure of municipal provision, such as in the UK and France, public spending reduction and redirection has created a widening gap between centre and periphery, social arts and flagship arts and arts amenity and cultural industry production activities. This is apparent even within local urban areas where a spatial and economic divide is in fact reinforced by cultural planning which focuses on cultural industries/tourism quarters, whilst adjoining areas lack community and cultural facilities or the means by which local residents may overcome the barriers to participation in both cultural and related economic activity, as in the case of Vienna (see below) and East London (Landry *et al.* 1997a, b, Mokre 1998, Evans and Foord 1999). In macro-economic terms there has been a social and cultural 'crowding-out' and a spatial concentration through economies of scale (mega-projects and complexes) and largely mono-cultural developments at the cost of more diverse, local cultural facilities and programmes (Evans 1999b).

How far urban regeneration was actually *cultural* policy led in these European case-studies is also questionable, since the cultural sector has not generally been involved at the planning and design stages of development to any significant extent. As the authors themselves observe in some cases: 'Bologna did not develop a vigorous cultural policy in response to urban decay' (Bloomfield 1993: 91), and 'Glasgow lacks an integrated cultural policy', and, tellingly, 'In part this is a reflection of an ambiguity in the relationships between culture and the development process' (Booth and Boyle 1993: 42). Glasgow is an oft-reviewed example of the arts and urban regeneration, both from a re-imaging and a competitive city perspective, including its perennial comparison and competition with Edinburgh. This was played out over the location of the new National Gallery of Scottish Art between 'big brash' Glasgow and 'effete old' Edinburgh. As

Sudjic – who himself joined the ranks of cultural envoys as Director of Glasgow's 1999 Festival (see above) – comments:

> In the post-industrial world a national museum has come to take on the national significance as a car factory or airport . . . the bargaining chips that a new generation of entrepreneurs desperately fight over, markers to prove their ascendancy over their competitors. More than trophies of civic pride they are seen as the job-creating building-blocks of local economies.
>
> (1993: 5)

The 'ace card' played by Glasgow was the offer of £10 million of EC regional aid (an amount not matched by Edinburgh which was ineligible for such assistance) if the gallery was sited there: this is an example of supra-national cultural intervention, and one little to do with arts planning but carried instead on the back of regional economic development criteria as discussed in Chapter 7 (Evans and Foord 2000b), and in Mommaas and van der Poel's words: 'the city as a kind of commodity to be marketed' (quoted in Bramham *et al.* 1989: 264).

Hot on the heels of press coverage devoted to Bilbao's new Guggenheim franchise, then Berlin and Rem Koolhaus's casino-outpost in Las Vegas (modelled on the Venetian Grand Canal!), the city of Liverpool has begun to woo this Foundation offering a £60 million package of public (Lottery, European) and private funds for a building to house more of the Guggenheim collection. A (World Heritage) site close to the Merseyside Tate Gallery and Walker Gallery is proffered by this city still trying to shake off a negative image and losing battle with its north-western city competitor, Manchester: 'aware of how Liverpool had so dismally failed to capitalize on the "Mersey Sound" during the 1960s, Manchester began to market its popular culture during the 1990s, for example the Greater Manchester Visitor and Convention Bureau was launched not at a major hotel, exhibition or convention centre but at the Hacienda, the city's premier music club' (O'Connor and Wynne 1996: 84). Moreover, Lyon in Southern France, along with more than 60 cities around the world, from Rio to Recife, is pursuing the Guggenheim Foundation and its footloose collection and brand for a further satellite museum to complement its major convention and transport developments. Back in the home of the original Guggenheim Museum (Plate 8.1), a $850 million proposal (involving Gehry again) for a new Guggenheim includes a library, educational facility, theatre, skating rink and a park floating above four existing piers on Manhattan's East River: 'the museum not only as exhibition space, but [also] as pedagogical institution and urban attraction, the old Guggenheim fused with the Rockefeller Center' (Ryan 2000: 91). Meanwhile 'Bilbao babies are being born everywhere' is the comment on Gehry's interactive Experience Music Project (EMP) in Seattle. Built as a homage to local hero Jimi Hendrix, this aims to capitalise on the thousands of visitors to Hendrix's grave in Renton, south of Seattle, Washington State, as Graceland has served as the shrine to Elvis. This may ensure its viability in contrast to the National Museum of Popular Music which languishes unvisited in Sheffield, West Yorkshire (see above and Plate 7.4, Chapter 7). City location alone is not sufficient to generate interest – symbolic and vernacular associations are needed to overcome the arbitrariness of the new, as well as inherited cultural facilities (Lynch 1972). Whether aspects of popular culture can

Plate 8.1 Guggenheim Museum, New York

successfully be museumified, e.g. sport, pop music, is also questionable where reduced to collections of artefacts, memorabilia and recordings which are obtainable and better experienced elsewhere.

Glasgow is now in its third phase of culture-led regeneration, which commenced in the 1980s with its *Glasgow's Miles Better* campaign (1983). This was followed by hosting the national Garden Festival in 1987 and the *European City of Culture* in 1990 (Booth and Boyle 1993) and the continued expansion of cultural events, e.g. *MayFest*, Jazz Festival, arts facilities, e.g. Museum of Modern Art, despite budget cutbacks in education and museum services, and most recently as host of the 1999 *Festival of Architecture Design and the City*, which has included further building conversion projects such as the Lighthouse Architecture Centre – the Grade I-listed Glasgow Herald building – celebrating its architect Charles Rennie Macintosh, and other Scottish and international designers. The site of the Garden Festival stood redundant for many years, in the manner of other late twentieth-century EXPOs and mega-events, and piecemeal development of the Clyde regeneration zone is predicted to produce an architectural zoo of unrelated buildings in the absence of a master- or cultural plan for the area. The population of the city also continues to decline and drift outwards, however, and Scottish inbound tourism, a key factor in new project viability and regeneration programme investment, is also decreasing year-on-year, with the prospect that cultural facilities will not survive and funding programmes be curtailed (after 1990 some arts facilities closed less than a year after *City of Culture*, e.g. the Third Eye Centre). As Hewison posited: 'will the benefits to the city's centre have been felt in Easterhouse or the Gorbals. Will Glasgow still be a cultural centre after the circus has moved on?' (1990: 176). The unsustained nature of the 1980s' cultural investment programmes, caught in public service rationalisation and budget constraint, also reflected the discomfort with the nature and beneficiaries of the cultural programmes and eco-nomic activities that were in fact generated. The diversion of local cultural policy and industry strategies into regional and even international cultural tourism strategies is now the major dilemma for both local amenity and cultural development and one which requires a more local economic response, as Bianchini suggested:

> the challenge for the next decade will be to go beyond narrowly consumption-ori-ented strategies, and the ultimately destructive 80s zero-sum game of competing for limited pools of inward investment or tourism revenues. It will be necessary to develop more locally-controlled production systems, be they in . . . manufactur-ing . . . or in cultural industries like film, fashion and design.
>
> (1991b: 12)

The urban regeneration process which itself has shifted in direct relation to global capital, geopolitical and supply and demand movements – e.g. the over-supply of office space; new technology in design and information flows, growth in residential and leisure–retail development; South East Asian economic and political crises, etc. – con-tinues however to look to opportunities for value-added and 'quality' to major development projects, including design, animation, public realm (e.g. landscaping, public art) and high-art temples. Size increasingly matters in these grand designs – wit-ness the latest mega-project in the centre of Vienna. Here a *museumsquartier* is underway, claiming to be the largest cultural construction area and one of the ten largest cultural districts in the world. Occupying the former Imperial stables, behind the Museums of the History of Arts and Natural History, nearly twenty separate arts organ-isations will be located in this cultural quarter which is expected to 'attract tourists and make money, a cultural centre for the neighbouring district and give new creative impulses to the city' – in the words of the scheme's project manager: a 'Shopping Mall for Culture' (www.museumsquartier.at). Less high profile and monumental cultural developments also coexist or rather coincide since they seldom have any connection with these *Grands Projets*. In Vienna since 1995 an EU-funded URBAN project (ECU32 million 1995–9) has focused on 77,000 dwellings in the *Guertel West* zone, which is densely built and occupied, lacking in amenities (1 square metre of green space per inhabitant) and over one-third of whose residents are non-Austrians (e.g. Turkish and East European immigrants), with double the city's rate of unemployment. One element of this programme of urban renewal seeks to develop a new social and cultural public: 'to provide possibilities for the public representation of different lifestyles, cultural needs and achievements' (Mokre 1998). A particular project, the 'Mile of Youth Culture', consists of avant-garde art projects, youth/multicultural, and retail and restau-rant facilities. Even this community development programme has economic imperatives, however. The programme is expected to be self-financing and attract people from all over the city as well as visitors. This conflicting 'partnership' (public–private) regime has already seen a Turkish youth arts project *Echotek* fail because of unrealistic financial expectations and this scenario is played out throughout cities undergoing urban regen-eration where social development through culture is subjected to the same criteria and rationales as major visitor-based, flagship schemes (Evans and Foord 2000a). However, the many major art institutions housed within the capital-intensive *museumsquartier*, like their counterparts in other cities world-wide, continue to attract direct subsidy, sometimes over 75 per cent of their annual budgets (e.g. Tate Gallery, London; Evans 1999g). More acute than even the economic rationale, the 'extremist/populist' right-wing Austrian government, which has used lawsuits to silence outspoken artists and scientists, has withdrawn funding from two existing cultural groups (*Public Netbase* and *the depot*) that were to occupy premises in the *museumsquartier*. Both were meeting points for cultural and political resistance to the political regime, both have lost their place in this new cultural quarter, officially 'to make space for other as yet undeveloped innovative activities' (M. Mokre, Austrian Academy of Sciences, Vienna, personal communication, 26 September 2000).

The major cultural houses therefore exist largely on parallel lines from the often neighbouring community arts, cultural development activities, and the source of much new art and creative activity, whether so-called avant-garde or non-legitimate and 'ethnic' arts. In the case of Tate Modern on Bankside, London, which opened in mid-2000 as one of the country's millennium *Grands Projets* (Table 8.3), the adjoining artists' colony occupying a former industrial building at the same time lost its home as property rents and high-value usage capitalises on this transforming cultural quarter. The immediate success and acclaim that greeted the gallery has been partly overshadowed within this institution by the concurrent poor attendances at the original Tate Gallery (renamed 'Tate Britain') across the Thames at Pimlico, despite new building works and exhibitions there. The as yet unanswered question is how far such major arts houses can attract a *net* increase in visitors, or whether saturation can in fact be reached (as is evident in static and even declining audiences for much performing arts; Evans *et al.* 2000 and see Chapter 5). In the case of an earlier Tate Gallery satellite in St Ives in Cornwall, the location and siting followed an established visual arts tradition there (galleries, studios) including artists such as Barbara Hepworth, Ben Nicholson and Patrick Heron, and like Bankside the reuse of an existing utility site, in this case a local gasworks. This gallery is promoted as a successful example of culture-led economic development (including European Regional Funding of 25 per cent of the capital cost), in what was an economically depressed region with few prospects or growth possibilities even within the existing tourist market (Arts Council 1994). Local resistance was apparent with local people preferring other amenities, such as a long called-for swimming pool, and questions over why a modern art collection was to be sited there at all. That the gallery would charge for entry (unlike its London main collections which are free, ensured in the case of Tate Modern by additional central government grant-aid) added insult to injury. Neighbouring towns and villages, several some of the poorest in Britain, have seen little benefit from this national gallery outpost and here like so many other places the arts and economic regeneration are highly concentrated in both a small number of individuals/enterprises – established and incoming – in the local area, not least the cultural institution itself, with high leakage of economic and cultural benefits out of the impact area altogether.

The scale and location of major cultural facilities also presents commercial and retailing possibilities which dictate the design, programming/curatorship and management culture under which they operate and relate to their key stakeholders – national and city cultural ministries and agencies. This retail phenomenon is now ubiquitous – from the 'ACE café with a museum attached' (coined for the V&A Museum makeover in

London; Evans 1995c) to I. M. Pei's Louvre extension and underground shopping plaza/entrance *Carousel du Louvre*, drawing perhaps on one of the first leisure–shopping experiences under one roof, the Bon Marché department store from where modern techniques of painting display were first derived (Williams 1982, Rearick 1985, Cowen 1998). As Ryan observes: 'the new museum may attach itself physically to the old (part parasite, part life-support system) and the contemporary nature of such institutions seems irrevocably to revolve around new photo opportunities, so-called Star Architecture and magnified scenarios for shopping' (2000: 90). Paris, Berlin and Vienna of course have a history of 'masterplanning' on a brutal scale (see Chapter 2) which directly or indirectly provided the historic cultural quarters now serving as national heritage and symbolic sites for visitors, and which in the case of Vienna helped create the heritage island on which the contemporary makeover (*museumsquartier*, see above) is to be located. As Robins claims: 'Urban regeneration reflects a more acceptable face of rationalism, and fails to come to terms with the emotional dimensions of urban culture' (1996: 88). Ellmeier and Rasky respond thus: 'today the tasks of city planning also include compensating for differences and creating necessary community in order to allow the city to function at all. . . . If inhomogeneity becomes visible, if the idea of the homogenous national or city culture is no longer tenable, then the city, the urban space, becomes important' (1998: 80). However the evidence over the past twenty years in cultural cities and major sites and their emulators, is that urban space still retains its homogeneous state, with the compensatory and uncommon by definition and design, effectively marginalised in both spatial and symbolic terms.

With hindsight it can be argued that we are now in the third phase of urban regeneration in terms of cultural policy and development – the earlier period before the liberal planning and private sector-led phase peaking in the mid-1980s saw the community arts and social action movements engage with urban policy and growing unemployment (particularly 'structural' and youth), manifested in the growth of community/arts centres and emerging cultural industries practice. This second phase of arts and urban regeneration coincided with the embracing of 'private–public partnership' and the arts regenerative role by cultural agencies, in the overt adoption of the economic importance of the arts rationale (as discussed above). The current phase exhibits aspects of both previous periods – the social exclusion agenda, access to the arts, neighbourhood renewal (Shaw 1999; see Chapter 9) on the one hand, and the private sector financing, small business (SME) development and creative industry initiatives on the other. Partnership is a more central and symbolising aspect of this latest version of *governmentality* (Foucault 1991, also Foord cited in Evans and Foord 1999, 2000a), including the responsibilities of the citizen within this social contract, but regime theory which previously distinguished US growth coalitions from the more socio-political European approaches to urban governance (Stoker and Mossberger 1994) is now more convergent in this current era. Culture is a universally common, even requisite, theme and component in major site- and area-based regeneration programmes, and arts interests now occupy part of these urban partnerships, although as Fanstein argues (1994), it is not just *who* is involved but *how* some actors enforce their objectives that matters. Local governance and power is seldom stable or equal, which naturally dictates the nature and beneficiaries of the 'culture' that receives both direct resources and consideration in project design and development plans. Representation within the arts also reflects the cultural elites, political preferences (e.g. aesthetic/design, creative industries versus community arts, multicultural versus pluralism) and the critical role of intermediaries in the development process, as brokers between planning, resource allocation and the 'locality' (Evans and Foord 2000a, b).

To an extent, the patchwork of city and urban financing programmes converges or at least purports to conflate these social and economic rationales for culture within the urban policy agenda. However whilst some critiques see the late twentieth-century post-industrial city as now no longer linked to the nineteenth-century tradition of arts and the pursuit of cultural homogeneity (Ellmeier and Rasky 1998), the concern for amenity, social exclusion and an economic creative industry base has clear resonance with the Victorian rational recreation (and earlier industrial) movement and the long-established relationship between culture and commerce (Casey *et al.* 1996, Hall 1998, Cowen 1998). The tensions between cultural diversity, nation-state and dominant (European) cultures is of course a persistent concern and source of conflict not only at the nation-state level (e.g. Francophonie, Islamic), but also at the regional level where the tension between *cultural capitals* – fulfilling their role as cosmopolitan and international city – and the regional government's notion of 'identity' often involving a rewriting of history and mono-cultural image is evident as already highlighted in Barcelona (Catalonia) and Montreal (Quebec). How far the multicultural, pluralist and 'identity' debates have seriously impacted on urban culture and regeneration policy and processes is not clear – there is however little sign that the institutional hegemonies, including the EU, World Bank and UNESCO, responsible for both arts and heritage and urban economic policy programmes have embraced or reflect these concerns and the attempts to widen cultural policy and more democratic planning. Perhaps the key question to be asked, therefore, given the experience and evolution of cultural policy in the urban regeneration era, is how far this experience has informed current policy and practice, how the role of intermediaries reflects policy objectives (where these are articulated) and where policy is 'located' in terms of professional, institutional and cultural interests (Evans and Foord 1999). Cultural planning in one important sense, drawing parallel with community planning approaches, assumes an important position in widening and democratising the policy and resource allocation processes, both through its spatial and environmental focus and in its lesser reliance on art form and legitimate cultural practice and institutional rigidities.

Cultural cities of the 'south' and Westernised urban planning

As has been discussed in earlier chapters, urban cultural morphology owes much to the inheritance of pre-industrial cities and both classical and pre-Colombian influence, largely 'non-Western' or at least informed by non-Western cultures. The urban renaissance and its post-industrial reinvention may therefore have some of its roots in Western Europe (although much exaggerated and re-imagined), but in the globalisation era of late capitalism, the form and function of cities of culture increasingly follows universal lines (Hall 1977). Although the spread of cultural consumption and production may be global economically, as Hobsbawm maintains, 'globalization isn't a universal process that operates in all fields of human activity in the same way' (2000: 62). This is so in the case of say politics, and is mediated by environmental factors such as geography, climate and history. Different cultural practices also lend themselves to transference more easily than others: 'Traditional culture spreads through a European model that has been adopted globally and therefore globalized: a concert program in Osaka, Chicago, or Johannesburg will present the same kind of repertoire: European classical music. This is not true of literature because of a very powerful limitation on globalization; namely language difference' (ibid.: 122). This is evident in the continued development of traditional opera houses (Figure 3.1), theatres and concert halls in non-European and 'non-indigenous' nations and the museumification of artefacts and collections in societies where this negates their cultural value and significance (Clifford 1988, 1990). Popular culture is however more syncretic, and Hobsbawm draws the distinction between so-called high and popular culture because 'the latter is shared by everyone, including those familiar with high culture, but the opposite is not true. . . . This is why the global icons come from popular culture' (2000: 123). This perhaps understates the significance of subcultures and alternative cultural capital, as opposed to the more commodified and 'accessible' (i.e. supply led, 'cool') outpourings of pop industry culture.

Perhaps the Asian country that has so embraced Western culture and post-modern design and, of course, contributed to its commodification through media technology and trans-national expansion (e.g. Hollywood, musical equipment – Sony to Yamaha), is post-War Japan. The paradox between this visible consumer culture on one hand and traditional Japanese restraint and understatement, the work ethic and loyalty to employer and family on the other is evident in Japanese cities and their 'centrelessness, neon-saturated streets, temporary looking buildings, simulation zones (shopping centres, love hotels, amusement and virtual reality centres, cinemas) . . . both sites of consumption and sites/sights to be consumed' (Clammer 1987: 47). Japanese contemporary culture can be described as one of excess and visibility, a 'society of spectacle' (Debord 1983), and this is seen in the popularity and growth of theme parks and the atypical success of the mega-event and EXPO in Japan (e.g. Osaka in 1970 attracting over 60 million visits, and the planned 2005 EXPO in Aitchi; Nakata 1998), in contrast to other countries where resident and visitor attitudes and responses are at best mixed. The European spatial model of cultural planning is less apparent here than in Western and colonial cities. In terms of the arts and cultural industries, however, the capital Tokyo, as in other countries, dominates Japan nationally with 20 per cent of all museums and theatres, 60 per cent of performing arts companies and 68 per cent of film and video production based there (Yamada and Yasuda 1998).

In Japan and other Asian cities, a new middle class is also emerging as it did in Europe: 'open to the new and anxious for urban sophistication, they are the ideal consumers of the cultural products of the new global market. . . . However they are also profoundly conservative in social terms . . . a hybrid of Westernised modernism and nationalist tradition' (Hanru and Obrist 1999: 12). Ibrahim draws a distinction between the European (urban) renaissance (of the fourteenth to seventeenth centuries) and the secularity of the Enlightenment, with the Asian renaissance which has its foundations in religion and tradition (1996). This is less apparent however in the case of India's Bollywood, the Kung Fu movie industry of Hong Kong, or the cultural assimilation seen in Bhangraland, London and other diasporas in European and North American cities. Moreover, the speed of development and of global capital and cultural flows in East Asia as in other fast-developing regions (e.g. Latin America) has meant that culture is 'by nature hybrid, impure and contradictory. As new cities are built, and existing cities expanded, renovated and transformed, signs of different cultures are emphasised in order to celebrate globalization' (ibid.: 10). Furthermore, as Wu points out, foreign investment in urban development represents the global dimension of place-making, whilst in terms of the reshaping of the urban landscape 'internationalisation not only brings alien lifestyles, but also provides a way to interpret their symbolic meanings' (2000: 1364).

What place *planning* has had in the cultural development and design of these cities is not clear, so even where masterplanning is exercised either by international urban designers/developers and/or city and national government, this has been a far cry from the more considered planning for communities, amenities and forms of community and cultural expression that would give residents and workers some sense of ownership and stakeholding in their environment. When financial crashes brought these financial city-states and their dependant states to their knees, it was of course these communities that suffered most from the fallout and unsustainable property and share values. The Westernisation thesis (King 1991, Sklair 1991) that views developing world cities as passive victims of consumer goods through 'coca-colonisation, western films, music and multinational enterprises' argues that 'western cultures are slowly dominating urban lifestyles, and leading to homogenisation across the developing world' (Potter and Lloyd-Evans 1998: 116). As Potter and Lloyd point out, however, such consumer culture whether product, 'place' or participation-based is not available or accessible to all where internal inequalities are marked and the social divide ('exclusion') is growing. This is not however solely a developing country phenomenon – few so-called advanced capitalist cities are not also divided on social, spatial, economic and cultural grounds, and the profile of cultural consumers and audiences for the subsidised arts and media (e.g. Internet access) (and as seen in Chapter 5) is as concentrated in higher income groups as it ever was in modern times, with levels of local amenity deteriorating in favour of larger scale and centralised cultural flagships and commercial entertainment complexes. Social exclusion/inclusion now resonates in Europe and North America, to which the creative and knowledge industries it is hoped, will alleviate and empower through *Information Society Technology* (Werthner *et al.* 1997, EC 2000), just as the poverty-elimination goal of donor-states requires engagement with World Trade and global economic rules, including 'opening up' to international tourism, new technology and the cultural industries (Landry 1998, 2000, Evans and Cleverdon 2000).

Another, apocalyptic perspective on the teeming, uncontrolled city syndrome (applied primarily to developing but less so Western cities) applauds the 'courage and endurance of people in the slums . . . admire and wonder at their capacity for adapting, for building their own shelters, for creating life for themselves, for finding a livelihood somewhere in the city economy' (Seabrook 1996: 5/6). This is also reflected in city cultures identified with the *favelas* of Rio and the 'carnival spirit', the shanty towns of the Caribbean, the townships of South Africa and 'black' ghettos in American cities (e.g. Harlem; Younge 2000) and of course their export to and fusion with mainstream Western culture, from rap and reggae to salsa and soca. 'Street' culture may be a catch all term (also Fyfe 1998), but the reality is that these communities have nowhere else to go, have few if any local cultural or community amenities or outlets other than a shared church or community hall, and no access to established arts and cultural production facilities. As Deckker writes on Brazil where, after the 1970s' demographic explosion, civic life was effectively destroyed: 'the most characteristic evidence of culture now is not the façades of theatres but innumerable television aerials and even satellite antennae on even the poorest *favela*' (2000: 184). It has been engagement with the mainstream cultural industries of the West that has provided the dissemination and commodification possibility for a few, but with little or no impact on the communities from which such creativity and talent emerges. Celebrating (and patronising) the 'local' where cultural melting-pot meets powder keg is also resisted by Massey who 'warns against romanticizing the conception of a community concentrated in space' (1994: 163–4, Tomlinson 1999: 157). However, whilst globalisation has threatened ethnic groups in terms of the irresistible acculturation process and 'creolisation' through participation in different cultural practices, consumption and codes (moral, cultural, universalist), as Jacobsen also observes: 'An ironically reinforcing bond between local identities and international normative patterns leaves the state on the sideline. . . . Thus, ethnic groups have secured, at least theoretically, international support in their jockeying for cultural recognition and political influence' (2000: 22).

Whilst the international post-modern ('po-mo') 'style' (*sic*) of building and urban design has been evident in lesser developed country (LDC) cities, often under the guidance of Western architects in the manner of colonial and post-colonial influence, as the pace of development slows down accelerated by the South East Asian economic and property crisis in 1997, more vernacular solutions to their social and environmental climate are being pursued, such as in Vietnam and Malaysia. As Massey *et al.* also point out: 'There has been an attempt by certain Islamic countries in recent years to design and build a future which does not follow in every detail the model of development exemplified by the West. . . . Certain films have been banned, and some kinds of music; the arrival of western cultural influences has been carefully monitored and controlled' (1999: 120). However, non-terrestrial media and other forms of global cultural transmission (including international tourism) render these attempts fragile, as Eastern Bloc and other overshadowed neighbours have experienced (e.g. Central America). Whilst official and public culture is on display in downtown and historic centres, more 'hidden' consumption is also to be found, as Seabrook discovered in *Sunday in the Cinema*:

> in the cinema in a little *soi* or side street that leads off one of the busiest roads in Bangkok. The entrance is at the end of the alley, discreet; the only sign of it is a semicircular concrete awning, which once advertised the films when the cinema may have been more respectable. They can't even announce the films now because they are not really films at all, but fragments and discarded footage from Japanese, German, Hong Kong and American porn. . . . This seedy, run-down little cinema will surely not long withstand the urgencies of development in Bangkok, already overshadowed by a high-rise condominium, and occupies land that is far too valuable to be left to this strangely innocent answering of human need.
>
> (1996: 264–5)

International intervention ('aid') in the form of development agencies such as UNESCO and the World Bank also promote a model of economic development

through culture and the built environment, which looks to Western-style cultural consumption and production, and the reliance on a cultural elite and intermediary – both
local and global. For instance, Quito in Ecuador is attempting an approach similar to
Barcelona's to rehabilitate the central section of the extended historic centre of the city
(Rojas 1998), whilst Zanzibar the largely Muslim island community off the coast of
Tanzania is set for a major (£1 billion-plus) tourism and infrastructure make-over,
including the Stone Town, a newly inscripted World Heritage Site, based on Puerto
Banus, an inauthentic *po-mo* marina development along from Marbella on the Costa del
Sol (Evans 1999c). The focus on world and symbolic heritage sites in the cities of both
developed and developing countries requires that a balance be struck between local and
national imperatives – qualities of life, economic and physical access, minimising gentrification effects and the imposition of 'staged authenticity' in terms of the heritage that
is conserved – and by whom and for whom is urban culture to be interpreted and maintained? Shackley and others advocate greater application of a form of cultural planning,
and the imposition of pricing mechanisms in heritage areas where 'Large visitor numbers, poor interpretation, little available information, crowds, congestion and pollution
effect the quality of that experience, a quality which can unfortunately only be maintained at a high cost' (1997: 205). Who draws up, implements and enforces such plans
and controls (e.g. pricing, development) is an equally important question and one
which must start with the inheritors and resident communities that have often stew-
arded heritage sites but which are typically losers in the masterplanning process (e.g.
Palestine, Mayan), and in the land-use and development aid distribution (Evans 1999c).
As Nasution maintains: 'To restore their traditional culture and grand monuments
Asia's first local heritage advocates . . . turned to former colonists for advice . . . but who
misunderstood the complexities of cities that are not just living, but teeming with
life . . . their local counterparts and cultural aficionados sometimes failed to translate the
ideas of urban rehabilitation into local realities' (1998: 28).

The juxtaposition of commerce with culture, alongside or even in place of public culture ('realm') in the form of cultural venues, facilities and monuments appears more
intense in New World cities undergoing modernisation than in the old cities. This is in
large part due to the short period over which modernisation has occurred compared
with old industrial and world cities, and in some cases their leap from primary to tertiary
and post-industrial stages of economic development. This is particularly manifest in
terms of monumental edifices: 'complexes of offices, shopping malls, entertainment centres and international hotels. Thus Hong Kong has Times Square and Pacific Place,
Kuala Lumpur has Sunway Lagoon, Singapore [see below] the Great Wall City, Beijing
has the New Dongan Centre . . .' (Hanru and Obrist 1999: 11). The 450-metre-high
Petronas twin towers in Kuala Lumpur encapsulate this extreme of place-making
through the competitive flagship, always eventually outdone (as the tallest building in
the world) in this case by the Shanghai World Financial Centre under development, and
by 2003 a 2000-foot-high cylindrical building on a small footprint in Chicago.
Designed by the same office that developed Canary Wharf in London Docklands and
Battery Park, New York, the towers featured in the Hollywood film *Entrapment* (1998).
In one scene Sean Connery rises from slum dwellings to the skyscrapers, an image which
led to the film being banned by the Malaysian Prime Minister (who claimed that the
slums were in southern Malacca). As Rykwert notes, however, this was not a defence of

artistic authenticity, but 'what really provoked [the Prime Minister's] ire was the read-ing of the twin towers as an image of social inequity, since, of course, Kuala Lumpur has plenty of slums to show' (2000: 227). Nearby to the towers, a Malaysian market has been re-created to provide a taste of authenticity amongst what Koolhaus coined the 'Generic City'. Many Asian cities have a policy for the systematic re-creation of 'history' and the refurbishment of 'indigenous' culture along the line of Disney and Las Vegas, it being one reason for the cities' inhabitants being attracted to such places abroad, and consumed without any sense of irony, as well as their replication at home (e.g. EXPOs, see below). Here, however, 'the result of such initiatives is the disappearance of real his-toric areas, flattened to make way for hyper-"real" simulacras of tradition' (Hanru and Obrist 1999: 12).

Another example of competitive urbanism in this region is Singapore, located almost on the equator and perceived as a pivotal point between East and West. Already the international air transport hub and convention city, a *Renaissance City* tag has been applied by the city government as part of an explicit importation of a Western-style entertainment strategy. The *Global City for the Arts* is premised on Singapore becom-ing an investment base for leading arts and entertainment enterprises in the region; the theatre hub of South East Asia and therefore a prime entertainment destination for vis-itors: 'a cosmopolitan city plugged into the international network where the world's talents and ideas can converge and multiply' (STB 1996: 9). A major development is planned, the *Esplanade-Theatres by the Bay* – a $250 million project that comprises a 1,800-seat concert hall and 2,000-seat lyric theatre next to a modern Marina Bay hotel and retail complex where the Singapore River widens out past the high-rise business dis-trict (Figure 8.1).

This Esplanade was supposed to house smaller studios and performance spaces for local groups – 'intimate Asian performance and Chinese opera' – but these plans were eliminated early on. Arts practitioners expressed concern that the Esplanade, with its mega-structures and high rentals, will be amenable mainly to blockbuster events such as foreign pop concerts and Broadway shows and be less accommodating to smaller, local, experimental and non-profit productions (Chang 2000: 824). In the words of one observer: 'a salubrious venue for top performing groups from the developed world as they cycle through Asia . . . while having no benefit for Singaporean experimental art' (Kong and Yeoh, forthcoming: Chapter 4). Again this development process lacks either a cultural plan or consensus over which cultural and economic benefits are likely to accrue from such an inward investment strategy, and importantly which cultural activ-ities and potential might be excluded, and what kind of identity Singapore's administration seeks to project and why.

In the powerhouse industrial city of Shanghai since 1990 (when investment liberal-isation opened up to the West), major urban development schemes have combined foreign capital and design with local as opposed to socialist controls on planning and land-use. Globalisation in this case has also empowered local levels of planning and governance, but also elites – political and economic – at the cost of centrally planned economic and therefore urban development programmes. The models of urban devel-opments in this case follow the international style and imperatives, as Wu observes: 'Foreign architectural firms are involved in urban design and planning. Located in the heart of the city is the Shanghai Centre a vast multi-use complex comprising 472

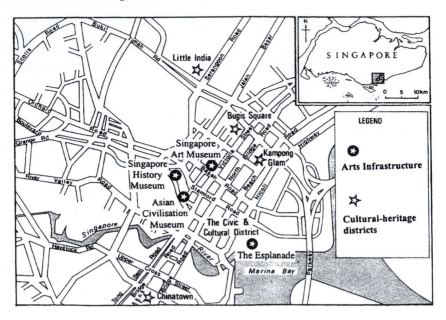

Figure 8.1 Singapore's Central Area: selected arts infrastructure and cultural-heritage districts
Source: Chang 2000: 822

luxury apartments, 25,000 square metres of prime office space, a theatre, a trademark exhibition atrium and deluxe 5-star hotel' (2000: 1365). This city-within-a-city was designed by a British firm and built by a Japanese contractor, whilst Singapore's Esplanade is US- and UK-architect designed and engineered. The influence of Western design and capital also extends to diplomatic intervention (with architect and building contracts a valuable export trade), for example in the case of Beijing's Forbidden City. Here the governors of this great piece of city-making looked to a major cultural complex development just outside of the city gates to host opera house, concert hall, main theatre and performance spaces, the 'world's biggest' (able to hold 10,000 people). The competition-winning British architects were told three times that they had won, then three times that they had not. With French government intervention this commission went to a French firm that proposed a more grandiose design – a *pomo* glass oval dome – recalling Napoleonic scales. As the losing architect put it: 'the building as icon rather than the cultural building as part of an urban complex' (Farrell 2000: 32).

The Malaysian politician Anwar Ibrahim, writing on the *Asian Renaissance*, sees the response to acculturation in nurturing an Asian *esthetique* rather than retreating to its diverse and rich cultural heritage: 'in recent times one has witnessed the overwhelming diffusion of Western or Western-influenced cultural products. . . . Not only has Asia to fortify itself against the possibility of negative cultural bombardment, [but also] it has to be able to make a positive and lasting contribution to a New World civilization which is just and equitable' (1996: 97). He therefore sees an opportunity for Asian culture and its diasporic influence as 'a powerful counter movement to the tendency towards homogenization, the cultural reductionism that comes with globalization. . . . Only creativity and imagination would provide Asian societies with cultural empowerment, not only to withstand the new and more subtle forms of domination, but [also] equally to offer the world their own cultural output' (ibid.: 98). How far cultural development is *planned* of course rests with the forms and systems of local and city governance; notions of cultural rights, and the extent to which these are present in Asian as in other urban societies is questionable. (Deputy Prime Minister Ibrahim was imprisoned for nine years by the Malaysian courts in a show-trial instigated by the Prime Minister in the infamous politically motivated sodomy case.) Malaysia ironically looked to the UK in the mid-1990s for the development of a national cultural policy, importing a team from the North West Arts Board to formulate its cultural plan and art form-based policies. This is a pattern that has been repeated in re-emerging countries from Croatia to South Africa (Landry 2000) where British arts consultants have been engaged to help develop cultural policies (often supported by national and international funds) and urban masterplanners and designers deferred to for major development schemes – a sign of their lack of confidence and regard for their own ideas, visionaries and cultural base. As Ibrahim observed: 'The great irony about Asia is that its great thinkers and works of art and literature had to be discovered by the "West" before they could reach a wider audience among Asians themselves' (1996: 98), and it seems that this need for Western validation extends to the planning of culture and cities as well.

How far cultural planning is driven by global and competitive pressures and the need to participate in (and therefore facilitate) international touring circuits and the distribution of artistic and cultural products, rests in part on the degree and robustness of cultural integrity and distinctiveness that societies exhibit and can maintain (King 1991). As Cohen argues in the contemporary era, 'the most prominent examples of cultural fusion in the arts do not come from global centers but rather from the world's periphery; they represent primarily an attempt at localization of global stylistic trends – the fusion of Western artistic styles or forms with local third or fourth-world cultural elements' (1999: 45). This dynamic exchange can serve not only to bring international and other cultural forms and practice to a local audience, but also to enable local artists to reach wider audiences and gain recognition for themselves, their groups and even for the (national/ethnic) cultures that they may represent. As Cohen adds: 'The artists thus play an interstitial role, striving to bridge the disparate worlds between which they are suspended, without, however, losing their local voice and identity' (ibid.: 45). Examples are best known in world (e.g. 'African-rooted') music(s) and literature, as well as in visual arts/crafts, fashion and, most ubiquitously, food.

The role, inheritance – tradition, heritage, spirit – and status of the artist and arts group will also influence urban design and planning in terms of more obvious activities such as public art, production and exchange, and the relative freedom that states allow in artistic creation, collective consumption and social action. The wider benefits claimed for both a more consultative city planning process and investment in cultural infrastructure also affect the likely success of participation and practice, for example reduced crime rates which, it is claimed, are directly associated with design quality, amenity and location decisions. In the USA quotes from city police chiefs attest to this view: 'We believe that arts activities can generally help reduce street crime. Both in those areas of

Boston which have regular street cultural activities and in our theatre districts, there tends to be less crime'; and 'I do believe the greatest interest and participation in cultural events in San Jose is a factor in the low crime rates we enjoy' (Kreisbergs 1979). Animation and the importance of the public realm, as well as safe, lighted private areas versus shuttered shop fronts, closed malls and darkened alleys – 'out of hours' (Bianchini *et al.* 1988), are therefore cultural planning issues as much as cultural programmes and facilities themselves. For instance in Barcelona between 1981 and 1997 over 140 urban space projects were completed, but mostly in the form of *plaza dura*: small, hard-surfaced squares and piazzas. The dominant public space aesthetic in the city 'belongs to the tradition of no trees'. These new squares are designed to be outdoor living rooms, not gardens, 'involving a public architecture of intimacy, one that brings people together in an experience of confidence and trust' (Worpole *et al.* 1999). More recently there has been an emphasis on encouraging institutions and private companies to create small parks and gardens in the centre of this densely populated, heavily built-up city. This is a long way from the 'Edge City' (*Glossary of a New Frontier* Garreau 1991: 443–59) where a park is in developer-speak either a 'Passive Leisure' or an 'Unstructured Open-Space' environment. Even masterplanning architects fall into this trap. Richard Rogers in the Urban Task Force Report (DETR 1999) claimed that 'open space is the glue which binds together buildings . . .'. As a building-designer, this perspective is perhaps not surprising. However someone should point out that whilst *sniffing* glue might take place outdoors, it is the 'space' that comes first (and last, as buildings become derelict, obsolete, fall and are removed; Bohrer and Evans 2000: 148). As Sten Göransson of the Department of Landscape Planning at Alnarp University more generously put it:

> the green urban elements distinguish and give character; they divide and structuralize, they bind together and create wholenesses; they facilitate orientation; they have a contrasting and a softening effect; they create a human scale; they reflect cultural and natural history; they symbolise and represent (e.g. nature, park, countryside); they show culture, art and architecture and they are important for the visual image and for the public image of the city.
>
> (quoted in Worpole *et al.* 1999)

Barcelona was awarded the 1999 RIBA Gold Medal, the first time a city rather than an individual architect had been so honoured, 'partly to send a message to Britain's politicians'. Josep Acebillo architect and former director of urban projects in Barcelona criticised the British reluctance to involve local people in regeneration projects, stating that 'if Margaret Thatcher had been mayor of Barcelona, the city's public realm would be nothing', whilst Barcelona's mayor cited the fact that 'crime rates had dropped from 25% to 5% in 10 years, whilst in 10 years of law and order in London it rose one-and-a-half times' (Maragall, quoted in Fairs 1999: 1).

Barcelona, Glasgow and other self-styled cultural cities cannot stand still or remain complacent if they are to retain their cultural capital status and levels of visitor and cultural economic activity. This requires regular addition to both cultural facilities and programmes. In Barcelona this is typically shaped by a visionary plan, in the words of designer Enzo Mari: 'a real city should include a certain utopia in the vision of its future' (Barcelona Future 2000: 4). This manifests itself as a proclaimed capital city of knowledge which will be marked by a 'renewed spirit of enhancement of creativity and where the commitment to culture will inspire the main decisions on the economic and political spheres' (ibid.). Another global event, this time the UNESCO-sponsored 'Universal Forum of Cultures' in 2004, will be hosted by Barcelona as part of its reclamation and regeneration of the River Besos between the city's urban nucleus and Sant Adria. Since the injection of European, national and inward private investment arising from the Olympics and regional development, the city has seen the new National Art Museum of Catalonia in the remodelled National Palace of Montjuic; a new Music Auditorium which together with the National Theatre of Catalonia completes a cultural axis that has opened at Les Glories and the new Gran Teatre del Liceu. Not satisfied with its cultural offering, Barcelona's ten-year city plan benefits from general policy initiatives such as a combined Public–Private Agency for the Support of Cultural Enterprises, a Metropolitan Council of Arts and Sciences with plans to promote audiovisual and film production. Projects underway include the Theatre City on Montjuic, the Central Library of Barcelona at El Born and the enlargement of the Picasso Museum. Barcelona perhaps presents us with the classic control case of how far the cultural city project can be sustained over the longer term and how far it can effectively create a comparative advantage that can survive regional and international competition, fickle visitor markets and the regional investment regimes (i.e. post-dependency). In many respects the widening spatial dimension to the city–region has extended this strategy and opportunity rather than an over-reliance on the inner core and historic centre and sites. How far Barcelona is and will emerge a more creative city is less clear and harder to judge whilst the benefits that may accrue to residents, the unemployed and the lower echelons of service workers will be on test here as they are in the urban regeneration and place-making underway in emerging and post-industrial cities elsewhere. As Scott suggests: 'Provided that the right mix of entrepreneurial know-how, creative energy, and public policy can be brought to bear on the relevant developmental issues, there is little reason why these cities cannot parlay their existing and latent cultural-products sectors into major global industries' (2000: 209).

A tale of two cities

In contrast to the regional-led urban renaissance of Barcelona, the Thatcher and Mitterand eras provide two *world city* examples, in particular the differing approaches to the regeneration of major subregional areas in London and Paris and their respective regimes. London and Paris also represent particular old cultural capital and touristic cities, given their similar scale of visitor activity (more than 20 million a year) and diversity (Pearce 1998 on Paris, Evans 2000c on London). Their approach to regeneration and place-creation has followed different political and planning solutions, including the link between cultural facilities, tourism and transport provision. Archetypal regeneration areas that dominated both spatial and public–private partnership programmes during the 1980s are Docklands in East London and the two Parisian *grands travaux* of La Défense and La Villette (Table 8.2). Although La Défense was started in the 1960s and Docklands only in the late 1970s, 'unlike its contemporary, the disastrous London Docklands, [La Défense] was developed along carefully planned lines, using a mixture of public and private backing to generate prosperity and new life' (Stungo

1994: 18). The short time-span foreseen in Docklands in 1980 of ten to fifteen years contrasts with La Défense where the Établissement Public D'Aménagement de la region de La Défense (ÉPAD) was established in the mid-1960s with a thirty-year life, which has been extended for a further ten years (ÉPAD 1993).

Both development agencies – ÉPAD and the London Docklands Development Corporation (LDDC) – were centrally funded and appointed with the aim of being self-financing once infrastructure investment was in place. Following debts of FF680 million in 1968, ÉPAD showed a profit by the mid-1990s, whilst the LDDC always relied on government funding until being wound up in 1998. However, significantly the Paris project shares power equally between government and incumbent local authorities, whilst the latter were disenfranchised in the ministerially appointed LDDC in 1981, with statutory planning powers removed from local councils to the development agency, as with other early models of Urban Development Corporations (UDCs) imposed by central government in other cities (e.g. Merseyside; DCC 1987). This move deliberately aimed to minimise the statutory planning process and what was perceived as crowding-out by the public sector (in land-use, employment and investment) in favour of a fast-track, liberal development regime backed by investment incentives such as a ten-year holiday on local property taxes ('rates') in a designated Enterprise Zone. ÉPAD how-ever has worked from masterplans dating from 1964, and in the early 1980s the Parisian planning authorities, alarmed by the loose sprawl of the city to the west, devised a mas-sive plan to redevelop the east of the city and return Paris to its traditional tightly knit, high-density plan. In marked contrast, Docklands offered a planning-free environment, allowing developers to take the lead: 'Docklands started as a story of hope; a dream of opening up the area to meet the needs and aspirations of the East Enders who have lived there for generations. Once hijacked by the private sector developers in league with a new market-led government-sponsored approach, it rapidly turned into a nightmare of deregulated planning and massive over-development' (Coupland 1992: 160). In terms of the effect on cultural provision, this lack of strategic planning and the demise of com-munity-based arts projects saw several theatres and museums struggling and lying dormant, whilst the new Canary Wharf Office complex programmed arts events, effec-tively underpinned by substantial public money (Docklands Forum 1989, Evans 1993a). The same developer (Olympia & York) programmes free events and entertain-ment in its Battery Park complex in New York and adjoining policed park, play and public art area (Plate 8.2) subsidised by its prestige tenant, the World Financial Center. As Zukin asks, do these 'false gathering places . . . create the new urban legibility?', noting that 'New Yorkers who admire the public spaces say, "It doesn't look at all like New York"' (1996: 54). This approach, which ignores planning and local needs, also makes little or no reference to existing arts and entertainment provision and the impact on arts organisations, where in London Docklands: 'in the 1980s Developers required to provide arts facilities as part of planning gain, often did so without consulting pos-sible future users. As a result, new facilities were created that were inadequately designed for use by arts organisations' (Horstman 1994: 4–5). The property crash that put both the flagship Canary Wharf development (and O&Y's other mega-office projects in New York, Battery Park and in Toronto) and the neighbouring new multipurpose London Arena into financial crisis epitomises the short-term nature of a market-led reliance on local arts provision where: 'The notion of "regeneration" adopted was

Plate 8.2 Battery Park, New York

primarily property and land-based' (Bianchini and Schwengel 1991: 219). Through a land-use and arts-planning vacuum, this area of London had less public arts provision than it had before the eighteen years of substantial public investment in the Docklands started, despite early warnings: 'The docklands contains the most appalling mismatches between sites and buildings, needs and means, money and quality, aspirations and achievements' (Wolmar 1989: 5, also DCC 1987, 1990, Colenutt 1988, ALA 1991, Brownhill 1990). The shortcomings and the failings in both local democracy and planning and reliance on a high-leverage property-finance formula saw Docklands' viability undermined and efforts to develop cultural activity and mixed usage wound down. Within two years of the LDDC's 'Arts and Tourism Plan' (1989), both programmes and staff support were rationalised and private development schemes which offered the prospect of arts amenities and facilities proved to be short-lived.

One of the barriers to development and therefore key to the success of both areas – transport access – provides another contrast. The French RER (high-speed suburban railway) placed La Défense only four minutes from the Étoile (Arc de Triomphe), and with Euro-Disney at its eastern terminal further new Metro lines and stations, plus increased (double-decker) train capacity, and a TGV station were created. The investment in public transport access to La Défense was however integrated with a policy of providing pedestrianised areas and facilities: 'It is enough to walk through La Défense to see that here the pedestrian is king. The esplanade has done away with the car and strolling is once again real pleasure' (ÉPAD 1993: 4). The high degree of pedestrianisation allows a wide range of public and indoor entertainment, exhibitions, festivals, e.g. National Jazz Competition, and plans for an omnimax cinema, the *Colline de l'Automobile* and the City of the Image, in addition to the *Grand Arche* itself, which hosted one of the key Bicentenary celebrations and which has become one of the city's most visited attractions. The support of public art, sculpture parks and the nearby park district which houses the Opera Ballet School and Théâtre des Amandiers has created in this particular example of regeneration an urban policy that is not incompatible with a living and working environment now driven by commerce – a critical mass requiring the integration of design, planning and 'urban culture'. This cultural initiative won the 1989 French Award for Patronage of the Arts and this experience and planning approach is not limited to the development of La Défense to the west of Paris, since: 'The plans for the eastern city are a model or urban design-led regeneration . . . [which] emphasizes the linkages in Paris between public patronage and modern architecture, urban design and cultural investment, and public education that seem nonexistent in London' (Punter 1992: 81). Bearing in mind the inter-city competition between world cities and emerging cultural capitals, London's position, despite its comparative advantages, rests increasingly on its physical and cultural environment, otherwise: 'Cities of Paris and Frankfurt which take trouble over their environment and their public transport, and ensure that there is variety including culture in their central business districts, are increasingly likely to draw commerce from London' (Sherlock 1991: 161).

Attention to the public realm, mixed-use and cultural activity is not confined in these cities to downtown and central core zones. In the north-east of the city, *Parc de la Villette* was developed from reclaimed land and derelict sites and industrial buildings on the outskirts of Paris. Formerly the location of a livestock market and slaughterhouse from the 1860s serving Paris and linking the Ourcq and Saint-Denis canals, the park is approached via canal, metro or car/coach and bus, and covers 136 acres. As well as public art, covered walkways (ground and raised), squares with cafés and play areas, the park contains a grand hall/exhibition centre (1500 capacity), a 6,400-seat theatre arena and a museum/entertainment quarter containing an omnimax cinema, science museum with planetarium, aquarium attracting high school/educational usage (the museum is very hands-on/experiential), families, traditional park users and visitors. The performance element, in addition to the hall, a former industrial building, is supplemented with outdoor events, including giant film screenings. The development received considerable national funding (*Grand Projet*) and like La Défense is now one of the city's main attractions for locals, Parisians and visitors alike – no mean feat given its unlikely urban fringe/industrial wasteland location and competition from tourist honey pots. In its first year alone the park attracted 3 million visitors and receives over 5 million annually, a sum equal to the Eiffel Tower.

Meanwhile in London, the failings of the limited capacity and slow Docklands Light Railway (DLR) system have joined the ranks of London's transport folklore, while essential underground and cross-rail links were repeatedly delayed by the UK government's reluctance to invest public money and the private sector's cold feet about funding something that the government appears not to believe in (Bashall and Smith 1992). The building of the extension of the Jubilee Underground Line, which cost over £3.5 billion but finally linked Central London with parts of Docklands and East/South East London, was first mooted in *1945* but completed only in late 1999, just in time for the Millennium Exhibition site (Dome) at North Greenwich (see below) served by a new station. Like modern transport that opened up access to higher scale and collective arts and entertainment venues a century earlier, public transport and integration with existing systems and modes has been instrumental in enhanced levels of cultural activity, as Bilbao, Lyons and Paris have demonstrated. Within a month of the Tate Modern opening in a partially converted Bankside power station served by new and expanded stations on the extended Underground line, over a million people visited the gallery, three times the expected number with 4 million visits against its first annual forecast of 2.5 million (CELTS 2000).

This new gallery has also hosted a major comparative exhibition, 'Century City: Art and Culture in the Modern Metropolis', exploring, as Hall (1998) has done at great length, the relationship between creativity and the city, taking nine cities representing at critical times 'crucibles of innovation'. Selected cities include Lagos, Bombay (Mumbai), Moscow, Tokyo, New York, Rio, Vienna, Paris and London. However, like Hall, the causal effects between cities and urban culture, and artistic and innovative milieu is neither clearly or convincingly articulated, nor is the relationship – good and bad – between the planning and development of cities and culture fully recognised. In part this reflects the tension between art and architecture, as Sudjic comments: 'Artists may provide the preconditions that create a cultural climate in a city, but it is the architects who actually build and shape them' (2001: 10). In part this also reflects the use of the city as curatorial backdrop on which literally to hang and promote contemporary art: 'the city as the biggest cultural story in town' (ibid.; cf. Hanru and Obrist 1999, Design Museum – Architecture Foundation 2000).

One cannot help but contrast the endogenous attitude towards the city that these examples present. The French have a permanent exhibition in the City of Paris (Pavillon de l'Arsenal) with major exhibitions celebrating *la ville* (e.g. Pompidou 1994), which is visited by thousands of people each year. Other cities increasingly incorporate city planning – past and future – in permanent and temporary exhibitions, such as Glasgow's 1999 Festival of Architecture, Design and the City; in Barcelona (1999) and during Santiago dè Compostela's *City of Culture 2000*. London's attempt at establishing architecture centres and the establishment of an Architecture Foundation have however foundered in the uncomfortable divide between art and architecture, between architecture and planning, and between architecture and itself. Despite the obvious need for integration in urban policy and planning, London's response has rested heavily on the enterprise of individual boroughs, not least since they represented significant populations and discrete cultural and heritage locales: 'Each of its 33 boroughs has the population of a small or medium city. But many of these boroughs have arts facilities considerably inferior to those of most cities. Since the demise of the GLC . . . London has lacked a strategic level of local government. This has impaired arts planning' (Arts Council 1993a: 115). As Lichfield therefore suggested: 'Local authorities should reassert their primary role; assume the initiative in planning for renewal. This should be done through local area-based initiatives, rather than what is seen as reliance on less responsive national and regional grant mechanisms' (1992: 4).

London, like other Old World cities, presents a schizophrenic planning and cultural paradigm. Its continuity and ability to generate and absorb change is in contrast to both better 'planned' and culturally resourced competitors. This historic paradox is also apparent through its milestone exhibitions which have spawned key cultural facilities and quarters, namely the Great Exhibition of 1851 and its successors (Tables 3.1 and 8.1), the post-War Festival of Britain in 1951 and in 2000 the British Millennium Festival – what became New Labour's nightmarish *New Millennium Experience* at the Dome in Greenwich. (The Dome project was instigated by the former Conservative government and operated by the New Millennium Experience Company on behalf of the government's Millennium Commission.) How far the inheritance from this reclaimed site, initially to be sold to the Japanese Nomura group for a *Dome Europe* theme park, might emulate the museum island of South Kensington and the Festival Hall on the South Bank remains to be seen. Its ambitious new operators claimed it would create 'the first urban entertainment resort offering the best of European entertainment, culture and cuisine'. However this sale offer was withdrawn at the last minute as the Dome struggled to maintain its solvency and rescue its decaying public and media image. A 'sobering' (*sic*) thought is that this Japanese investment corporation was also the largest owner of licensed pubs in Britain, the attention of another form of theming. What this does represent, however, is the return to the great exposition-as-entertainment event bridging site-based regeneration with efforts at thematic cultural celebration, a counter-intuitive and both politically and economically risky return to more homogeneous eras, to the values and unity of the past. Indeed, aside from the numerative–biblical definition of the Millennium, figuratively this also refers to 'a period of good government, great happiness and prosperity' (Evans 1996a).

Festivals, expos and the mega-event

> The expo is to the city what fast food is to the restaurant. It is an instant rush of sugar that delivers a massive dose of the culture of congestion and spectacle, but leaves you hungry for more.
>
> (Sudjic 1993: 213)

From the position of local area city planning policies which together with infrastructure and other environmental improvements in many respects are the 'unseen' aspects of cultural amenity planning, perhaps the most visible examples of public culture in the late twentieth century and which raise issues of cultural planning, public choice and democracy are the contemporary mega-events and cultural festivals. Cities have again embraced these politically and economically high risk ventures, as Stungo observes 'not since the nineteenth century has architecture been used so consciously to promote civic, indeed national, pride' (2000). Their location, scale and content, the rationale for their funding and impact on local, existing and aspired-for arts and cultural facilities is therefore considered further as the *fin de siècle* passes into the new Millennium in a self-conscious frenzy of cultural events and buildings. Two iconic manifestations of what is a primarily urban cultural renaissance are the festival or 'EXPO' event, and the *Grand Projet Culturel*.

At the outset, the evolution of the contemporary public festival as symbolic and visitor attraction – 'instant heritage' – presents a *problematique* if their purpose and sustainability is of concern beyond the calendar cycle of ever-growing cultural feasts. The festival listings and review sections devoted in newspaper supplements and guides; the tourist board promotions and theme-overkill all suggest an offering that is neither sacred nor profane (Falassi 1987), but a formulaic device for tired venues, competing town and city promoters and the performing arts circuit, complementing off-season hospitality and tourist itineraries. The EXPO-style cultural festival and mega-event have now joined the ranks of nationalistic and destination promotion devices for which major public and private sector stakes are gambled, despite the fact that 'the urban visitor attractions field is littered with the debris of over optimism, over rated projects and written-off loans' (Middleton 1994: 88).

Getz's generic definition of a festival – 'a public themed celebration which is concentrated in time and delivered with a clear purpose' (1991a) – whilst functionally correct also tends to lack a clear purpose in the case of the contemporary festival. For instance, Hall distinguishes religious from cultural festivals, including milestone (e.g. centenary, anniversary) events (1992: 22). However the commodification of major festivals (e.g. Edinburgh) such as in the UK where only 38 per cent of arts festivals are run on a voluntary basis (Rolfe 1991) contrasts with more traditional festivals which still retain more of their original purpose and indigenous involvement and their sacred and profane roots. This can be the case even where festivals are large-scale cultural events such as the carnival *Mas* in Trinidad (Mason 1998) and Rio, and their recreation in Toronto's *Caribana* and London's Notting Hill Carnival. These latter festivals combine local, participant and tourist, but belie months of planning, workshops and craft production and rehearsal before the events themselves (Owusu and Ross 1988). Significantly they also take place in 'contested spaces', often in conflict with state governments and official festivals and touristic events. Historic festivals such as those held in Venice, Valencia, even the Paris 'Autumn' and Nîmes' and Seville's Férias, also retain their sense of ownership, whilst not 'excluding' visitors, and serve as opportunities to celebrate place and reaffirm cultural and civic pride (Evans 1993a, b). The regenerative potential of festival locations, as well as the events (or series) themselves, has also been an important component of urban revitalisation programmes and developments in Europe, Australia and North America, and the awareness created through the hosting of mega-events, notably sporting competitions, has been a universal panacea for developing country and advanced state alike. This has also been the case irrespective of indigenous relevance, such as Seoul's Olympic Games, where many of the games were alien (and the sponsor Coca-Cola's catch phrase 'Coke Adds Life' translated to 'Coke brings your ancestors back from the dead'!; Evans 1993a).

Festival sites integrated with mixed-use urban development schemes have also been celebrated and replicated as models, from Baltimore's Festival Harbor Place (cf. Southampton's *Ocean Village*, Law 1992; Toronto's *Harborfront*) and the time-limited Garden Festivals promoted in Britain as a regional regeneration and inward investment strategy (PACEC 1990). The European *Cities of Culture* have also been used to promote urban development and cultural tourism, attempting to reconcile unity with diversity (ethnic, regional) and focusing on the festival as part of the Europeanisation project (see Chapter 6): 'when we became aware of the city again as a place of culture, style, and artistic excellence . . . when industrial production was less of a boast than nice squares and art galleries' (Jones 2000: 5). This promotional exercise has now been replicated in the Americas, with the first *City of Culture* held in Merida, the state capital of the Yucatan in South Mexico in 2000.[1] The festival's role in animation and rediscovery has also been seen in English industrial heritage locations such as Little Germany in Bradford and Gabriel's Wharf, Coin Street in London, often using temporary buildings and sites in the period between reclamation and redevelopment. Their regenerative potential has therefore not been lost on areas of urban decline: 'The hosting of mega-events is often deliberately exploited in an attempt to "rejuvenate" or develop urban areas through the construction and development of new infrastructure . . . road and rail networks, airports, sewage and housing' (Hall 1992: 69). This was seen in Montreal's EXPO '67 as 'an excuse to pyramid dozens of public projects including the new subway system, highway and 745 acres of parkland' (Peters 1982, in Hall 1992: 71) and more recently in Seville's EXPO '92 and the surrounding Andalusian region, and Barcelona's 1992 Olympics, through European regional development grant-aid (Evans, 1993b, 1998b). As Sudjic remarks, 'Seville's plans [were] an ambiguous mixture of old-fashioned pork barrel politics . . . involving a massive diversion of state funds by politicians anxious to secure re-election . . . with sophisticated attempts to overcome the problems of economic backwardness' (1993: 31).

The hallmark or mega-event almost by definition and scale often transcends any planning, cultural or even economic rationale and assessment – although all of these are to a greater or lesser extent claimed as justification by the host city/government through international competition and domestic consumption, creating in Horne's view 'a fabricated public culture that purports to be the culture not only of the rulers, but [also] of all the people' (1986: 184). They are overtly political and given the scale of investment and infrastructure required to achieve viability also require a degree of national consensus and minimalisation and marginalisation of resistance and protest. Hall lists specific protests against the hosting of Olympic Games from Mexico 1968 onwards (1989: 95) and even the unity promoted through the Montreal EXPO '67 masked underlying conflict: 'Quebec separatism was only the most conspicuous; growing ethnic and regionalist diversity, the women's movement and Native assertion also challenged the existence of any one set of symbols to fit all' (Kroller 1996: 6). Here cultural symbolism was created, challenged and revised with unprecedented intensity: the bitter debates over flag, anthem and the EXPO '67 logo, for instance, tied up parliament for weeks – these symbols often became the opposite, namely reflections of fragmented and conflicting identities. Urban regeneration, derelict land reclamation and the development of a landmark scheme were the prime conditions laid down for the location of the official Millennium Festival in Britain, thus continuing the thirty-year line of world fairs in the 'fourth period – the city of renewal' (Hall 1992: 29). The British Millennium Festival as a national celebration and invitation to the world, located in an already eminent cultural tourism capital, therefore faced both the politics of place creation and the re-creation of national unity (Irvine 1999; also see below). As Bonnemaison states, the hallmark event 'functions like a monument, supporting and reinforcing the image of power, whether religious or secular' (1990: 25. Here 'secular' is ironic given the Judaeo-Christian calendar used to justify this millennial mega-project (and others), which was largely a secular offering, financed by gambling (i.e. Lottery) proceeds (Evans 1996a).

The Great Exhibitions

The renowned Great Exhibitions and early World Fairs (see Chapter 3) were held in another age. However they still provide a hint of the formula required to succeed and maintain visitors after the launch year. The largest exhibitions (10–12 million visitors a year) all had the features of sprawling sites, a wide range of categories, 'produce' and very vocal government participation (Benedict 1983, Ryder 1984, Greenhalgh 1991).

The late nineteenth-century exhibitions had semi-permanent facilities that were reused at least four times in succeeding years, thus maintaining the same level of visitors, leading to permanent venues that spawned Earls Court, White City and later Wembley Stadium. The creation of a permanent facility or landscape is therefore essential in establishing a sustainable EXPO site (e.g. Crystal Palace, Eiffel Tower, and Royal Festival Hall). Category A Universal EXPOs[2] are normally not allowed more than once every ten years because of the huge cost of staging for the host and for exhibitors. British Garden Festivals as temporary sites are less successful examples as the long-derelict festival sites of Glasgow, Liverpool and Ebbw Vale in Wales attest. These were reinvented in Britain during the 1980s based on the German model of *Bundesgartenschauen*, which began their modern form in the late 1940s as part of the rebuilding of post-War Germany (Gooding 1995). Their poor sustainability and after-use record in the UK rested largely on their over-reliance on a leisure–property formula. The choice of derelict and contaminated sites (as with the ill-fated Millennium Dome – a British Gas site) leads to high capital costs and benefits are only likely to be achieved in the long-term, if at all. Of the 1,310 acres of the Seville EXPO site, only 20 per cent is currently in use, with a theme park mothballed awaiting refinancing. Lisbon's Category B EXPO '98 was not expected to recoup the £1 billion capital investment until 2009 at the earliest. Even Lisbon's smart new waterfront is a curious edge city that is not successfully integrated with the fabric of the place as a whole. Hannover's EXPO 2000 operators were advertising the sale of the site and facilities even at the peak of this event, which like its year-long equivalent in London (Millennium Dome) has been distinguished by a lack of attendances, high costs and severe criticism, including demonstrations by local and environmental groups on its opening (Irvine 1999). Despite the high capital costs and questionable after-use of such event-led regeneration, international exhibitions continue to be sought by national and regional governments. Table 8.1 lists the EXPOs and major Fairs in the post-War period.

The feasibility and pursuit of political prestige and credibility driving the competitive festival process also requires consideration of both social impacts and community involvement in such a major urban regeneration project not least since they generally emerge outside of the normal land-use planning horizon and process. The evaluation of 'host' impacts of major events is, however, less considered than economic measurement (Ritchie 1984, Smith 1991, Hall 1988, 1992, and Owusu and Ross's socio-cultural history of the Notting Hill Carnival 1988). Community motivations and the issue of authenticity in event tourism are investigated in Getz (1994), Mayfield and Compton (1995) and Uysal *et al.* (1993). However the evolution and planning of a mega-event is by its very nature a top-down process and outside of the control and scope of individual or even a forum of community representatives (including local businesses). The political imperatives driving such urban and tourist developments 'create a political and economic context within which the hallmark event is used as an excuse to overrule planning legislation and participatory planning processes, and to sacrifice local places along the way' (Dovey 1989: 79–80). Mechanisms to engage and empower local groups start first at the partnership level: a place on the board. Second, 'community visioning' processes can be employed as part of the community planning exercise, since, repeating Teitz: 'public determined facilities have a role in shaping the physical form of cities and

Table 8.1 Post-war EXPOs, fairs and UK garden festivals

Year	Exhibition/festival	Category	Attendance (millions)	After-use facility
1951	Festival of Britain		8.5	Festival Hall and South Bank, London
1958	Brussels World Fair		41.5	Atomium & Heysel Exhibition Centre
1967	Montreal EXPO	A	50	Art gallery, Biosphere
1970	Osaka, Japan	A	64.2	mixed developments
1974	Spokane, USA	B	3.8	opera house, convention centre
1982	Knoxville Energy EXPO	B	11	convention centre
1984	Liverpool, north west England	garden	3.4	(redundant site)
1985	Tsukaba	B	20	Science City
1986	Stoke, northern England	garden	2.2	housing, leisure
1986	Vancouver EXPO	B	16	convention centre
1988	Brisbane	B	8	convention centre
1988	Glasgow, Scotland	garden		hotel, exhibition centre, part-redundant site
1990	Gateshead, north east England	garden	3	housing
1992	Ebbw Vale, Wales	garden	2	housing, leisure
1992	Seville	A	20	infrastructure
1993	Stuttgart	C		public park
1993	Taejon, Korea	B	10	high-tech exhibition centre
1998	Lisbon	B		park, business park, oceanarium
2000	Greenwich, south-east London	Millennium Experience 'Dome'	4.7	Sale of the Dome to the Japanese investment bank Nomura for $100 million aborted (cost of site/structure £758 million)
2000	Hannover	A	20 (estimate)	Housing, park exhibition centre (extension), roads and rail upgrade
2005	Aichi, Japan	A		infrastructure

Sources: Allwood (1977), Benedict (1983), PACEC (1990), Greenhalgh (1991), Evans (1996a)

quality of life within them' (1968: 35). Community benefits of mega-events are also argued by Hall (1992: 82, also Burns and Mules 1989), including the less tangible but no less real impact on cultural identity, civic pride, community development, and, for large events, even 'psychic' benefits. As Hall also notes, however: 'The social dimensions of hallmark events are unevenly distributed through a community in the same manner as the direct and indirect economic impacts of an event' (ibid.: 82). This worsening ratio of costs to benefits is most acute in the crisis which surrounds large-scale events, in the development and construction of facilities and infrastructure which is often 'fast-tracked' through planning procedures and where the evaluation of the social and economic dimensions of the event through a public consultation process remains incomplete (ibid.).

The transformation from a living and working urban tourist destination to the hosting of a major festival, event or exhibition therefore raises some of the most complex and conflicting forecasting, planning and political issues that arise from visitor-led urban regeneration initiatives, not least their social and distributory effects, since as Hall states: 'it should be recognised that social impact evaluation will ask the difficult question of who benefits? A question which goes to the very heart of why cities host hallmark events in order to improve or rejuvenate their image and attract tourism and investment' (ibid.). As Ritchie and Smith have observed, the latter requires consideration at the development stage rather than as an a posteriori exercise as viability is questioned and blame apportioned: 'cities considering the staging of such a mega-event must anticipate a significant rate of awareness and image decay and take steps to counter it' (1991: 3). The integration of environmental impact assessment, carrying capacity modelling and community consultation exercises also offers a more sophisticated evaluation of the opportunity costs and benefits of such an urban mega-event, but the existence of a cultural planning framework would enable this process to consider such temporary invasions and their aftermath over the longer term and in respect of a wider range of social, economic, land-use and other environmental factors.

Grands Travaux and Projets

Like the mega-event, cultural planning in terms of local amenity and assessment of need and preferences are seldom considerations in the mega-architectural *Grands Projets* that have served as national political–cultural statements, whether inherited from city fathers or promoted by contemporary government – city, regional or national – and patrons. In Paris where 'the Pompidou Centre was decided against all planning authorities, whose discourse or speech was "no more institutions" and "no more Paris institutions". Nevertheless, Pompidou decided to go ahead' (Girard 1987: 10). A similar decision was taken by Mitterrand (and Lang) over the Opera Bastille, 'which related either to an historical tradition or an intuition, or vision, of the monarch, [which] could not by definition be rational' (ibid.) – or be based on a cultural democratic 'plan'. The French system is said to be a 'cultural monarchy', 'where he pleases, the minister in office defines his options and takes his decisions in the fashion of a sovereign, according to the principles of "enlightened despotism"' (Wangermée 1991: 35). Boylan stresses the dominant role of successive presidents from Pompidou onwards rather than culture ministers (1993). He argues that presidential self-aggrandisement has been more powerful than culture ministers, who have largely been insignificant short-term holders of the post except for the first, André Malraux, and the socialist Jack Lang during the Mitterrand presidency. What many of these *Grands Projets* have in common is an exorbitant cost, often turning out to be overdue and substantially higher than first budgeted (some overrunning by over 100 per cent), and a defiance of public planning or choice. Over £3 billion in capital investment has been made in this case (Table 8.2) with over 600 provincial *Grands Travaux* projects costing in excess of a declared £200

Table 8.2 Capital cost of the *Grands Travaux*

Project	£ (millions)
Musée D'Orsay	169
Cité des Sciences et de l'industrie de la Villette	569
Parc de la Villette	140
Cité de la Musique à la Villette	97.8
Grand Louvre	526
Institut Arab Monde	30.7
Grande Arche (La Défense)	113.4
Bercy – New Finance Ministry (to clear space for the Grand Louvre)	375
Opera Bastille	293
Bibliothèque de France	549
Centre de Conferences Internationales	263
Renovation of various national museums	105.2
Total	3,231.1

Sources: Biasni (1989), Comedia (1991b: 54)

Note: during the 1980s the cost of these projects represented an average of 0.2% of the state budget

million, and by now these totals will have been exceeded by ongoing and additional spending, such as the final phase of the *Grand Louvre* project.

To what direction these architectural and cultural monuments look is hard to judge conclusively. However it is difficult to ignore their historic and classical associations and aspirations, whether civic, monarchic or nationalistic in origin, since: 'Most of these symbols of our new era would have been intelligible to a top-hatted or crinolined Victorian as they are to today's trainers-wearing masses' (Stungo 2000). Unlike today, the private patronage that funded and initiated many of the civic cultural institutions and facilities was seldom centrally orchestrated (in some sense the projects were carried out in competition or even in spite of central influence). As Stungo goes on to say: 'there seems something imposed rather than organic about today's civic pride' (ibid.). Baudrillard perhaps put it in its post-modern condition when on a lecture visit of London in 1997 he remarked that 'there was to be no Millennium since the history of the late twentieth century had already been written'. What underlies this discomfort, this staged authenticity is part of the wider effect, supposedly, of globalisation – through the convergence in urban design and economy and therefore in acculturated consumption. King goes further in this, viewing the institutional cultural policies as essentially 'anti-culture', or, in Handler's words: 'everyone wants to put [their] own culture in [their] own museums' (1987: 137):

> The extent to which states (or towns and cities) do *not* have their own historical museums, do *not* have self-conscious 'cultural policies', do *not* have historically-informed conservation policies, are *not* concerned about cultural homogenization, national identity and westernization, is the most accurate and telling comment on the uniqueness of their cultures and sub-cultures; the degree

to which cultures are self-consciously 'different' is an indication of how much they are the same.

(King 1991: 153)

Not only the serial replication of urban culture and consumption, but also the mobility of what were previously largely rooted locales for collections and companies has also seen the arts and artefacts as footloose. For example, in the Yorkshire industrial city of Bradford, the local authority's cultural policy (Arts Council 1991) originated in 1979 and was always tourism led, although local regeneration initiatives focused on community festivals and animation. Responding to a loss of 63,000 jobs during the 1970s, the City Council revived a cultural tourism strategy originally launched in the mid-1960s which concentrated on the industrial heritage, the Brontë connection, the rural landscape and the vague 'arts in the districts'. Outcomes were not quality of life or cultural opportunity for residents but the increase in visitor numbers, building a tourist base (e.g. overnight accommodation) and encouraging business relocation, and generally to raise the profile of the city outside of Bradford. Cultural flagships were part of this regeneration: the National Film & TV Museum, Alhambra Theatre, several heritage centres and museums: 'landmarks of the industrial revolution. . . . These large-scale recycling projects have turned into spectacular pieces of urbanism, more Cecil B. de Mille than town planning' (Sudjic 1993: 185). This included exploiting the David Hockney connection with the city in the gallery at Salts Mill, Shipley, and a long-delayed move of the South Indian collection from the V&A Museum in London to a proposed converted Manningham (Lister silk) Mills at an estimated capital cost of £60 million. This move was presumed to be a welcome gesture toward the large community from the Indian Sub-Continent resident in the city since its migration to serve in textile mills, now cleaned and swept to house imported collections and displays. Most of the collections proposed to be transferred were in fact Hindu and Jain, predominantly of human form deity figure representations and potentially highly offensive to Bradford's predominant ethnic minority group – Muslims of Pakistani, Bangladeshi and Punjabi origin. What scope for contemporary Pakistani culture and youth arts expression or production was to be offered by this version of a cultural economic strategy, again is not clear. The V&A quietly dropped the Bradford project in the summer of 1995. The V&A proposal although unfulfilled was one example of this competitive urban cultural challenge, the 'footloose' museum and art collection with cities fighting over their new 'home' (*sic*) and regional outposts hoping to meet both distributive aims in arts policy terms and at the same time help meet criticisms of galleries and museums with most of their collections hidden and mothballed:

> The proprietors of almost any halfway respectable art collection, financially viable opera company or solvent regional orchestra now find themselves in the same privileged position once occupied by Euro Disney and Nissan. They are assiduously courted, flattered and bribed by every ambitious city eager to make its mark . . . and set up in their back yard.
>
> (Sudjic 1993: 4–5)

In the unseemly battle for relocation of the Tower of London's Armoury collection between two other Yorkshire cities of Leeds and Sheffield, the latter would have appeared a more appropriate 'vernacular' home (with its steel/metal crafts industry inheritance), but Leeds won the contest in 1996 with a commercially financed 'theme-museum' which required sufficient returns from 750,000 visitors a year. Within three years annual visitors did not reach 400,000 in total (including free admissions) and by 1999 only 250,000 visits and with £20 million of debts the government stepped in to bail out and restructure this particular *unplanned* regional *Grand Projet*. The international branding of the museum itself is seen most recently in Bilbao, Northern Spain and the development of a Guggenheim Museum – a thirty-year franchise from the New York 'original'. Using cultural tourism as a tool of re-imaging – Bilbao an industrial and polluted city nicknamed 'Orificio' locally – the expansion of airport and road and rail networks has also featured in the city's regeneration, including a new airport terminal (capacity for 2.5 million) designed by Santiago Calatrava; a new underground/metro running alongside both riverbanks, designed by Norman Foster; and the Intermodal Station at Abando, linking high-speed train, bus, metro and car parking as part of a mixed-use residential and commercial development, designed by Stirling and Wilford. The Guggenheim satellite museum located in a much improved harbour zone and designed by US architect Frank Gehry is described thus: 'a titanium monster which resembles an intergalactic ocean liner grounded on the shores of the [River] Nervio . . . the real reason Bilbao is on the tourist map and has moved to the top of the cultural tourist's must-visit list, a hot spot for 21st century tourism' (Barrell 1998: 3), and as if to pacify the international gallery-going visitor: 'there's thankfully no sense of the city being overwhelmed – you're unlikely to encounter another foreigner outside of the museum' (ibid.). Basque reaction was less enthusiastic: 'immediate and much of it negative . . . infuriated by the secrecy surrounding such a large project . . . and estimated to cost 400 million pesetas a year, to them the money would be far better spent tackling directly the problems of the Basque Country' (MacClancy 1997: 2). On its opening, a security guard was hurt in a bomb planted by the Basque terrorist group ETA (since Guggenheim Bilbao, now more active) – whose culture is on offer here, in a tourism-led regeneration strategy adopted by the city authority, as in Leeds and Bradford and countless European, American and Asian cities? The cultural tourist could naively be presented as a political neutral in cultural politics that have been fought out before their arrival. Returning to Bradford again where the combination of the 'unique' (*sic*) and manufactured with the 'familiar' is the strategy adopted by the city's Arts and Museum Officer:

> I imagine a retired couple in Bonn reading information as to where they might go for a cultural experience anywhere round the world, and that there is at least a good chance they might come to Bradford, because of its *unique attractions*, because of the *attractions manufactured* by the will of those involved in public and private sectors, and with the sure and certain knowledge that they would find the sort of facilities they would expect in their *home town*.
>
> (Arts Council 1991, cited in Evans 1998a: 13)

Back in the home of the modern *Grands Projets*, the core–periphery, flagship–amenity dichotomy is writ *Grand*. From one highly self-interested viewpoint:

> These new instruments of culture born of France's major construction projects in Paris and in the provinces are so many open houses which rally behind a certain idea

of the city, this collective lifestyle under which, for the first time, some 2,500 years ago the Greeks sketched the outlines of what we now call democracy.

(F. Mitterrand, quoted in Biasni 1989: 5)

Whilst from another viewpoint:

whatever their value as architectural set-pieces, they are not the much-vaunted harbingers of a proclaimed urban renaissance. On the contrary, like circus games, they direct attention from the inexorable erosion of Paris and the brutal neglect of its suburbs.

(Scalbert 1994: 20)

The scale and cost of the French *travaux* whilst substantial, has not been unique, or of course restricted to the capital, Paris. Like the *Maisons de la Culture* of the 1970s, mayors and city halls of regional cities have created new and renovated cultural facilities, from Grenoble, Rennes, Lyon to Marseille where a $1.2 billion five-year programme of development of housing, industrial (*Euromediterranee* international business centre) is underway, including upgrading of former village centres. This revived European urban confidence is leading to the (re)creation of self-styled 'city-states', which aim to secure international competitive edge through investment in public realm – new art galleries, libraries and museums are therefore an essential ingredient to this civic mix. This is not limited to the more Catholic, Renaissance countries, for example in Rotterdam, The Netherlands and Uppsala, Sweden where large library and urban renewal projects incorporate *museumparks* with higher education facilities and housing.

Major cultural building programmes (if such a term can be applied, in practice these were not 'planned' in any organisational, spatial or strategic sense) are evident in Britain following the new Lottery-fuelled arts, heritage, sports and Millennium 'landmark' projects. Here over £2 billion of Millennium Commission and other Lottery funds has been awarded to national and regional projects to 'mark' the coming of the new millennium, and to shore up the physical cultural infrastructure that had been in decline in some respects since not only the Victorian era, but also the 1970s' decline in public capital spending. (In a study of local authorities in England, the backlog of maintenance and repairs of arts and recreational facilities was estimated to be over £102 million compared with a capital budget including new projects of only £20 million in 1996/7; Evans and Smeding 1997: 16–18.) Table 8.3 provides an example of these particular *Grands Projets* that have common themes – those of urban regeneration, science and technology, the environment, and a general tendency towards *edutainment*. These exclude the designated British Millennium Festival at Greenwich peninsula, notionally the 'Home of Time' (on the zero degree longitude Meridian line), which at capital cost of over £750 million received £628 million of Lottery funds towards the infamous and temporary 'Dome' (see above). Since such schemes under the government's own Lottery rules could not be solicited, few of these new projects, particularly new developments on new sites, are expected to attract sufficient visitor numbers where not already located in accessible locations. The absence of any planning framework for what has been the largest public investment in cultural facilities for several generations risks serious problems for many of these schemes and a less than welcoming reception from the public which has had no involvement in the allocation of the resources or the location and design of the facilities themselves. As cultural activity and facilities accumulate unplanned in this way, the cultural map is redrawn and spatial relationships change accordingly, inevitably and continually undermining previous expectations and demand.

Table 8.3 UK Millennium Commission-funded projects awarded £15 million and over

Project	£ (millions)	Opening date
Tate Gallery of Modern Art, Bankside, London	130	spring 2000
Millennium Stadium and Rugby Museum, Cardiff	134	1999
Earth Centre Environmental Research Centre, Doncaster	100	1999 (phase I)
Millennium Point Multimedia learning centre, Birmingham	113	autumn 2001
The Lowry Waterfront Theatre and Gallery, Salford	98	spring 2000
British Museum Great Court, London	94	autumn 2000
Bristol Wildlife Cinema and Science Museum	97	spring 2000
Odyssey Project Science and Sports Centre, Belfast	90	autumn 2000
University of the Highlands & Islands, Scotland	86	autumn 2001
Millennium Seed Bank, Kew Gardens, London	81	summer 2000
Maritime Museum and Landmark Tower, Portsmouth	79	spring 2001
Eden Project plant research, St Austell, Cornwall	77	spring 2001
Millennium Link Forth Clyde and Union Canals	78	spring 2001
Our Dynamic Earth Environmental visitor centre, Edinburgh	71	1999
Wales Millennium Centre, Welsh Culture Theatre, Cardiff	70	spring 2002
Library, business and heritage centre, Norwich	60	spring 2001
International Centre for Life, Newcastle	58	summer 2000
'National Discovery Park' multimedia centre, Liverpool	54	winter 2001
National Space Science Centre Planetarium, Leicester	46	spring 2001
Hampden Park Stadium, Glasgow	46	1999
National Botanic Garden of Wales, Myddleton	43	spring 2000
Manchester Millennium Quarter city centre regeneration (post-IRA bombing)	41	winter 2001
Steel Industry Exhibition Centre, Rotherham	37	spring 2001
Glasgow Science Centre	31	spring 2001
'The Deep' Aquarium and Research Centre, Hull	18	winter 2001
Total	1,832	

Source: Millennium Commission (direct source/communication, 1999)

The familiarity with which many of these neo-classical/post-modern edifices would in Stungo's view be shared with the Victorians (see above) provides a clue as to their essentially classical form and conception, notwithstanding modern technology and materials, and as Davey *et al.* observe, the effect of these imposing structures is long established:

> Buildings made primarily for arts and recreation have always had an element of the impervious. The great recreational types we inherit from antiquity, the theatre, the

circus, the stadium and the amphitheatre are all inward turned, as their present-day successors are . . . which sit in unhappy conjuncture with matrix of spaces and life that surrounded them.

(1993: 4)

This has also been true in the case of some post-War municipal arts districts that lack both a sense of place and a user-friendly design scheme. As Barker observes: 'The zoning of the arts into megaliths like the Barbican and the South Bank created a bleakness you do not find in the spec(ulative)-built West End or in old, reused structures' (1999: 31), a situation which their expensive and problematic upgrades and re-masterplanning have sought, unsuccessfully, to improve. The new culture-houses – galleries, museums, libraries, arts centres and other civic structures – have of course supported an illustrious roll-call of international architects, a special cultural milieu whose signature buildings are sought as much as artistic directors and curators. Their international scope is evidenced by the fact that much of their work is located outside of their own country (e.g. Hadid, Rogers and Meier; Blaser 1990), but they also represent a tradition from the earlier twentieth-century utopian architects epitomised by Le Corbusier, Frank Lloyd Wright and Ludwig Mies van der Rohe whose power and influence far exceeds that of the jobbing *arkhitektron* (masterbuilder) of classical Athens or 'design-and-builders' such as Sir Christopher Wren. Meier's pinnacle of the 'high museum', the Getty Foundation's new art museum which has rehoused most of the collection from its Malibu 'roman villa' (*sic*), requires the visitor to pre-book a parking space with a so-called rapid transit link (in reality a slow ride) to the six-level 1,200 underground car park. On a busy weekend visitors can experience an hour's queue to get down the hill at closing times. As Sudjic notes, this 'brings home starkly how different life is in a motorised metropolis from the two-hour pedestrian precinct view of most European cities' (1993: 135). The designer's role in planning for arts facilities, which are often predetermined, is limited (as of course is that of the artist or performer), but given their symbolic and spatial importance greater consideration for place, the vernacular and impact on the locale might be expected. As Immanuel Kant wrote on the essentially human purpose of architecture: 'the beauty of a house or a building (be it a church, palace, arsenal, or summer house), presupposes a concept of the purpose which determines what the thing is to be, and consequently a concept of its perfection' (1790 in Beck 1988: 230). Kant of course was neither architect, builder nor planner. One of the late twentieth-century's modern architects, the Brazilian Oscar Neimeyer, is associated with museums and civic structures that are acclaimed as works of genius, although their interiors often do not fulfil their promise. His Museum of the Founding of Brasilia, the ultimate *new city* (and declared World Heritage Site by UNESCO in 1987 because of its importance in twentieth-century architecture and planning), could be mistaken for a mausoleum, located underground with no provision for interpretation in a bland space. Brasilia was inhabited as the value of public realm was subjugated to global consumerism and the modern movement's adoption of separated functions (*Charte d'Athene*; Jencks 1996, Rykwert 2000). As Deckker observes, this 'concept of urbanity did not really embody an adequate representation of civic space . . . the cultural facilities of the utopian *superquadras originais* do not form a cohesive whole: the city simply has no institutions where people can meet and enjoy the public realm. . . . There are almost no cultural events in the city at all' (2000: 189). As he goes on to say:

> The lack of a central cultural area deprived the city of a focus of cultural life; the National Theatre is totally isolated . . . the Cine Brasilia (Niemeyer 1960) is widely acknowledged as the most comfortable cinema in Brasilia, but it is intimidating to walk to it at night across its landscape space from the adjacent *superquadras*. There are no bars or cafes to provide urban life, let alone refreshment, in either place. The general public prefers the shopping malls with cinema complexes and 'fast food' outlets, a simulacrum of urban life. (ibid.)

His recent new art gallery in Niteroi forms a visual landmark overlooking the bay towards Rio, but again in a monotonous public but inaccessible space (Plate 8.3). The land and facility usage in this functionally poorly designed gallery remains unresolved, rendering this cultural 'landmark' sterile and inanimate, and like the new Guggenheim: 'Bilbao looks great now, but in 30 years will it prove less flexible?' (Thurley, quoted in Irving 1999: 28). The new Gallery of Modern Art in Barcelona (Meier again), for example, sits uncomfortably in a poorer district of the city – the Raval formerly known as the Chinese quarter – overlooked by terraced housing partly removed to accommodate this new gleaming edifice (Plates 8.4 and 8.5). The deliberate location of this *Grand Projet* in a deprived and marginalised neighbourhood is now a common regeneration mechanism (lower land costs and availability also facilitate this, of course), but the sight of local people sitting alongside (and never inside) this building, as opposed to the museum and adjoining cafés/bars and 'public' square, ghettoblaster and large dogs in tow, seems both problematic and uncomfortable. This development has

Plate 8.3 Art Gallery, Niteroi, Rio de Janeiro, by Oscar Neimeyer (1997)

Plate 8.4 Inside the Museum of Contemporary Arts (MACBA), Barcelona (1998)

Plate 8.5 Plaza, Museum of Contemporary Arts (MACBA), Barcelona (1998)

succeeded in attracting a museum publishing house, university faculty and design centre, however within the city and the locale itself the Raval is perceived as a less successful artistic zone with its aspiration of becoming a Parisian-style *Marais* or Barcelona's SoHo – few galleries and artists have moved into the area despite the efforts of city and cultural administrators. As Ulldemolins observes: 'the success or failure of cultural quarters will be interpreted as a result of conservative strategies of risk-taking action by art merchants and as a consequence of the symbolic value that the quarter has been accredited by the local community' (2000: 19), and these two actors cannot be assumed to have either recognisably consonant value-systems or similar reactions to the external creation of a cultural flagship operating in a global context (i.e. international cultural tourism and art-house and gallery markets).

Exceptions of modern architecture and more 'open' design, including an essential concern for the public realm, access and safety, are however apparent, despite the dominant post-modern architectural style and conservatism. This includes adaptation of classic modernist examples such as Mies van der Rohe's McCormick House in Long Shore Drive, Chicago (1951), one of only three houses he built in the USA. This has been saved from demolition, dismantled and transported to a new site in Illinois to form one pavilion as part of the new Elmhurst Art Museum. Mies himself had written in 1943 in *A Museum for a Small City*: 'the first problem is to establish the museum as a centre for the enjoyment, not the internment of art. In this project the barrier between the work of art and the living community is erased by a garden approach used for the display of sculpture' (quoted in Wislocki 2000: 18). Arts facilities within park and garden settings (as noted in Chapter 4) retain a symbolic and urban design solution to the closed nature of many cultural buildings. Another urban example and perhaps one of the most significant of the earlier Parisian regime was the forerunner to the *Grands Travaux*, the Pompidou Centre. Opened in 1977 and designed by Renzo Piano and Richard Rogers, the Pompidou 'showed how modern technology could offer ways of integrating a huge gallery into the heart of an ancient city, at the same time, enhancing the life of both' (ibid.: 5). In another example, Niels Torp's *Aker Bryggee* in Oslo, successfully integrates recreational, commercial and domestic life, with theatre, cafés, galleries, sporting facilities incorporated into shops, flats and office complex. Late modernist architectural design has therefore demonstrated in several towns and cities that 'imagination has been used to counteract the isolationist and reductive tendencies of buildings for recreation to involve them in general life in a multi-dimensional way' (ibid.). Whilst the Pompidou Centre, re-emerging from a £55 million refurbishment has undoubtedly impacted on the Place Beaubourg within which it is sited and is likely to continue fulfilling its symbolic and touristic role, its artistic/functional purpose is less valid since less than 20 per cent of its 25,000 daily visitors actually enter the 'art museum' facilities themselves, preferring just to meet, hang-out and walk through (cf. *Kiasma*, Helsinki). The Pompidou, like so many public art centres and museums, serves in large part as just that, a culturally legitimated 'amusement park' and culture café (Heinich 1988, Evans 1995c). Museums represent a particular litmus test of community culture given their longevity and role in bridging representations of the past with the present and hope for posterity and the future – the *Museum Time Machine* (Lumley 1988). Once a lodestar by which the citizen could navigate the uncertain depths of cultural value, the museum today is, as Giddens writes in *The Consequences of Modernity*, 'no longer certain of its role, no longer secure in its longevity, no longer isolated from political and economic pressures or from the explosions of images and meanings which are, arguably, transforming our relationships in contemporary society to time, space and reality' (quoted in Irving 1998: 26).

To take this further, Bennett writing on *Community, Culture and Government* from a cultural policy perspective rejects the oppositional position that places museums and communities on one side of a divide as part of creative 'bottom-up' processes of cultural development, and the state or government on another as the agents of external and imposed forms of 'top-down' cultural policy (1998: 202). Drawing on James Clifford's work (1997) and the idea of *museums-as-contact zones*, the contemporary museum (and equivalent 'centres') that many of the late twentieth-century people's palaces represent are seen to have departed from their nineteenth-century predecessors, the *museum-as-collection* which represented 'passages from colonial periphery to metropolitan centre', to a museum which 'relocates the object as the site for a process, often bitterly contested, of the negotiation of meanings and values between different cultures' (ibid.: 203). How far the scope, scale and *critical curatorship* offered by many of these culture-houses, new and repackaged (*old wine in new bottles*), meets this challenge of the cosmopolitan city and cultural development is likely to depend on the strength and openness of the curator him/herself in: 'orchestrat[ing] a polyphonic dialogue between the different voices and values emerging from the multiple constituencies that comprise the culturally complex structure of contemporary civil society' (ibid.: 204), and also on the degree of cultural planning that contributes to their formation, location and development. This is a far cry from the nationalistic, universalist nature of the early museum collections and colonial-inspired Great Exhibitions discussed above. However this vision has yet to be realised outside of the refined atmosphere of the curator's thematic versus chronological style of interpretation (Serota 2000).

France and post-Lottery Britain present perhaps special cases in *Grands Projets* and cultural building shopping lists, but many emerging cities and those undergoing re-imaging, such as Barcelona, Frankfurt, and Latin American and South East Asian cities, have indulged in major mixed-use and waterfront developments. A third special case few would argue against is 'new' Berlin. The return to Berlin as the capital of a reunified Germany has witnessed perhaps one of the largest public building programmes involving international architects and public as well as corporate developments (e.g. Mercedes Benz, Sony). However the impact on cultural provision in East–West Berlin in this transition has been a painful one. As mentioned in Chapters 2 and 3, post-War reconstruction in Germany saw the development of hundreds of theatres and culture-houses and halls, whilst newly divided Berlin in 1948 saw a number of new private and public buildings under the direction of the city architect/planner Hans Scharoun, such as the *Philharmonie* sited on the empty and devastated area of the ghostly Potsdammer Platz (redeveloped now as a mixed-use entertainment complex), the Prussian State library, and in the 1960s Mies van der Rohe's National Gallery in the Cultural Forum complex. As Taylor reminds us:

> As a result of its isolated position, West Berlin could survive only with the constant help of Bonn – economic, cultural, moral. . . . Financial and other concessions were introduced in order to attract people, including those in the culture industry, to settle there, while cultural morale was boosted by guest performances from actors and musicians.
>
> (1997: 364)

In East Berlin, the government did preserve and repair the Staatsoper, the Altes and Bode Museums on *Unter Den Linden* (see Chapter 3), driven as much by a desire to gain international prestige and publicity as by genuine cultural conviction, with the familiar Marx-Engels-Platz serving as the new centre for ceremonial and state occasions. The East German leader Walter Ulbricht was also instrumental in the redevelopment of the 2-kilometre-long avenue (Stalinallee, also known as Karl-Marx-Allee) flanked by seven-to-ten-storey apartment and office buildings, including shops, restaurants and leisure facilities (Taylor 1997). Today, the dismantling of the Berlin Wall and the arguable duplication in cultural provision when set against costly priorities of unification has meant that 'state sponsorship of culture also has to be fought for . . . if a subsidised cultural enterprise was deemed surplus to requirements the authorities would not hesitate to withdraw support' (ibid.: 391). The Schiller-Theater in Charlottenburg was one such victim despite (and as Taylor argues, because of) its progressive tradition. Dismantling of two city-state regimes also meant that cultural and educational institutions were also rationalised, with this dual cultural legacy, but any from the East had to be sanitised, 'decommunised' before considered worthy of ongoing support. The reconstruction of the Reichstag (to Norman Foster's design) also echoes Fritz Lang's futuristic vision of Berlin in 1926 (Richie 1998). The rush to reinstate and re-image the artistic heritage over the rubble ('heritage'?) of the Third Reich has also seen new and renovated buildings to house over 150 collections in the old museum island, from the Marlene Dietrich museum in the revamped Potsdammer Platz to Daniel Libeskind's new Jewish Museum, with numerous art galleries and arts centres occupying reused buildings. Although as discussed in Chapter 6, the cosmopolitan artists' colony is being displaced here as in other cities, which is being divided again along capital land-use fault lines. What is most ironic in the dismantling of socialist cultural provision in Eastern Europe is that the physical models of integrated workplace/industry, with housing, social and cultural amenities including holiday facilities (Evans 1995b: 70–1), are being replicated in the 'West', for instance fourth-generation business parks with 'low densities heavily landscaped and [an] ecologically sophisticated environment. Social and leisure facilities on site, public transport . . . and housing provided as part of the package' (Doak 1993).

This arts and civic building spree is not confined to European cultural capitals (and from 2005 the EU's annual *City of Culture* award will be renamed *Capital of Culture*), for instance Montreal, Canada where from the mid-1980s the city was emerging from an effective freeze on spending on all new cultural facilities imposed by the new Liberal Government, and the painful and costly memory of past mega-projects (EXPO, Olympics). Over six years C$440 million was invested in cultural facilities in Montreal, although this was not part of either a cultural plan or real assessment of 'need' (other than the aspirations of their proponents and resident companies). Further projects include a major extension to the *Cinemathèque québecoise* (C$50 million) and resurrection of a 900-seat concert hall and waterfront redevelopment modelled on Barceloneta. Each project has a particular story that determined its location and rationale. However the availability of public funding was based on the pursuit of city-imaging and employment growth in the arts and cultural industries that studies at the time highlighted (see Chapter 6). Table 8.4 lists the prime projects receiving support during this period and distinguishes those funded at city, state/provincial and federal levels (purely Québecois projects not qualifying for federal cultural funding).

The political issue for this linguistically and therefore culturally 'divided city' is the territorial aspect which a particular location infers. With an east–west city divide between the Anglo and Francophone communities, and allophone groups occupying 'neutral' enclaves in between these contested areas, the locational decision brought with it this

Table 8.4 Major investments in cultural facilities in Montreal, 1988–93 (C$ millions)

Facility	Government of Quebec	Government of Canada	City of Montreal	Private sector	Total cost
Museum of Fine Arts	33	33	0.9	27	94
Canadian Center of Architecture	4	4		57	65
Biodome	57		6		63
Museum of Contemporary Arts	37.2				37.2
McCord Museum of Canadian History	3	3		24.5	30.5
'Place des Arts' Arts Centre	25.7				25.7
Pointe-à-Calliere Museum of Archaeology	9	12	6.6		27.6
Museum of Humor	5.5	5.5	2.5	7.5	21.1
National Monument Hall	8.8	8.1	0.1	0.7	17.7
Biosphere Environment Center		17			17
Pierre-Mercure Music Hall	3.3	3.2	3.2	4.1	10.6
Insectarium	1.1	0.6	0.2	0.7	5.6
Theatre d'Aujourd'hui	2.2	3.4	0.2	0.1	5.9
Hotel-Dieu Museum		2.2	0.1	2.3	4.6
Dance Agora	2.9	0.5		0.6	3.9
Rideau-vert Theater	1.5	2.2		0.6	4.2
Bonsecours Market Exhibition Center			3.7		3.7
Saint-Gabriel House	0.2	1.4		0.8	2.5
La Licorne Theater	0.3	0.2		0.5	1
Writer's House	0.3	0.2	0.1	0.5	1.1
History Center of Montreal	0.3	–	0.3		0.6
Total	195.3	96.5	23.7	126.7	442.1

Source: Latouche (1994)

particular political dimension: 'a most delicate task in a city where ethno-cultural and socio-economic fault lines have the bad habit of often criss-crossing one another and changing over time' (Laperrière and Latouche 1996: 13) and therefore: 'Most of the cultural facilities their location and their program can best be explained through a vision of the Montreal urban scene as a divided and conflictual landscape. Few had anything to do with any urban regeneration strategy' (ibid.). One response in the form of a more pluralist approach to diversity (as opposed to multiculturalism/marginalisation and assimilation through acculturation) has been the provision of arts venues and projects that celebrate the fusion and interaction between various cultural forms and traditions, such as the Theatre of the New World, located between the linguistic zones of the city (Plate 8.6).

As these authors also note, confirming the value of the smaller and middle-scale facility:

> By world standard many of these new cultural facilities are relatively small [compared with the French *Grands Travaux du President*]. . . . The idea seems to have been to create relatively small facilities with as large an audience as possible while the traditional large institutions have had to content themselves with playing the role of local host to events coming from the 'larger' world.
>
> (ibid.: 19)

The last comment is pertinent to the flagship-building strategies adopted in most towns and cities, because whilst some performing companies (e.g. Sadler's Wells Royal Ballet relocating from London to Birmingham in 1990 as the Birmingham Royal Ballet), artistic directors and curators are wooed to other places, the creative content and resident artistic capital is increasingly absent – receiving houses where before there were producing houses (e.g. theatre); touring exhibitions and collections versus in-house or

Plate 8.6 Theatre of the New World, Montreal (1999)

new work; touring circuits of dance, drama and bands, whilst 'home-grown' is not considered or able to access the larger venues. This was recognised in the Arts Council national review leading up to a *Creative Future* (1993a): 'The arts and media in Britain are in crisis. Scarcely a day goes by without press stories of theatres closing, grants being cut or audiences declining; of a lack of good innovative work in all art forms, of the absence of a sense of direction, purpose and adventure' (Arts Council 1991: 1). Showcasing of new work (e.g. crafts, artists, young companies) does of course take place, but this is a very minor and often token use of cultural spaces and resources, such as ticket discounts, childrens' and schools concerts and performances, with arts in education and community arts work relegated as the poor cousin in professional involvement and facility programming (see Chapter 5). A decline in *cultural capital* cannot therefore be offset by a burst of spending on *physical capital* alone. For instance, the international eminence of English theatre, long-rooted, is not guaranteed despite the recent investment in restoration and new performing venues: 'theatre now faces a greater crisis of confidence than at any time since the war. The regional repertory system, where actors, directors and designers once learned their trades, and the source of much of the product which later filled touring theatres and the West End, and enriched film and TV, has been in slow decline' (Longman 1999: 7).

Conclusion

The self-conscious city of culture has long and varied antecedents although today none can claim the centrality that arts had in either society or in the urban morphology of pre-industrial and original renaissance cities. The dual notions of the cultural economy, as Scott put it, between 'the commercialization of historical heritage, or large-scale public investment in artifacts of collective cultural consumption in the interests of urban renovation' (2000: 5) tend to both be present in these post-industrial versions. Examples are seen in the conversion of 'heritage' industrial buildings to art houses (e.g. Tate Modern, London; Baltic Flour Mills, Gateshead/Newcastle (Plate 8.7); *museumsquartier*, Vienna) or the siting of new facilities as part of historic quarters (e.g. contemporary arts museums in Barcelona and Santiago, Spain (Plate 8.8); Musée de la civilisation, Quebec). Their relationship to one another, rationales for their location, resourcing and prioritisation present a key tension in terms of cultural planning, facility design and the identities that cities and other urban locations may choose to project. Whilst on one hand the clustering and 'privileged locus of culture' is well established in the economic geography of culture: 'this process has deeply erosive or at least transformative effects on many local cultures' (ibid.: 4). The effects of urban regeneration that draw on one or both of these re-imaging and economic development strategies persist however in ignoring the spatial, social and uneven distributive impacts (e.g. employment, economic leakage, visitor flows), even when this is repeatedly claimed as their prime benefit and goal; and the narrowing in the range of cultural activities and experience on offer, i.e. homogenisation. The mega-event whether overtly cultural or more broadly based (e.g. EXPOs) represents the extreme case in practice, ironically despite their attempts at distinctiveness in theme and design. Coupled with nationalistic overtones their claims at influencing urban regeneration have been overstated, not just in the failures that have bequeathed their host cities

Plate 8.7 Baltic Flour Mills, Art Gallery conversion, Gateshead, Tyne and Wear (2000)

Plate 8.8 Galician Contemporary Art Gallery and People's Museum, Santiago de Compostela (2000)

financial burdens and redundant, difficult sites – from Lisbon to Montreal – but even those associated with successful revitalisation such as Barcelona which was already undergoing major regeneration and redevelopment before the Olympic bid had been won: 'if you want fruitful regeneration a party is perhaps not the place to start' (*Building Design* 2000: 13).

Evidence in terms of the mismatch between areas receiving and benefiting from urban revitalisation through culture and related forms of cultural consumption, and those most in need socially and economically – the many (majority of) towns and cities that are not able to sustain the critical mass, passing trade and compete with the 'monopoly powers of place' (ibid.: 5) – all points to both a lack of planning, of cultural planning even where culture is the prime element, and a crisis in local governance. This therefore reflects the regimes and power-play existing in the development and competitive-city process, and in particular the role of intermediaries that mediate and broker the global with the local. As Sassen points out in *Global City* (1991), impacts are rooted in the local because this is where the power relationships and integrations of globalisation are seen and felt: 'In this view local communities are seen as the essential receivers and transmitters of the forces of globalisation' (Richards and Hall 2000: 3). This is at best a passive role, however, since they are not the owners of either plans or the producers of much of the culture that finds itself the centre of attention.

It would be short-sighted however to write-off the cultural city and creative city advocacy, not least since its multiplication makes it a universal phenomenon in the evolution and contemporary analysis of cities, nor to over-generalise and conflate the impact of those cities and experiences that provide better models or possible guidance for the less successful and a response to the more deleterious effects to which writers such as Robbins, Harvey and Bianchini have drawn attention. Much of this criticism is located fundamentally in the power (political, economic/ownership) over place and aspects of freedom over cultural expression. This suggests that post-industrial urban regeneration has parallels with conflicts over land-use and the tensions between local communities, economies and governance and between cultural diversity and heritage, which are a feature of many developing and second-world countries (e.g. Indonesia, Mexico; Style 2000) and their collision with national–global pressures for efficient production and the imperatives of so-called free trade. In particular given the focus of this book, the extent to which forms of planning, how far the experience of culturally informed planning and the importance afforded both the soft and hard arts and cultural infrastructure have been elements in cultural and city development remains a key question. To an increasing degree, a comparative framework is provided by the growing number and longevity of cities undergoing the regeneration and repositioning of their economic base, their changing landscape and the shifting lifestyles and aspirations of their residents, both transient and incumbent. In some respects this demands a more holistic and interdisciplinary approach to the amorphous notion of cosmopolitanism, global cultural effects and the measurement of continuity and change in post-industrial society. This, I would argue, is assisted in the juxtaposition of culture and planning which meets some of the limitations of traditional economic and cultural geography, of cultural and policy studies and the various studies of the 'urban' – closer to what Soja seeks, in theory if not in practice, through a more productive synergy between critical cultural studies and geopolitical economy (2000: xiii). Despite their concern for human geography and the study of human culture (Tuan 1976), however, the proponents of both post-modernism and the post-metropolis seriously lack either a real feeling for the power and effects of creativity, cultural development or the arts' metaphysical and symbolic, as opposed to their observable social effects and physical form.

Notes

1 Merida, known as the 'white city', was the former Mayan city of Itza before the Spanish Conquest. Under the edict of Carlos V of Spain in 1542, all colonial cities had the strict grid plan imposed, organised around a central plaza that housed the main buildings of control/power, with zoning for the colonialists and the subjugated. The centre was destined for the Europeans/Creoles, whilst to the west of the city two suburbs were to be occupied by Mayans, and one to the east was for the Atzcapotzqalco Indians who were brought into the city by the Spanish invaders. Later a northern suburb was created to house the 'negroes and half breeds'. These suburbs – small townships – had their own native authority and representative town council under an Indian chief appointed by the regional governor. In time the centre gave over to the encroachment of these 'suburbs', which grew outwards taking their indigenous residents further from the centre and which as a result lost its geometric street layout and spacious form (exhibit, Cuidad Museum Merida, Yucatan, Mexico, 1999).

2 Categories: A, more than 200 hectares; more than 130 participating countries; each country given a plot of land on which to build their own pavilion; B, more than 80 hectares; more than one hundred participating countries; the host nation provides exhibition space; and C, International Garden Festival.

9 Planning for the arts
An urban renaissance?

The versions of cultural amenity and cultural economic planning presented here from both the historical and contemporary urban perspective have been viewed and analysed at the local and micro-level to the city–region and national assessment of the distribution and valuation of the arts and cultural industries. Whether primarily driven by employment and economic policy, wider social or specific urban and cultural policies, the planning of the arts in their benign and boosterist states can be argued to have a particular place in city formation, development and renewal. The sacred and the celebrated are both reflected in the material culture, 'performance' and heritage that cities possess through their arts and culture, as Tuan maintains: 'Past events make no impact on the present unless they are memorized in history books, monuments, pageants, and solemn and jovial festivities . . . on which successive citizens can draw to sustain and re-create their image of place' (1977: 174).

As Chapter 1 explained, cultural planning and the provision of arts and related amenities in differing spatial and experiential dimensions might be expected to be virtually bypassed and made obsolete by the globalised forms of cultural dissemination and consumption, and by the increasingly privatised realm in which society recreates and participates in artistic and related leisure pursuits. The models and patterns of civic and national cultural development, in pursuit of glory and universal recognition, are apparent however in 'old' industrial as much as 'new' and emerging cities and regions. This is notable through the grand architectural statements and the necessity for institutional culture-houses – opera, theatre and concert halls, arts centres, galleries and museums – to compete and compare with those long-established in the so-called developed nations (e.g. Beijing's new opera house), whether or not they have indigenous relevance or domestic appeal. As Tuan again observes, the civic leaders of the new cities and modern city-states were 'required to speak with a loud voice. Strident boosterism was the technique to create an impressive image . . . with munificence such as large-scale public works and the subsidization of art [since] the boosters could rarely vaunt their city's past or culture' and therefore that 'symbolic means had be used to make the large-nation state seem a concrete place' (ibid.: 174–6). Whilst the institutional and national cultural centres and events retain a residual value and importance, despite in many cases their declining popularity and narrow class base, it is the everyday lived cultural practices and experiences which signify, to borrow Williams's phrase, 'common culture'. As Willis argues:

the new temples of High Art . . . may enjoy some corporate popularity, but as a public spectacle not private passion, as places to be seen rather than to be in. The prestige flagships are in reality no more than aesthetic ironclads heaving against the growing swell of Common Culture. Let's follow the swell.

(1991: 13)

Willis also suggests less reactively that some of these mainstream cultural institutions should also be focal points (cf. 'arts centres' in Chapter 4) and should facilitate partnerships and collaborations with local arts and cultural activities and networks. For example, the development of local libraries and museums through more animated and accessible forms of interpretation; arts in the community and education; and the use of interactive technology (e.g. *Digital Dancing*, Dance Umbrella, UK) could be seen to offer a bridge between the sterile high and popular culture dialectic and he suggests a more cultural democratic approach, again echoing Williams (1981, 1983):

The recent successes of certain museums and art galleries in appealing to a wide range of people and communicating with new audiences, and the continuing success of many libraries in providing an ever wider range of symbolic materials, rest not upon extending an old idea to new people, but on allowing new people and their informal meanings and communications to colonise . . . the institutions.

(Willis 1991: 12)

Whether the expansion of cultural facilities – new, relocated and renovated – witnessed in the late twentieth century has signalled an attempt by weakening nation-states and regions to regain their symbolic and economic power, or whether city-states have looked to the past civic fathers, merchants and industrialists to regain or reposition their image and local economy, the map of public cultural provision has not been so worked-over since the Victorian and industrial eras, the earlier urban renaissance or indeed the classical periods from which civic art and culture largely draws its inspiration and form. This counter-intuitive development exhibits aspects of all of these, not least the prosaic embrace of the cultural economy in new and old modes of production, from multimedia and information communications technology to live performance, design and handicrafts/designer-making.

Cultural tourism and the widening motivation for city and historic area visits – leisure, educational, family ('visiting friends and relatives'), business and convention – also demands a range of services, facilities and experiences which include the live and visual arts, museums and animation through festivals, cultural itineraries/routes, events and pleasure zones. The expansion – ethnic and geographic – of migrancy and settlement, the returning diasporas both temporary and permanent, have also created a movement of peoples who bring to and draw on cosmopolitan cultures. The consumer of public culture therefore includes the resident, worker, visitor and investor, with the politician and mayor as much as the artist/promoter and marketeer seeing a role in advocating and providing for the cultural city. The continuing growth in tourism and in trade exhibitions, conferences, conventions and centres also contradicts the futurist scenario confidently predicted for their substitutes in virtual travel, video-conferencing, ICT and e-commerce. This human geographic phenomena has directly and indirectly fed the new and repackaged cultural venues and activities; however the extent to which they can be sustained artistically and financially has yet to be seen, and likewise whether the growing circuit of competitive *cultural capitals* can survive and flourish.

Notwithstanding the transient mega-event and *Grands Projets*, and outside the central cultural zones and islands of culture and entertainment – towns, districts and 'urban villages' have seen, albeit unevenly, a resurgence in civic pride in their local and municipal culture. The revamping of town centres and recognition of the new cultural and crafts economy, can optimistically be seen to represent significant elements of the local economy and landscape. Town centre revitalisation strategies, for example, have attempted to deal neatly with the demands of modern urban living based on cultural consumption and participation by seeking to overcome barriers and develop more sustainable urban policies linked to transport, employment, housing, leisure and their spatial relations. This has widened the role of local authorities and local business associations in town centre management, animation and urban design schemes and at a regional level this has been reinforced by linking the city plan to the concept of *strategic town centres*. The attention given to town centres and larger 'urban villages' rather than the more glamorous and powerful city centre and downtown zones, and whether established, in decline or emergent (Haringey 1991, Davies *et al.* 1992), is both a response to the out-of-town and urban fringe drift in leisure, retail and housing development (predicated almost wholly on car usage) and to the advantages of a central location. These include the benefits of local critical mass, cross-trading and public transport links, which such centres can offer, including of course the location and development of cultural facilities alongside education – schools, colleges and specialist facilities, e.g. design, ICT and producer services. This network of town centres is key to strategic planning in traditional metropolitan regions defending themselves from the edge city and out-of-town shift. These are seen as having potential to 'provid[e] a sense of place and focus for communities . . . increase accessibility to a range of services, extend economic activity beyond daylight hours, and sustain and enhance . . . community and cultural features' (LPAC 1993: 21–2). A sign that the market recognises their viability again is seen in the opening of multiplexes, leisure–retail centres and refurbished civic arts facilities in towns, rather than sub/urban fringe and green-field leisure 'parks', and in the reversal of the population decline in inner urban areas. This is a vision presented in the series of studies which led up to the UK Urban Task Force's *An Urban Renaissance* (DETR 1999) and manifested in city economic and population growth, from Manchester to Denver (Gratz and Mintz 1998), with the opportunity for the development of brown-field sites and industrial buildings and a return to the revitalised and supposedly enriched inner city (Worpole and Greenhalgh 1999).

Those looking for the quick fix – and this is a basic limitation of political terms of office, development finance and some community expectations – is that the regeneration process that inevitably starts from an entrenched, fragmented interest and land and building usage base, is a long-term process. The mega-development schemes, whether event, flagship or mixed-use projects, pushed through normal consultation procedures or with promises of benefits which history suggests are unlikely to be fulfilled (e.g. Garden Festivals, EXPOs) offer a faster route and concrete evidence on behalf of their creators and boosterist promoters. Given the complexity and wide-ranging concerns that cultural planning and more consultative planning approaches demand, time is probably the most valuable resource but one which is given least credence in regeneration and development planning horizons. These are often driven by time-limited public funding and performance indicators and the bluff and threat from footloose commercial financiers and developers, including mobile cultural entrepreneurs. The realities of resource constraints, political and executive capabilities, and incrementalism that dictate land-use and economic change is seen, for example, in the North London Borough of Haringey (population about 200,000). The borough contains a post-War town centre (the optimistically named Wood Green) 'distinguished' (*sic*) by a failing post-War 'Shopping City', divided by a major road, and a regional bus garage. Whilst Wood Green was 'short of the kind of "Fortress City" described by Mike Davis (1990) in Los Angeles, questions of access, surveillance and control [were] all too present' (Jackson 1998: 188). The local council convened a multidisciplinary Urban Design Action Team (London Borough of Haringey 1991) that sought to generate solutions for its declining base, poor urban design and linkage to an adjoining Alexandra Palace and Park (*People's Palace*, see Chapter 3), which was also in financial crisis. Few of the design solutions forthcoming from this exercise were enacted, however the borough also undertook an economic strategy study for the arts and cultural industries in 1993 that looked to the development of cultural quarters in the town (Evans 1993d). Nearly ten years on from this visioning process, the town has invested in small-scale urban design improvements, opened a twelve-screen multiplex showing a mix of blockbuster, Bollywood and art house films, and now hosts a major cultural industry facility, The *Chocolate Factory*, including the local arts council and media training, a university art and design campus, and small cultural enterprise production. This example is unexceptional, largely incremental, but provides evidence of both the time-scale and steady development required for cultural planning to work through in a complex and dynamic urban location and which, it must be accepted, is now the norm.

The early examples of town centre malls typified in North America had of course been originally associated with the pedestrianisation experiments first seen in Kalamazoo, Michigan in 1959 (Brambilla *et al.* 1977) and cities' pursuit of traffic-free areas for their central business districts, with similar solutions seen in Latin America and in Europe such as the hypermarket and retail 'park' (*sic*). However as Goss maintains (and this has also been claimed for the nineteenth-century US suburb; Sennett 1970), shopping malls began to 'reclaim for the middle class imagination, "The Street" – an idealized social space free, by virtue of private property, planning, and strict control, from the inconvenience of the weather and the danger and pollution of the automobile, but most important from the terror of crime associated with today's urban environment' (1992: 24). The same view can be directed at the cultural flagship and complex that forms the centre of urban regeneration schemes, particularly for night-time usage. However arts and town centre complexes that retain significant open, public realm areas, including foyer events and performances (Plates 9.1 and 9.2), can and do overcome the essentially privatised and ultimately retail-driven controlled environment which also reverts to a barren, unsafe (e.g. shuttered shops/doorways) and wind-swept site when closed. The commodification of these spaces is however a tempting strategy once public usage is established, in part to maintain their revenues and lessen reliance on public subsidy evident, for example, in the redevelopment proposed for the South Bank Arts Centre in London and the museum retailing operations and

Plate 9.1 Free performance at the Royal National Theatre, South Bank, London (2000)

Plate 9.2 Free performance at the town square, Guadalajara (1998)

their proximity in New York (MOMA) and Paris (Louvre, Les Halles). The privatisation and commodification of public cultural spaces also represents a failure of planning itself, since the effect on independent retail, workspace and specialist services in adjoining areas is negative, in a similar manner to touristic and heritage zones (which is in fact what they have become), and the market-orientation of these institutions incurs serious opportunity and transaction costs to their core mission and organisational culture (Evans 1995c, 1999g, 2000b).

Cultural planning and public goods

The now-recognised shortcomings in the planning and land-use system point not only to dilemmas in the political economy – for example, defending the rationale for public/merit goods, the mixed economy, mixed-use of buildings, the public realm and space, and recognising pluralism and diversity rather than assuming convergence/assimilation – but also to the need for the widening and (cultural) democratisation of the planning function and plan formulation itself. As the Brick Lane Community Development Trust commented from East London:

> Participation must continue beyond the planning stage of developments to their implementation and management. . . . We did not want simple results like planning permission granted or denied, or one-off planning gains, like money or a community centre . . . what we wanted was to join in designing strategy for the economic development of the area over the next 20 years.
>
> (ten Kate 1994: 16)

Where a simple market mechanism is applied even benignly (the 'invisible hand'), that is, where no public/merit good argument holds, there can be a rapid and irrevocable unravelling of the cultural production chain. This begins with the removal of the stepping stones in the hierarchy of cultural provision and building and site uses, and the break-up of artistic communities and workplaces that have evolved through such inter-reliance of skills, expertise and facilities evident, for instance, in crafts and designer-making, and which leads to a consequent loss of synergy and critical mass. Once this has happened, in the case of cities undergoing 'development gain', it is questionable whether the reversal of such trends or their re-creation as some form of planned formula can be achieved, at least not without considerable reinvestment which alone cannot guarantee the conditions under which a cultural milieu and particular landscape of cultural activity can again thrive. The reassertion and in some sense redefining of *public goods* (or 'non-private' benefits; O'Hagan 1998) is also required in part to defend the concept against the arguable failure of recreational services and facilities to meet wider equity and participatory tests and in part due to the absence of community cultural planning, as well as the false consciousness offered by consumerist culture (or what some economists would still term 'market failure'). This is not confined to specific activities, services or facilities, however, since as is also claimed: 'The city itself indeed, may be regarded as a public good of a most enduring kind . . . [public goods are] the staples of a basic quality of life' (Worpole and Greenhalgh 1999: 16). As Zukin warns, however, the democratisation of urban planning and greater concern for the visual and physical environment has not tended to be inclusive or consensual (with parallels to conservation and environmental movements) and, as I have presented here, it is seldom adequately integrated with wider planning and urban design formulation: 'The notion of art as a public good also raises problems about a city's ability to maintain its identity as a culture capital despite demands to share the benefits cultural strategies bring' (1995: 155).

Local cultural strategies: the arts and social exclusion

A more recent attempt to focus again on the local area in terms of cultural planning also suggests a return to a concern for amenity provision within a local environmental context. Whether this signals a reassertion of the value of social (amenity) arts *per se* is doubtful. However the linkage made in Britain and Canada between the *Arts and Neighbourhood Renewal* (Shaw 1999; also SAC 1992, Landry 1996, Jeanotte 1999) at least draws on the more integrated approaches to planning at a local level, and the contribution that cultural development may have to particular urban problems of social exclusion, poverty and economic decline. The social exclusion agenda being taken up by the New Labour Government in Britain, following its French and wider European and US foundations, is also mirrored in Australia (Guppy 1997) and Canada (Jeanotte 1999), again an indication of convergence in social and cultural policy and a counterbalance to the greater emphasis afforded the economic value of the arts and cultural industries. Several studies of the social impact of the arts leading up to this policy concern and linkage (SAC 1995a, b, Comedia 1996a) therefore respond in part to the economic impact studies and rationales which proliferated in the 1980s (see Chapter 6).

So whilst the creative industries' policies that many developed and developing countries have embraced have gained further momentum (DCMS 1998, Landry 1998, World Bank 1998), capitalising on the so-called social market and the egalitarian potential of *Information Society Technology* and the cultural 'knowledge' economy, a concern for social exclusion and access to arts activities and institutions has also required the reconsideration of local arts provision within a neighbourhood and wider socio-economic and environmental policy agenda (Shaw 1999). This focus is not entirely divorced from the cultural economy since this is seen to hold out opportunities for areas of poverty, unemployment and related decline (e.g. in health, housing and environment) through small-scale cultural production, education and training, and community enterprise, as well as opportunities for celebrating cultural diversity ('Rich Mix'; Evans and Foord 1999, Foord 1999, Worpole and Greenhalgh 1999) – such as the support of ethnic arts and festivals. It should be said however that tensions between notions and realities of national(ism) and integration/assimilation, multiculturalism and pluralism have not, in European or other cosmopolitan cities, been seriously confronted, let alone resolved in either cultural or social policy spheres (Evans *et al.* 1999). The 'other' still exists largely outside of the built environment, public amenities, and legitimate (subsidised) arts and cultural facilities. As Christopherson makes the distinction between 'genuine ethnic culture' and 'that which is manufactured for sale' (1994: 414), the exclusive separation of the private and public also conflates a wide range of market practices and opportunities and understates the powerful hegemony and intermediary brokerage that dominate public cultural and regeneration strategies, as I have discussed in respect of arts and urban regeneration and cities of culture. Jackson therefore suggests that 'notions of consumer citizenship need to be carefully situated in and socially differentiated. . . . Rather than assuming that commodification and privatisation are inherently undemocratic and reactionary social processes' (1998: 188). The cultural democratisation phenomenon offered by popular media, electronics, communications technology and subcultural formations therefore applies to many of those groups which have traditionally existed outside of the mainstream and cultural resource base.

In Britain the government's broad social exclusion policy agenda, to which all ministries were required to respond (Social Exclusion Unit 2000), has looked in particular at neighbourhood- and housing estate-based urban (and rural) programmes in targeted areas with high socio-economic deprivation levels, and in local amenity terms this has been promoted through the framework for *Local Cultural Strategies* (DCMS 1999). These, significantly, seek to integrate cultural activity with other aspects of the local environment, including economic development, transport, health, environmental quality (LA21), education and, importantly, statutory land-use plans. The Cultural department's *Guidance* (DCMS 1999) indicates the range of planning and consultative processes that a local cultural strategy might encompass (Figure 9.1)

In terms of higher levels of coordination and consolidation, e.g. scale hierarchies of provision, city and regional planning, these local area plans would form part of a *Regional Cultural Consortium* (i.e. city-wide) through which local/regional cultural planning issues would be resolved, such as duplication and gaps in provision, strategic and local facility provision, and related infrastructure needs, both cultural (e.g. art form, diversity, resources) and environmental (e.g. transport). A Cultural Partnership is thereby established 'to promote cultural issues, and to develop a [city]-wide approach to cultural matters' (CSP 1999: 14). As an indication of the range of interests represented in this new model, the *Cultural Strategy Partnership* includes bodies responsible for Heritage, Libraries and Archives, Film, Sport, Tourism, Lottery, Parks, Museums, the regions' voluntary sector, inward investment and city promotion, and government regeneration programmes (LAB 1999: 3). To quote one city authority's cultural planning brief: 'for the first time a statutory power/duty for local authorities to . . . promote the social, economic and environmental well-being of their communities' (Sunderland County Council 2000: 1.2.1). The way in which this power develops could be of great importance in the area of cultural services and an integrated strategy should enable local authorities to approach the issue with confidence. When coupled with the integrated approach offered by the local borough plan (Unitary Development Plan; see Appendix I), this approach moves closest perhaps to the notion of cultural planning outlined earlier, particularly if the Community Plan is designated a duty within which the resident community is genuinely able to participate and influence. Without a strengthening of local governance and more attention being paid to the distribution of power – encapsulated in the conflictual process of 'governmentality' (Foucault 1991, Barnett 1999) – and without greater control of public choice and resource distribution, local cultural plans will however never move from the 'plan' state or will prove ineffective against the established hegemony and paternalism which has typified planning for the arts in the past. As Bennett argues, culture itself should be thought of as 'inherently governmental' so that 'culture is used to refer to a set of practices for social management deployed to constitute autonomous populations as self-governing' (1995: 884; also Barnett 1999: 371).

The 'urban renaissance' may also be seen as the reinterpretation and rediscovery of the local rather than through the negative notions of 'structural pessimism' (Byrne 1997) or as a compensatory response to *degeneration*. The arts and cultural industries, contrary to the views of King (1990, 1991), can distinguish themselves by restoring identities (Hough 1990) as well as local economies, in an eclectic urban society conscious of not only just the traditional, but also of other cultures (and lifestyles), whether

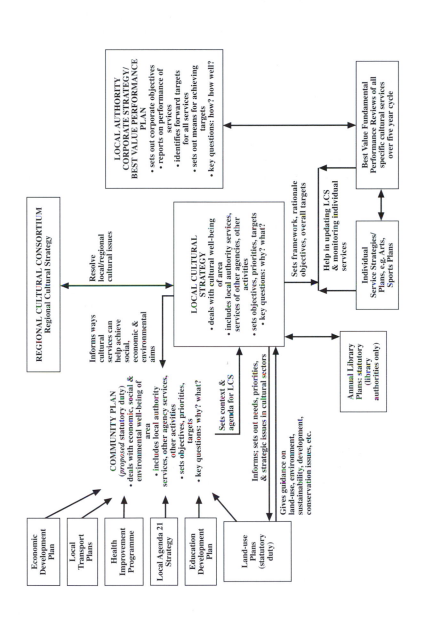

Figure 9.1 Linkages between strategies and plans

Source: DCMS (1999 'Exhibit Five')

also local or experienced globally through the mass media, and by exchange/fusion in all its forms. The multicultural and pluralist state of contemporary *world* cities suggests that neither a reductive public culture (Horne 1986) nor reactive and serially replicating civic arts planning are equitable or adequate for their purpose. Cultural planning based on identified local needs and profiles would need to reflect the cultural make-up of local, borough and even subregional areas and thus extend to contemporary and 'common' culture as well as to traditional art and heritage.

Culture or heritage?

A reconciliation of heritage interests – the built environment, artefacts, performing and visual arts 'classics' – and living and working culture interconnected through a common history and cultural development, would therefore appear to be essential for a more sustainable cultural policy paradigm. Whilst conservation areas, listing of buildings and heritage status protect façades and 'sites', no such protection is afforded artists (with few exceptions), cultural production or the creative end of the cultural economy and mix of building uses. The World Heritage Convention (UNESCO 1972) itself established a formal obligation for states to adopt a general policy to give the cultural and natural heritage a function in the life of the community (Article 5a) – it is up to each Member State however to define these 'properties of outstanding universal value' (Article 1) and put forward proposals for World Heritage Site designations. There are few cases of such community planning (and management) in practice, or examples of cultural resources within which museums and heritage sites play an important part in the regeneration of communities (Newman and McLean 1998: 149).

This is not a specialist heritage debate however as the over-concentration of visitor and therefore economic activity in a very small number of European and other heritage sites indicates (see Chapter 7). The heritage movement, dominated by international agencies and conservation experts, is now challenged with the twin demands of the modern movement (twentieth-century architecture) being considered as part of the 'universal patrimony' from Miami (Art Deco) to a Manchester high-rise housing block, and the pressure from developing countries seeking to attain World Heritage Site status for the first time, from Spanish Town (Kingston, Jamaica) to Stone Town (Zanzibar, Tanzania). In many respects this is skewing the cultural planning of towns and cities in developing and re-emerging states where heritage tourism in historic zones coupled with development of Western-style museum, gallery and culture-houses exhausts scarce culture budgets and investment. This is also transpiring in established urban areas undergoing heritage and museumification, such as Maritime Greenwich in London and Old Quebec (declared World Heritage Sites in 1985 and 1997 respectively), where the everyday is overshadowed by the imperative of marketing and image-creation and the sterility that heritage zones demand (Evans and Smith 2000). In Greenwich, for instance, this situation has little or no resonance with what is a multicultural and in parts deprived residential population, or to the particular cultural activities that the population actually participates in and to which it aspires (London Borough of Greenwich 1998). The heritage management plan in this case has no wider reference to either contemporary urban culture, the resident community or its coexistence (English Heritage 1997). The definition and delineation of *who* the community is also often fails to reflect the displaced or those to whom heritage attaches but who may no longer be resident in the locality itself – heritage 'development' (*sic*) is also regrettably responsible for the enforced displacement of communities or crowding-out through not only high land/property and tax costs (Evans 1994, 1999c, Rojas 1998, 1999), but also insensitive planning legislation.

The inheritance of cultural and other amenities in our towns and cities also reflects the experience of both political consensus and control evolving for instance since the Victorian and equivalent rational recreation eras. These dual ideologies reflect the attempts from the eighteenth and nineteenth centuries onwards to reconcile class conflicts through civic, urban arts and recreation provision (Yeo and Yeo 1981, Harris 1994). Marx importantly distinguished between 'heritage', which encompasses all historic and style periods, all social formations without exception, from 'tradition', which is only a component of the former – the 'wealth of ideas consolidated in the public mind, which requires a choice, acceptance and interpretation of the heritage from the point of view of certain classes, social layers and groups' (Andra 1987: 156). Tensions caused through selection, choice, re-evaluation and cultural change would be a natural but difficult aspect of this process. However Stark more constructively offers a vision of a future balance between the arts of the past, the present and the future:

> Our argument . . . is that the culture to which we should aspire in the [twenty-first] century is one that balances more carefully the availability of the best of the world's artistic heritage . . . with investment in the production of the arts by professionals and in facilities and opportunities for the population of all ages, cultures and traditions to enjoy, explore and celebrate the arts through active participation.
>
> (1994: 3)

'Non-plan'?

As early as 1926, a caption in Fritz Lang's silent film *Metropolis*, which presents the legend of the Tower of Babel in terms of the construction of the modern city (based on Berlin), warned 'Those who toiled knew nothing of the dreams of those who planned.' The notion and fear of *planning* used here says nothing of the infrastructure, access and 'rights' to cultural expression recognised in the post-War reconstruction and welfare state era (Henry 1993: 15–25, Sinfield 1989) and therefore of the resources required for a balanced civic and urban existence. The investment in human capital, the opportunities for the acquisition of skills and cultural experiences, both professional and amateur, and the creation and maintenance of places where these experiences and interactions can take place, whether formal, designated or private and informal, are arguably essential components of both cultural democracy and sensible resource planning. This is the case whether the prevailing rationale is amenity- or production-based. The libertarian presupposition seems to be that *planning art into being* is an unavoidable outcome of arts and cultural planning or a plan *for* cultural activity. As is known with hindsight, the association of socialist planning with propaganda and censorship, and the over-prescriptive Fabian town planning that has been blamed for the high-rise and brutalist public (and private) schemes which litter cities and suburban fringes, keeps fresh the resistance to planning in any sense. In Britain, the *Non-Plan* was launched in the late 1960s by planners and architects (Peter Hall, *World Cities*; Reyner Banham *Theory and Design of the Machine Age*; modernist architect Cedric Price, 'Fun Palace'; and writer Peter Barker, *Arts in Society: The Other Britain*). Thirty years on this group sees the need to relaunch the *Non-Plan* manifesto (1999), believing that growth that happens without too much prescription is best (and vice versa) and that 'innovations are what determine the growth of cities, not planners' decision' (Barker 1999: 30). Quoting the timely *The Richness of Cities* (Worpole and Greenhalgh 1999): 'the only thing we can be certain of is the Protean character of cities, their resistance to top-down planning or prediction' (ibid.), but what is being rejected here is the reactive, paternalistic application of physical masterplanning, not that community planning and therein cultural development is redundant. The naïve assumption that, because a high-rise block is either full or empty/*hard to let*, or an out-of-town shopping mall and multiplex 'popular', this is the result of *public choice* (*sic*) between viable alternatives, or truly representative of the *public* at all, ignores the pattern of market failure. The notion that the so-called free market ensures choice and competition goes against the reality that increasing/defending market share, killing off or acquiring the competition, is the norm – not least in the leisure industries – as are the imperatives of commodification, scale economies and homogeneous/pastiche design (e.g. retail, leisure, branding), which are all the result of another form of 'planning', the corporate. A trans-national company, Heron International, opened the first of its chain of European retail–leisure developments in Madrid in late 1999 that claims to be 'the first development which is a "global brand", and reflects the unique characteristics of the locality through its design image . . . enforcing the global trend we are witnessing [as if this is a natural phenomenon] of retail and leisure developments which represent local community destinations' (Leisure Opportunities 2000). Such doublespeak would do the planners of *Metropolis* and *1984*, proud – like it or not, further Heron City's are planned for Barcelona, Lille and Stockholm, to be followed by Milan, Brussels and Lisbon.

Land-use planning and the recognition of infrastructure and production chain arguments are unlikely in themselves to guarantee greater provision, certainly not better 'art', let alone a greater cultural democracy: 'a society where people are free to come together to produce, distribute, and receive the cultures they choose' (Shelton Trust 1986: 111). However, without an equitable level of arts resources and the recognition of the *facilitation* of arts development through an adequate infrastructure such as venues for touring, production and education (school and informal), and an accessible hierarchy of facilities in order to achieve cultural development and enhanced participation, such provision is likely to fall short of national and local access and equity objectives. Planning, including establishing to some degree norms and minimum levels of provision based on existing facilities, activity and consumption patterns and local/regional 'needs', is one suggested approach. An arts and urban planning approach to cultural amenity and production may also engage, more naturally, a cultural policy and avoid the marked city centre/core and suburban/fringe imbalance of arts facilities, both spatially and population-based. The absence of such policies in the urban renaissance strategies of the 1980s and 1990s and the exemplars of regeneration suggests that the problems of sustaining the levels of public intervention required for many schemes and facilities, and extending such investment to underdeveloped areas, have been underestimated. This could also be presented as a failure of urban policy itself and notions of 'trickle down' and multiplier effects and the concept of public and private 'partnership' where 'the ideology is that of leverage, the use of minimal public resources to prime the private sector pump' (Shurmer-Smith and Burtenshaw 1990: 41). This mantra persists irrespective of political complexion, for instance in the promotion of the Public Finance Initiative (PFI) for public facilities, and the continued privatisation of public amenities. Planning has traditionally been silent (and powerless) in the financing of public and development projects. However in the nature of commercial development, privately financed public facilities essentially lose their public good protection – sooner or later returns are required that will burden their operators and inevitably influence access, pricing and programming and move towards commodification opportunities beyond the unobtrusive mix which a public cultural facility can support. As So and Getzels suggest (1988): 'Public planners must be able to take responsibility for understanding what the community wants and for shaping the future of the community's land uses to reflect those desires. Although strong believers in the free market may not be comfortable with this notion, it is at the core of planning' (quoted in LeGates and Stout 1996: 403). In practice, an overly pragmatic even corrupted situation prevails where economic development pressures and imperatives override community and environmental preferences. The corollary to this avoidance of public finance and borrowing in macro-economic terms is in theory lower taxation (or higher spending elsewhere) and therefore the greater freedom and disposable income for the individual (or at least those with a taxable income). Again this represents a transfer from public to private goods and a reversal of cultural equity and the benefits of collective cultural provision, which was the founding basis of much recreational and educational amenity in the nineteenth century and, ironically, their economies of scale (Jevons 1883).

A case for arts planning norms and standards?

A central concern of this book has been to examine critically the arguments and rationales in favour of planning norms in relation to the provision of arts facilities, which can be recognised and used in both town- and arts-planning contexts parallel to those long accepted for recreation and other amenities. In this context the examination of different models of arts planning also includes the related but wider notion of urban cultural planning, as defined in Chapter 1. Different methodologies used for amenity planning have been detailed in Chapters 4 and 5, which draw on spatial, 'demand', and other modelling and forecasting techniques. In particular the *hierarchy* and *production chain* concepts have been developed in relation to both local and strategic arts amenity and facilities, and in cultural production and distribution. Local area plans, whilst recognising and accepting arts facilities as essential elements of local amenity (and this has not always been the case in the past), have in some cases extended their scope to encompass cultural industries, production/services and other economic (e.g. tourism) development considerations.

Related to these are the underlying public/merit good rationales for such provision, although arguably private ownership and operation of arts and entertainment facilities and resources (including the 'closed' community) in no way preclude the need for planning. In this sense planning is 'neutral' since a balanced and sustainable distribution of arts, culture and 'entertainment' for a given area must encompass all provision and land-use, irrespective of the provider or owner. Matters of equity and access will continue to be the prime concern of social and arts policy (at central and local government levels) through subsidy, pricing and other mechanisms since planning alone cannot ensure or provide for this. However, the town-planning and related economic development system has a role through development planning and zoning in protecting and pro- moting public use of spaces and mixed-use of buildings (e.g. cultural quarters, conservation areas) and consequently mediating in land and property valuations – public recreational land, managed work spaces, live-work studios – as well the creative use of environmental planning in cultural provision itself such as in urban design, safety and protection of the public realm.

From an extremely limited base, and lacking in the quantitative norms of local pro- vision accepted for sports and other recreation facilities, the local economic and planning importance of Arts, Culture and Entertainment (ACE) has begun to perme- ate the town planning and recreation planning processes in some cities (see Appendix I). Rationales are largely environmental (built environment, conservation, transport) and economic, as befits the town planning function and concerns. The effective appli- cation of arts policy and cultural planning approaches in this context has therefore been one of pragmatism and, to a degree, damage-limitation, thus reflecting the particular era and political regime within which plans operate, and all too often these are drawn up amidst calls for greater 'vision' and partnership, whilst strategic (including arts) planning and the pursuit of cultural democracy continue to be largely absent (Tomkins 1993). The traditional borough-planning process reinforces this in that it is largely officer led and prescriptive in style. Interagency negotiations over site and area plans reinforce this bureaucratic approach and in most cases this has excluded the views of the arts and cre- ative sector and incumbent communities. The political geography of planning also continues to be ring-fenced, despite the highly artificial pattern of borough boundaries in terms of cultural participation, tourism and labour markets, and capital flows, includ- ing urban development itself, since as Smith writes: 'Most urban areas are legally defined by administrative boundaries, but these only accidentally reflect the range of everyday social intercourse' (1992: 107). This last point is obviously fundamental in cultural plan- ning in relation to regional- and structure-level planning of facilities and the relationships between participants, consumers and places of collective activity, as well as macro- and supra-national planning consideration. Scale hierarchies of cultural provi- sion, trans-national exchange and media as well as migration and *wanderlust* all bypass administrative boundaries, as Read writes, on the isolation of the artist, if not art itself: 'an island is only defined by reference to another land mass' (1964: 18). In terms of the infrastructure needed to support the flow of cultural consumption and experience, and in addressing the barriers to participation and expression still widely felt (despite some advances through, for example, community cable television and radio), micro-level arts planning policies cannot be seen as separate from regional and higher levels of plan- ning and ultimately cultural policy development. The attention paid to higher-level facilities, the temples and cathedrals to the arts which dominate cities of culture and their regional catchments, understates the contribution and distributory effects of the more participatory-based arts centres, media resource centres, and educational and community facilities that potentially serve a wider community. More importantly in planning and programming terms, all scales of facility and arts development (e.g.

outreach, arts in education, touring) should be seen as links in a pyramid of opportunity. This is needed in order to counter Horne's view of '"Art" [as] something that for many citizens is done *for* them and *to* them. The idea of actually going so far as to *make* art could seem impertinent. Indeed a great deal of art is presented in such as way that it is done against the citizens' (1986: 234). As Hewison responds, this should not be restricted to education and distribution, it also means access to the policies (and plans) of arts institutions and to their facilities and not merely seeing the public as consumers, passive constituents of the market (1990: 176).

Where attempts to create a coordinated and integrated policy or ideally a more comprehensive local cultural policy are reasonably successful – where they involve arts and community development – concerns widen to include matters of education and training, cultural diversity and operational issues relevant to cultural facilities and access. These require proper coordination at the planning level based on a knowledge of existing assets and resources (*cultural mapping*) and research into activity, usage, and community preferences and aspirations, even competition and change factors such as demographic, technological and comparative participation and case-studies. As Crouch also points out, 'amateur maps of popular-expert knowledge' are maps of popular culture in themselves, 'they represent cultural practices . . . representations of what locality means made partly in terms of leisure practices' (1998: 163). This *popular* geographic knowledge and comprehensive audit assessment can then be fed into the plan formulation, land-use designation process and democratic involvement in resource allocation and evaluation. For some time Healey *et al.* (1988, 1997) and others have advocated a more active involvement in 'collaborative' planning in the wider sense, and this is felt to be a necessity if a local response to globalisation is to attract both a consensus and sense of ownership of development and amenity – a recognition of what George Nicholson (a former GLC councillor) called *The Campaign for Messy Government* (1990):

> the old planning idea that populations were merely demographic statistics with easily identifiable needs for units of accommodation, transport links, and social welfare needs, is giving way to the realisation that communities are made up individuals, sub-cultures, interest groups and coalitions . . . with different life-strategies.
> (Worpole and Greenhalgh 1999: 38)

Further evaluative research is also required to monitor and assess the implementation and robustness of arts and related policies over the life of the 'plan'. This inevitably requires a more longitudinal frame of reference, but one that cannot seriously rely on either linear and standard social and economic statistics or land-use change – both are superficial and weakened by the complexities of cosmopolitan society and human agency: things are simply not what they seem or easily categorised. Bianchini also notes the 'dearth of comparative knowledge and research on the richness of policy-making experiences and traditions at city level' (1993: 207) and calls for 'new methodologies and indicators . . . to measure the impact of cultural policies and activities in terms of quality of life, social cohesion and community development' (ibid.: 212; also 1994: 16). Such approaches and measurements require a more sophisticated range of methods than the short-term or quantitative economic impact techniques traditionally used, but which are still the prerequisites for political and financial evaluation of 'success' and 'returns': 'The cost–benefit equation is easier when culture itself is turned into an industry' (von Eckardt 1982: 125). Attempts at creating a more holistic quality of life index incorporating 'culture' have been developed for instance by the 'green' *Think Tank*, the New Economics Foundation (Lingayah *et al.* 1997), applying social/enterprise audit techniques which have also been developed through *Fair Trade* and ethical investment assessment indicators. These have however tended to focus on the social impact of the arts and neighbourhood renewal (Shaw 1999) and to an extent reductive performance indicators of 'diversity' (ethnicity, disabled, gender 'quotas'; Hacon *et al.* 1998). At the same time the evaluation of the legitimate, professional arts and creative industries continues to look to economic efficiency and impact validation (Evans 2000b) – two worlds for one 'common' culture or two 'cultures' for one world? A more inclusive and sophisticated approach is also demanded since, as I and others have argued (Pratt 1997, 1998, Evans 1999a, Scott 2000), the 'new' cultural industries have been overlooked by traditional economic and employment profiles that have been shown to understate both the informal, hidden arts' economy and the fragmented but industrious work-practices which increasingly typify the urban arts and cultural *locale* (Evans 1990, 1999a). Montgomery therefore argues that:

> There is a very strong case for conducting primary qualitative research in order to establish, once and for all, a sensitive but rigorous methodology, to define the [creative industries] sub-sectors in ways which reflect their day to day workings, and to achieve fully reliable estimates of turnover and employment which can be reassessed at intervals in the future.
>
> (Urban Cultures Ltd 1994: 12–13)

Without such recognition and understanding, as O'Connor concludes writing on the relationship between the cultural and urban infrastructure: 'The inability of planners to place value on such activities in terms of intellectual and cultural capital has meant that cultural industries have usually been the first victims of the regeneration they helped to inspire' (1999: 24).

A central question when considering arts and cultural planning is 'what are the best spatial arrangements for cultural activities'. As Johnson goes on to ask: 'what criteria underpin the notion of "best"?' (2000: 15) The tension and increasingly the conflict between arts amenity and public goods, and the commodification of the arts through the cultural industries and consumption culture encroaching upon the public realm, presents a dialectic in both planning and resource distribution. Rationales and planning strategies therefore depend on the balance between, on the one hand, amenity, welfare provision and the support of creative expression, and, on the other, the cultural economy – local and global:

> If the primary concern is with stimulating the development of high or low cultural interests and activities in particular sections of the population, then the answer may [but not necessarily] differ from that arising out of a concern to maximise the efficiency of the cultural industries, or encourage interaction between different cultural forms, or to gain the widest possible exposure from the population at large.
>
> (ibid.)

Examples of all of these positions have been discussed throughout this book, where a clear convergence not only towards the economic and related regeneration rationale for culture is evident, but also through the quantification and value-system that seeks to exploit cultural resources and assets as part of a competitive creative city goal (Landry 2000).

What the creative city and industries' approach has failed adequately to consider however is the role and relationship of the arts and creative activity through the production chain and hierarchy of provision framework (see Chapters 4 and 5) with the integration of local amenity and provision and needs of the city's residents as participants, creative people and discerning consumers. Strategic economic and land-use planning in this context therefore continues to understate or ignore altogether the human dimension and interrelationships that make-up the urban situation, notions of citizenship and ultimately the city's status and viability. The ranking of key arts cultural industry sectors in comparison with other cities (Comedia 1991b, Landry 2000) also follows a largely quantitative (physical facilities, employment, financial value) measurement that ignores both the quality and diversity of provision, its reach and equity possibilities, issues of mobility, and any association of the strength of creative work. In short, these exercises measure inputs and limited outputs rather than any consideration of outcomes or notion of (sustainable) cultural development (Evans 2000b). National, sectoral and city comparative studies also replicate the league table analysis such as *Top Towns* (Focas *et al.* 1995) that are based on the number of standard cultural facilities in a district rather than participation and production which will take place in a wider range of venues and places. International comparatives of cultural spending and lifestyles, such as complex *human development* and *quality of life* indices (Daly and Cobb 1989: 410-455, UNDP 1995) and cost of living surveys, also tend to produce contradictory and reductive city and country rankings – a good cultural city may have a poor environmental ranking – implying that a city can be dull and healthy or creative but crime-ridden. The renaissance of Harlem in New York, which Younge refers to as 'The Negro capital of the world, long associated with urban deprivation and cultural richness' (2000: 1), conflates both a long history of immigrant settlement (seventeenth-century Dutch, nineteenth-century Irish, Italian, and Jews, twentieth-century African-American from the rural South moving North) as other industrial cities have played 'host' (e.g. East London). This gentrification and urban regeneration also occurs in the areas of divided cities located often physically close to the high value commercial and residential districts. London is often ranked as 'best' cultural city but the 'worst' living city and the most expensive in Europe; contrast also Rio (fun, *favelas*, frightening) with the planned (read sterile, isolated retirement home) administrative capital city of Brasilia; Sydney and Canberra, San Francisco and Seattle and so on. We cannot therefore talk of cities as homogeneous places in any sense, whilst their multicultural, rich-mix-manufactured images reside largely in the literature and myths that their image-makers choose to project. As Ryan asks 'The Ideal City and the city as site of humanity's crisis and degradation are seemingly contradictory themes. . . . In the [architecture] schools, Sodom and Gomorrah are seldom even remotely mentioned. Could it be that these ostensibly oppositional views of the city – Good City versus Bad City – are actually inverses of one another, needing each other as sparring partners?' (2001: 23/4).

The urban chaos–cultured city dichotomy (and its corollary, the rural idyll re-created in suburbia/edge town and city, country crafts and in so-called indigenous societies) is thus perpetuated through the image of the creative cosmopolitan city, and this is clearly an urban myth that planners, politicians and developers seek both to design-out, sanitise and plan against. Even in rural development a qualitative distinction was made some time ago: 'Village Industries should be started again but not in the old basis at all. They should be an integral part of the whole country development, not arty crafty revival of badly done crafts' (Mairet 1933). A hypocrisy is also witnessed in the reaction to the forms of popular leisure that do not fit easily within the cultural city paradigm or regular street life, as in suppressed youth cultural pursuits such as graffiti art and skateboarding. Whilst graffiti gained some recognition when it went 'inside' (galleries, books and film), elsewhere, such as in India, it is both celebrated and forms a vivid representation of popular culture and the streetscape (Edensor 1998, Dawson 1999). Raves and the associated acid house, garage and techno music, and the drug scene are another planning dilemma: 'As the chroniclers of pop pointed out . . . this was nothing really new. Moral panics have occurred before in reaction to subcultures of [usually working class] youth, which is often defined as a repressed category in society' (Rietveld 1999: 42). Planning in this sense cannot hope to be inclusive unless respect for freedoms, emerging and non-traditional (even oppositional) cultural practices, and consideration of their need for expression (e.g. safe places for rave and dance events) is paid, as with existing forms of cultural and recreational amenities.

In distinguishing between the political economic and *symbolic* spheres of the built environment, Zukin 'focuses on the representations of social groups and visual means of excluding or including them in public and private spaces. From this view, the endless negotiation of cultural meanings in built forms – in buildings, streets, parks, interiors – contributes to the construction of social identities' (1996: 43). This suggests that if cultural planning is to distinguish itself from the economics of land-use and a negative starting position in the development control process, these 'interpretations and interpenetrations of culture and power' will need to be better understood and the symbolic economy recognised, and as she goes on say: 'To ask whose city? suggests more than a politics of occupation; it also asks who has the right to inhabit the dominant image of the city' (ibid.). Public choice devoid of creative action, risk and innovation is also a recipe for stasis, and a reversion to the safe arts of the past and of the (silent) majority. Planning out the new through unquestioned conservation, urban design and rigid zoning can stifle cultural development, whereas confidence in commissioning also suggests that opinions pre- and post-new art can and do change, as Williams argued some time ago. Take the story of the *Angel of the North* (Gormley 1998), a state Lottery-funded monumental statue first greeted by local and national disdain at the model and drawing stage, and by resistance to its installation (a 'Stop the Statue' campaign collected petitions, phone-in polls were ten to one against), and then genuine acceptance and ownership as this major public art icon took shape overlooking the A1 in Gateshead, North East England. Perhaps the lack of faith and trust in municipal culture and the traditional distant relationship between the artist, *locale* and consumers/participants is a reflection not only of elitism and hegemony, but also the absence of a more common engagement with culture through prosaic town planning and the design and location of facilities in society today.

As explored in Chapters 4 and 5, one of the main challenges to arts development agencies and advocates in the planning dialogue is the absence of robust and useful def-initions of arts and other 'amenity' and appropriate standards that could be applied to the interpretation of arts and cultural facility needs, in land-use and town planning terms (Cullingworth 1979, also with Nadin 1994), coupled with a lack of any sort of defined cultural policy at borough or national levels. As in the town and country planning domain itself, politicians are often hostile to the very concept of 'planning' (or con-versely over-enthusiastic and prescriptive) or even to the adoption of formal policies in the arts field. A notable response to this is seen in the strategic approach of the London *Arts Plan* and Toronto *CityPlan* (see Chapter 5), both of which adopted specific urban planning concepts and terminology (i.e. derived from environmental/town planning), moving closer to a human geographic analysis, and the language of the *Arts Plan* was made more accessible to borough planning officers, including Environment Ministry civil servants and ward-based councillors as a result.

The strategic versus local distinction in borough and city plans also raises an impor-tant methodological issue of whether the arts should be treated as a stand-alone topic area or should policies to realise the benefits of arts, culture and entertainment be incor-porated with broader mechanisms proposed in the plan, such as town centre strategies, urban design, public transport, enterprise development and training, and within indi-vidual development sites and zones (LPAC 1990b). This question goes to the heart of the development of a local *cultural* policy, as opposed to a narrower and separate *arts* policy (Challans and Sargent 1991), since defining arts provision through the typical art form and arts facility approach has failed to integrate wider (and higher priority) policy and planning imperatives, notably economic development and in the areas of education, health, housing, employment, environment, transport and other social aspects of land-use and development. A unified urban regeneration strategy would require an approach that incorporates all of these areas of provision, a cultural policy would be therefore to look to the integration of arts and cultural need and opportunity within each area of provision, such as arts in education, arts and housing, arts and health, public art/realm, the arts and cultural industries, employment and training, and so on. As Bianchini sug-gests: 'Local authorities would have to overcome "departmentalisation" and move towards a more corporate, integrated approach to policy-making in order to implement a cultural planning strategy' (1991a: 39).

However, a weakness of even the more progressive examples of departmental and policy collaborations is their agency and council officer dominance – with little real external representation, either from artists or arts and community groups evident. As Landry points out: 'Critics also claim planners underestimate social dynamics, which are as significant as land-use or property services . . . those in planning from land-use to marketing are not creative enough' (2000: 1268–9). This mirrors the professionalisa-tion and bureaucratisation of both cultural and other public policy realms and decision-making structures (Coalter 1992, Henry 1993: 113) and what Laffin and Young (1985) referred to as the council officer as 'bureaucratic politician' and Dunleavy as 'ideological corporatism' (1980). This also reflects the fragmented specialisms within the town planning (and engineering) functions themselves (Davidoff 1965). A cultural policy that sacrifices cultural democracy, and particularly the voice of the artist, risks the worse aspects of municipal culture and planning so feared by the libertarians and a fail-ure to meet cultural need and diversity. For instance, even Birmingham's design award-winning city centre investment strategy has been questioned (Loftman and Nevin 1993). This was subsequently attacked by senior government ministers who accused the Labour-controlled council of underspending its notional allocations, e.g. capital and revenue investment on education and housing, in order to pay for major new cultural, exhibition and visitor attractions for reasons of personal vanity and prestige on the part of leading city politicians. In their defence, this also emulates the 'models' offered in the USA (to where Birmingham's politicians looked in the 1980s), but which like the mall and multiplex had yet to pass the test of time. This classic city-centre strategy, like so many others, also ignores the fact that 'central area prestige projects can alienate residents in suburban neighbourhoods if local facilities are not also provided' (Symon and Verhoeff 1999: 741). Such negative reactions and opportunity costs illustrate the risk of such singular strategies, a risk which cities such as Singapore have either not fully considered (or worked through in a planning sense) or are prepared to suppress. In this as in other cases of flagship-led city regeneration (see Chapter 8), one might conclude that a cultural policy does not yet exist, rather an economic and city centre strategy based on blind faith in the 'trickle down' effect. A specific example of this is provided in a comparison of two established arts groups in Birmingham (the Ikon Gallery and the black arts centre The Drum), both Lottery and European-funded building relocations (Evans and Foord 2000b) but with sharply contrasting treatment and solutions. The former was resited in the central business and entertainment district, the latter in a less salubrious and non-central location, and as was concluded this 'reveals the need for micro-environmental factors to be taken into account when planning urban investment for White and non-White audiences' (ibid.: 7230). In this sense the process and eventual location can strongly influence image, access/usage, markets and consequently the viability of cultural organisations, and this therefore reflects the hierarchical values ascribed to certain cultural practice over others, even where ostensibly within the same art form or genre and within the same city and cultural regimes.

Aspects of a divided post-industrial city are therefore emerging in the very examples of culture-led regeneration that have sought to respond to their declining industrial role and purpose. Whether these are building on existing land-use separation and socio-economic and ethnic–linguistic divides, or reinforcing the core–periphery, they point to a failure in the planning process which is not really planning at all, but rather a combination of *a posteriori* rationalisation of development – whether market or public programme led – and what Zukin sees as the worst-case cultural strategies that 'do not reverse the hierarchies of place that lead to competition for distinctive segments of capital and labor . . . [and] suggest the utter absence of new industrial strategies for growth, i.e. lack of local strategies that have any chance of success in attracting productive activity' (1995: 274). How far, say, Barcelona and its comparators are and will in the future enhance and distribute their cultural capital and development potential through high-profile redevelopment and cultural programmes is perhaps too soon to judge. As has been seen in the past, creativity and cultural equity do not necessarily coincide with place-making efforts and growth economies. Less robust examples such as Dublin, Liverpool, Glasgow, Bilbao and in North American and emerging cultural cities in developing countries seem less likely to sustain themselves or achieve (or even articulate) genuinely cultural development objectives. They arguably lack a comprehensive evaluation or cultural planning approach to their development, where 'the common element in all these strategies is that they reduce the multiple dimensions and conflicts of culture to a coherent visual representation' (ibid.: 271). Minimisation and marginalisation of cultural diversity is one common aspect of many urban cultural regeneration processes considered here; homogenisation of the built environment and its capital formation is another.

Public choice and zero base

At the micro-level, where resource allocation processes impact on both short-term and crucial investment decisions, a *zero-base budget* (ZBB) exercise by a city, town or by a regional or national cultural agency may seem a recipe for trouble. However, only such an exercise, even if initially hypothetical, may genuinely assess artistic and community need against resources and provision, and the formulation of plans. Without this we are otherwise stuck with the inheritance of past preferences – moral judgements, 'taste', public good externalities and paternalism – and which act as a block to responsive planning and living culture. As Roberts observed over twenty-five years ago:

> The justification of the choice of activities undertaken by [civic] authorities is little more than historical accident plus the concept of 'worthy' leisure . . . private enterprise has offered us the frisbee, grouse shooting and Summerland, while Epping Forest, the Festival Hall and our local tennis courts are by courtesy of the public purse.
>
> (1974: 10)

Public choice is an intrinsic aspect of cultural and other types of amenity planning and development, but without planning engagement, government and other distributory agencies are left to decide 'whether to accommodate people's preferences, or instead try and change what people want . . . the circumstances conditioning their choices are themselves determined within the political process . . . choices heavily influenced by structural, institutional and environmental factors' (Dunleavy 1991: 256–7). The latter route is the natural default and Sennett in his long view of *The Fall of Public Man* (1986) sees this 'dislocation having ruined politics by tricking us into believing that issues of power and the allocation of resources can be dealt with in terms of trust and warmth' (p. xvii). Starting with a clean sheet in the resource allocation process will of course be threatening to existing interests and beneficiaries, however the exercise itself may require and produce a more serious evaluation of commitments and interests whose status is otherwise unquestioned and which effectively block new initiatives and needs (simplistically, the 'high arts' – heritage and classics versus contemporary and popular culture). As Borja and Castells point out, reflecting the pattern of urban regeneration and regional development programmes in Europe and North America:

> Lack of resources means that in practice higher layers of government replace local government through sectoral programmes or individual projects. In other cases action is taken by the private sector, without being integrated into a coherent

urban programme. In yet other cases, a major area of the city and of inhabitants [is] simply left without any cultural facilities.

(1997: 113)

In the spatial and financial distribution of resources to public culture and facilities, as we have seen time and time again, the centre-core dominates in the physical location, proportion of resources and nature of high-arts activity, reinforced by the urban regeneration and flagship responses evident in old and new cities the world over. This is increasingly the case at a supra-national, national, regional, subregion and city levels, and even within local areas as pockets of gentrification, new residential areas, share administrative boundaries with economically and amenity-poor districts, and as higher-level cultural facilities and quarters occupy transformative zones at the cost of neighbourhood facilities. Cultural resources follow this concentration. For example in Paris, Barcelona and in London where 70 per cent of public funding of arts organisations in England is represented by only 40 per cent of national audiences at recipient venues, while the five national companies based in the capital receive nearly 50 per cent of total national arts funding but account for only 16 per cent of total audiences within the subsidised arts sector (Evans *et al.* 1997, 2000b). Despite this largesse, the audiences and performances of the national 'flagship' companies have been in decline whereas their ticket prices have been raised far above the rate of inflation, so they now effectively exclude most income groups from these erstwhile public/merit cultural services (Evans 1999e). The Royal Shakespeare Company's audiences have been in decline at their London base, but in Stratford-upon-Avon they have been on the increase. The touristic value is apparent here, and this perhaps indicates a disjuncture that national cultural planning needs to address. With tourists making up an increasing proportion of museum and gallery visitors, festival-goers and street markets, price discrimination (and tourist/bed taxes at the destination; AMA 1990, WTO 2000) is perhaps one mechanism that may ensure local access and cultural resources to offset the negative impacts of cultural tourism and crowding-out. Tourism plans that do not consider the contribution and the impact of a wide range of cultural activities and facilities (wider than the traditional tourist arts) and arts plans that underestimate the role and impact of tourist activity – who is the tourist and why do/might they visit this area? – suggest that tourism should also be considered as part of a cultural plan for an area and city–region rather than being restricted to hospitality, carrying capacity and the marketing of predetermined itineraries and visitor attractions. The stubborn support and expansion (*Grands Projets*) of 'classic' and heritage-based culture-houses and institutions, including many museums and galleries – within limited cultural resources, lacking public (i.e. tax payers') support, identification and participation and moving from financial and managerial crisis to crisis – suggest that they should, with few exceptions, be taken out of the 'cultural equation' altogether (Willis 1991). Their cost–benefit and merit good rationale should be recast within tourism and national economic impact criteria, leaving both common culture and cultural development free to develop within a more creative and distributive framework, including the market where appropriate.

Whilst public culture emulates the commercial sector which freeloads on the former, there is also plainly a widening gap between public arts/goods and private urban consumption – failure to address this risks arts policy and agency being passed over in the continued trend towards the polarisation of participation, socially, demo-graphically, culturally and economically; a narrowing audience base; and the focus on the 'fortress' home and leisure experience. As Chapter 1 pointed out, the popular entertainment and cultural expression chosen by those for whom legitimated art has no relevance, the ill-defined socially excluded (Le Grand 1998) and those 'absent' from arts audiences all have one thing in common, they are silent and disempowered when cultural plans and facilities are developed by the intermediaries and bureaucrats for whom the arts are, literally, a preserve. Equally, cultural expression that takes place out-side of approved and designated places, whether a warehouse rave, community or ethnic festival, or the post-market gatherings ('fairs') that regularly occur in the fringes of cities such as Rio, suggest by their large attendances that common culture thrives, but when institutionalised physically and through cultural bureaucracy its essential nature is diminished. Tolerance may therefore be as good a response as control and legitimation. As Zukin says: 'Planned or not, a culture capital thrives in the inter-sec-tion of the business, nonprofit and arts economies. . . . Even if cultural strategies of economic revitalization succeed, it is not inevitable that the economic value of the space overwhelm the cultural power of the symbols' (1995: 151).

Whether the so-called leisure society and associated mobility and communications media produce the spatial restructuring that convergent theorists claim (Zelinsky 1992), or whether the associated individualism also creates a new regionalism: 'land-scape mirrors popular culture' (Jordan-Bychkov and Domosh 1999: 320). Leisure landscapes can be elitist, consumer or essentially amenity-based and are increasingly informed and influenced by visual strategies (Zukin 1995). Culture in this sense, and as I have attempted to present here, has a spatial dimension. Castells (1977) defined spatial structures as the particular ways in which social structures are spatially articu-lated (Pickvance 1976), but whilst the separation of social, cultural and economic activities has been a feature of modern town and masterplanning in the past (Jencks 1996), 'the long argued distinctions between activity and movement, between land-use and transport, between production and consumption have begun to dissolve' (Solesbury 1998, see also Marx 1973). As Rykwert also recently observed, 'all action to do with planning and building is inevitably political' (2000: 245), and I share his disquiet with the notion today of 'space' as the neutral focus of creation, i.e. *art spaces, space planners* and *masterplanners* who foreground their buildings and treat their backdrops ('the environment') as the benign open spaces between them (Rogers *et al.*, DETR 1999). As Rykwert prefers, 'all worthwhile building . . . must involve the making of *places* . . . enclosures that people can inhabit and appropriate without doing themselves violence' (2000: 245). The notion of planning for the arts within a town and specifically the urban planning framework, as I have argued, places social and cul-tural needs at the centre of the physical planning process: 'Constant community participation and involvement are needed to shape our cities and to make them com-municative' (ibid.: 246). Arts development, cultural democracy and recognition of the urban social and economic context therefore equally underpin the cultural planning approach against social, economic and technological change that renders traditional town planning and prescriptive arts policy ineffective. As the *fin de siècle* passes, Eric Hobsbawm in *Age of Extremes: The Short Twentieth Century* writes on the triumph of the individual over a more organised and collective society, where 'roles were pre-scribed if not always written':

> The cultural revolution of the later twentieth century can . . . be understood as the triumph of the individual over society, or rather, the breaking of the threads which in the past had woven human beings into social textures . . . such textures had con-sisted not only of the actual relations between human beings and their forms of organization but also of the general models of such relations and the expected pat-terns of people's behaviour.
>
> (1995: 334)

If arts and cultural expression are really 'rights' and their provision and practice are to be of continued importance to society, if the divided city and region is to be recon-ciled and is not to be wholly dictated and shaped by external forces and a globalised political economy, the planning of the arts to meet the needs of sustainable community, economic, educational and cultural development and diversity, is likely to require a deeper understanding of these 'threads' of human relations in this fragmented scenario as the basis of a renewed urban tapestry.

Appendix I
Model planning policies for the arts, culture and entertainment: a borough survey

Introduction

One way that cultural policies can be incorporated with the environmental planning process is through the development of mechanisms and themes by which cultural planning measures may be included in periodic borough and city plans. In London the requirement of each of the thirty-three local planning authorities to produce ten-year Unitary Development Plans (UDPs) for their borough, provided the opportunity for the interpretation of culture in town planning proper (DoE 1992a). The term 'unitary' referred to the fact that these represented a single tier of statutory planning, both strategic and local in effect, in the absence of a regional tier of government and therefore city plan. These borough plans are subject to local inquiry and consultation before their adoption, and at the time: 'There are high hopes for the new regime – It should reduce the resources devoted to planning appeals. . . . The planning system should become simpler and more responsive . . . making it easier for people to be involved in the planning process'. This made the UDP the primary consideration in development control, effectively making it an enforceable blueprint for each borough: 'the approach shall leave no doubt about the importance of the plan-led system' (Cullingworth and Nadin 1994: 58)

The main policy areas from which arts policies and other mechanisms can be extracted from borough plans (UDP) follow the *arts infrastructure* analysis explored in Chapters 5 and 6. In addition to specific arts policy statements within planning policies – general or area-specific – boroughs also had the opportunity to identify the arts as part of their strategic 'vision', and therefore included in Part I (Strategic) of their Plan. Policies have therefore been analysed under the following categories, which were based on the *Model Policies for the Arts, Culture and Entertainment* (LPAC 1990a, b). Under each main heading, arts-specific subpolicies and mechanisms are analysed as follows:

Part I

1   **Strategic and/or local context**
    The arts are considered to be of strategic, borough-wide importance, not just activity and local or site-based (LPAC 1990b).

Part II

2   Economy and employment
- Subpolicy: maximise use of facilities through enhanced economic activity, employment, capacity/usage and market for the arts.
- Policy implications: in addition to employment generation potential, include 'fostering voluntary and cooperative organisations which support the arts and to encourage the provision of affordable business premises and widening the use of open space' (ibid.: 6).

3   Environment
- Subpolicies: replace existing facilities; planning-gain; per cent for arts.
- Policy implications: relate to town and strategic centres, the regeneration of redundant and heritage buildings and their enhancement; security, particularly in town centres and fringe areas and urban design and landscaping.

4   Image
- Subpolicy: designation of arts and cultural quarter(s).
- Policy implications: include the integration of design details, information displays and environmental initiatives to 'develop a coherent image of ACE activities and in particular cultural quarters' (ibid.: 7).

5   Accessibility
- Subpolicies: dual-use of facilities (community, arts and sports, education); public transport provision.
- Policy implications: include liaison with transport operators, police and council departments to improve safety and security and to ensure adequate night-time services; car parking provision; disability access to arts facilities, and pedestrianisation schemes and networks.

6   Infrastructure
- Subpolicies: promotion/safeguard of facilities; safety on public transport.
- Policy implications: encompass the reconciliation of 'supply, demand and need for facilities'; cross-borough liaison and planning; the retention of socially valuable mixed-use of land in central London and in strategic and growth centres.

7   Equal opportunity
- Subpolicy: access for the disabled.

8   Design
- Subpolicy: good urban design.

9   Arts, culture and entertainment
- Subpolicy: separate chapter in Plan devoted to this topic area.

In point 9 a case is made to demonstrate the importance of the arts, sufficient to warrant a separate chapter in the Plan, 'However as planners involved in the UDP preparation, we also recognise that this view may not be universal' (LPAC 1990c: 3). The perception of special pleading may also disadvantage the promotion of the arts over other competing resource or land-use claims, however nearly 25 per cent of London borough UDPs included separate chapters on the arts, including four outer London boroughs (which have lesser cultural facility provision, physically and per capita).

***

Whilst there was no restriction in terms of content, borough plans concentrated primarily on land-use, infrastructure and the built environment, and the implications for these of economic activity and social need. The extent to which planning authorities, both through internal integrated policy development and external consultation, translate planning policy and social and economic development in terms of the cultural economy and arts amenities, is therefore evaluated in a comparative analysis of each London borough UDP. In particular, the adoption of arts planning model policies, and the rationale for their inclusion, is summarised in Table I.1, providing an analytical framework of the thirty-three borough UDPs in terms of these prime and subpolicy areas. Each reference indicates a separate inclusion of the relevant arts ('ACE') policy, by chapter or section reference, in the UDP document. Abbreviations are those used in each UDP and in most cases indicate the chapter topic:

- REC: recreation
- ACE: arts, culture and entertainment
- STRAT or ST: strategy
- L/A/R: leisure/arts/recreation
- ENV: environment
- TRANS: transport
- DES: design.

The first column (1) indicates that the arts have been mentioned in Part I of the UDP (S, *Strategic*), as well as under *Local* (L) planning policy consideration – in fact all thirty-three boroughs mentioned 'ACE' as a strategic issue in Part I, as well as in Part II under local planning topics (columns 2–13).

Summary analysis

The extent to which boroughs have incorporated arts and cultural policies and considered arts infrastructure issues in their borough development plans can be gauged from the following matrix. In particular this shows the frequency that particular policy guidelines and mechanisms have been referenced under each main issue or topic area and those policies that have been more popular or more easily integrated into the borough planning process, as summarised below. The frequency of references and incidence of borough adoption of these is closely correlated, showing a clear bias towards those arts policies with obvious impact on the built environment and the land-use and development process, and vice versa. The above analysis does reinforce an outer–inner-city

Table 1.1 Policies for the arts, culture and entertainment – borough UDP analysis

Borough	Strategic or local context [1]	Maximise use of facilities [2]	Replace affected facilities [3]	Designate cultural quarter [4]	Dual usage of facilities [5]	Promote/ safeguard facilities [6]	Percent for art [7]	Planning gain provision [8]	Public transport provision [9]	Safety on public transport [10]	Good urban design [11]	Access for the disabled [12]	Arts and culture chapter [13]
City of London	S, L	REC14 REC4	REC12 REC10			STRAT11 REC11/ 13 RIV3	RIV6 REC10		TRANS1		ENV5 RIV1 RIV6	REC11 SOC3	n/a
Barking & Dagenham	S, L	STG5 AT4	AT3	AT5 AT6		AT3 AT7 BTC10	AT9	AT9	CHAP11	CHAP11	CHAP7	E4 AT8 C14 LAR18	CHAP10
Barnet	S, L		L5.1			L5.2		L5.3	M4.1			M4.2	n/a
Bexley	S, L			TAL18 BTC9	TAL16	TAL4 TAL17 and 19	TAL23		CHAP8		CHAP5	TAL3	n/a
Brent	S, L	STR27 ACE2	ACE1	STR19 EMP11 S40 WTC5 WTC8 HN3	HN24–6	ACE5 ACE6 CF3 HN22/3	ACE3 E40 E25 WTC16 HN14/18	ACE4	STR28 STR33		STR4 STR5 E2	E16 E17 DS16	
Bromley	S, L	B/L.3	B/L.2 L17	B/L.1–3		B/L.1			CHAP6		CHAP9	APP. IV C6	n/a
Camden	S, L	LC19 SLC2	LC16			LC16 LC17 SCE2 EC15 SKC1	LC23 SSH3	LC19	STR1 LC18	STR5	SEN4 EN7	LC6 LC21	CHAP8
Croydon	S, L	SP35	LR2 CC13	SP38 LR5 CC11 CC12	LR7	SP38 CC11 CC13 LR5 LR6	BE15	LR6 CC12	SP17	SP21	SP1 B19 and 20	BE29 BE30 LR8 CC20	n/a

Borough													
Ealing	S, L	CF18	CF10	E25	CF37	OL15 CF1 CF19 EI.vii S24	P2 DEV37 P2		T7 T12		H30	T11	n/a
Enfield	S, L				AR2	AR1 AR4	AR5		T1 T2		GD4	GD12	CHAP13
Greenwich	S, L	ACE3	ACE11	ACE5	ACE7 C17	ACE1 ACE10 T23 W3	ACE8 ACE9 C6	ACE2 ACE4 ACE8	M10		D6 W4	ACE6	CHAP10
Hackney	S, L	ACE6	ACE3	R1	ST39	ACE1 ST42		ACE2 ACE4 ACE5	ST28	CHAP6	ST4 CHAP2 CHAP11	ST6 ACE1 ACE8	CHAP10
Hammersmith & Fulham	S, L	CS13	CS1		CS13	CS2 E12	EN13				EN8	EN11	n/a
Haringey	S, L	LE13.3	LE13.5 LE13.6	LE13.4		LE13.1 and 2 LE13.7	LE13.8 ENV1.19	LE13.8	TSP4		ENV1 STC4	CFS8	n/a
Harrow	S, L				S22	R2 S22			T15		E6 E44	A1–9	n/a
Havering	S, L	LAR2		LAR4	LAR3	LAR2 LAR4 STR33	ENV20		TRN3 TRN13	TRN14	ENV1	ENV1	n/a
Hillingdon	S, L	R6	R5	S1	R6	R9 R10 LE8	R10 LE8				BE8	AM1 AM14	n/a
Hounslow	S, L	R6	C6.1		C6.2	ENV1.8 ENV1.11	IMP3.1		T.1 T.3		ENV1.1	ENV1.8 ENV3.1 C4.3	n/a
Islington	S, L	STRAT R1	R23	IMP14	ED11	R21 and 22 V4	D9		CHAP6	CHAP6	CHAP12	ENV11	n/a
Kensington & Chelsea	S, L		LR23 LR24 LR26		LR27 SC10	STRAT34 LR28 LR33	CD para. 3.29 LR32	LR31 LR32	TR3		STRAT35 CD27	SC15 CD42	n/a
Kingston-upon-Thames	S, L	STR16	RL3 KTC15	KTC10 KTC15	RL4 STR12	LTC15	UD para. 5.37		STR16/19 CHAP11		STR9/10	STR11	n/a
Lambeth	S, L	G45	AT1 B8	AT2 AT4 B8 B9		AT2 B9	AT3 E19	E11–19 ED 2 S16–21	CHAP5	CHAP5	CHAP6		CHAP11

Table 1.1 (continued)

| | | | | | Subpolicy | | | | | | | | |
Borough	Strategic or local context [1]	Maximise use of facilities [2]	Replace affected facilities [3]	Designate cultural quarter [4]	Dual usage of facilities [5]	Promote/ safeguard facilities [6]	Percent for art [7]	Planning gain provision [8]	Public transport provision [9]	Safety on public transport [10]	Good urban design [11]	Access for the disabled [12]	Arts and culture chapter [13]
Lewisham	S, L			GEN39	GEN36	GEN35							
Merton	S, L	L21	L22 (ii)	WTC.11 WTC.22	L6	L21 L22	EB.27	L.22	M.1–3 M.6 M.8 M.10	M9	CHAP3 CHAP10 CHAP11		n/a
Newham	S, L		LR1	LR3	LR2.2	LR2	LR4	LR2.3 LR4				LR9 LR2.1	n/a
Redbridge	S, L	LP.RL10	LP.RL10	LP.RL9		LP.LAR10					CHAP9	LP.CS6	n/a
Richmond-upon-Thames	S, L	CET5 STG9	CET1		CET2	CET1 CET3 CET5 CET6 HEP4 HEP15	CET4	CET4	STG11 CHAP11	STG11	CHAP4	CET2 (d)	CHAP7
Southwark	S, L	ENV3.2 C4.1	C4.2			C4	ENV2.6		TRANS. CHAPTER		ENV. CHAPTER	C5.1	n/a
Sutton	S, L	CLF21 CLF27		CLF29 STC5	CLF8 CLF9 CLF27	G/CLF2	CLF30	STC5	CHAP9		CHAP4	G/CLF6	n/a
Tower Hamlets	S, L	ST38	ART2	ST10 ART6 CHAP12		ART1 ART6	ST40 ART5 ART6 DEV19	ART5 ART6 DEV3 DEV19	CHAP5		CHAP1	ST10	CHAP8
Waltham Forest	S, L		LAR14	SHP6	LAR14 (i) GSC11	LAR14		LAR14 (ii)	CHAP3	CHAP3	CHAP2		n/a
Wandsworth	S, L	LR2	LR1			LR3, 11 and 12	LR11		CHAPT		CHAP TBE	LR2 LR3	n/a
Westminster	S, L			THE6	THE3	THE5 CA2	DES17		STRAT17 TRANS9		STRAT22 DES1		n/a

Sources: Borough Unitary Development Plans (LPAC 1990a, b, Horstman 1994)

divergence, but this difference is neither comprehensive nor a simple split on party-polit-ical lines. The key interacting factors influencing the penetration of arts policies in borough Plans have been the scope and scale of recent development and of land-use change in a borough; the relative importance of economic development – a function of unemployment, social/'areas of community need' and consequent regeneration initia-tives – and the concentration of existing cultural facilities (Table I.2).

Table I.2 Planning policies for the Arts, ranked by frequency in borough UDPs

Policy/mechanism (n = frequency of reference in UDP)	Ranking of inclusion	Percentage of boroughs (n = 33)
Promote/safeguard facilities	1 (n = 84)	100
Urban design	2 (n = 47)	94
Access for the disabled	3 (n = 45)	88
Percent for art	4 (n = 44)	78
Public transport	5 (n = 41)	85
Designate arts/cultural quarter	6 (n = 38)	64
Planning gain	7 (n = 35)	64
Replace affected facilities	8 (n = 34)	78
Maximise use of facilities	9 (n = 29)	70
Dual-use of facilities	10 (n = 28)	67
Safety on public transport	11 (n = 10)	58
Separate 'arts' (ACE) chapter	12 (n = 9)	27

All boroughs included policies for the *Promotion and Safeguarding of Arts Facilities*, whilst most made two or more separate references, and all but three had policies on *Urban Design*. The latter reflects the growing adoption of design policy guidelines as part of the development control procedure. Nearly all boroughs made references to the needs and requirements for *Disabled Access* to facilities and public areas, with five excep-tions, including the central (West End) Borough of Westminster.

Percent for Art and *Planning Gain* policies were also included by the majority of bor-oughs. Given the ten-year duration of the UDPs, those boroughs not developing planning-gain and per cent for arts policies in their statutory plans could be seen to be short-sighted given the continued drift of commerce and industry (as the source of development funding) to outer London. In all of these cases, however, boroughs (offi-cers and members) were reluctant to specify prescriptive policies that would constrain or deter would-be developers (private *and* public), preferring a more free-market and site-by-site planning review, in the tradition of the British discretionary and negotiable planning system (Sharp *et al.* 1992).

Two-thirds of boroughs had included policies to support the designation of *Cultural Quarters*, focused on specific locations and centres. Such policies were generally pro-posed where a cluster or critical mass of cultural facilities were linked to the public realm (town squares, pedestrianised areas, mix of uses, such as shops, cafés and so on). These plan policies included continued support of existing cultural quarters, particularly in town and strategic centres and prospective areas in redeveloped areas or schemes. In contrast, the absence of such designations in some boroughs where they could have been expected revealed a lack of policy and planning coordination.

Whilst most boroughs had also adopted policies for the *Replacement* (78 per cent) and *Maximisation* (70 per cent) of arts facilities within their UDPs, several outer and two inner London boroughs including, most surprisingly, the city of Westminster did not feel it important to include either policy. In such former suburban boroughs the shortage of arts facilities was used to justify this omission, however with only one public arts centre in each and declining commercial entertainment, e.g. cinemas, the fact that these large boroughs had the most to lose from the closure or under-use of their local arts facilities, has not carried weight. This *laissez-faire* attitude is also remarkable in Westminster which has the highest concentration of theatres in Europe, despite the 'fall-back' provided by the preservation Theatres Trust and the protection of theatre buildings (mostly heritage 'listed') from change of use. Finally, this survey shows that the majority of boroughs recognised the importance of *Public Transport Provision* in providing access to arts provision, though fewer adopted policies focused on the improvement of safety and environmental barriers to wider arts participation and visits. These included several of the more urban, inner-city boroughs, including the key touristic boroughs in the central core.

It is clear therefore that through the UDP regime, planning departments for the first time were able to propose policies that dealt with the physical and built environment, through the development control (granting of planning permission, design, access) process and the provision of hard infrastructure, such as disabled access, urban design and safeguarding of facilities, as well as public transport provision. These policies fulfil wider social policy objectives not limited to the arts 'special needs' and reflect the traditional concern of town planning. Less confidence has been expressed by planners, judging by their proposals in the borough UDPs, in the more proactive and operational areas of the maximisation and dual-use of arts and cultural facilities, promoting greater safety on transport and the public realm. The recognition of urban design and the mechanism of per cent for art (and to a lesser extent, planning-gain used to enhance public art and design elements) are now well established, if not universal. From discussion with planners, this represents the heightened awareness, amongst both the public and elected members, of architecture and urban design (and modernist and post-modernist debates) and a public and professional reaction to the post-War mass-building, particularly high-rise housing and offices and belatedly to large-scale shopping and leisure–retail malls and centres. The importance of 'good design' (quality of building/materials, public realm, aesthetics, vernacular) was also recognised by planning and arts officers in 'quality of life' and borough image improvement and also the expectation that European planning guidelines (Environmental Impact Assessment; see Chapter 7) would require design and aesthetic standards. The greater harmonisation between EU planning legislation was felt to result in an even greater plan-led regime, which had encouraged the wider scope and interpretation of borough UDPs, and which would hold good into the twenty-first century.

Appendix II

Extract: *Space for the arts* (GLA 1991: 8–9)

The following list of mechanisms for implementing Arts, Culture and Entertainment 'ACE' initiatives is not intended to be a catalogue of all possible measures. It is indicative of the kind of methods local authorities could usefully employ. Equally important, they do not exclude one another. To be effective it may well prove necessary to put together a combination of measures to meet the needs of a particular building, site or area. Among the broader aims of economic and environmental improvement is the specific aim to protect vulnerable community facilities and starter businesses and allow them to realise their potential benefits to the economy, the environment and the well-being of an area in the longer term. Some measures are 'planning led' while others are the responsibility of other local authority services.

- Designation of Arts Culture and Entertainment (ACE) activity quarters where land-uses in classes D(1) and D(2) ('arts and entertainment') and other related land-uses will be expected to be predominant.
- A 'percentage for arts' scheme laying down a minimum, to be used in negotiations, including planning-gain.
- Including provision for ACE facilities in planning briefs for redevelopment or refurbishment.
- A presumption against loss of existing ACE facilities to other uses.
- Listing buildings or groups of buildings, or declaring small Conservation Areas to include ACE areas.
- Establishment of Community Development or Social Property Trusts to provide capital and revenue support for ACE activities, especially a variety of small managed workspaces.
- Agreements under section 106 of the 1990 Town and Country Planning Act ('Planning Gain'/Community Benefit) to ensure an ACE activity component.
- Direct council or joint council and private sector initiatives to support or promote specific ACE activities.
- As part of local authority disposal of assets, the provision of council premises for ACE activities at turnover or profit related rents. (Merging the planning and property portfolios might help this.)
- Maintain a register of private and public short life properties suitable for ACE activities.

- Develop an arts plan and where appropriate a leisure and tourism strategy for the Borough, together with neighbouring/subregional authorities and the marketing of these.
- Negotiate with public transport operators to improve services in association with ACE activity development.
- Ensure schools and especially colleges provide courses for skills needed by local ACE activities and that liaison is established between schools/colleges and potential employers.
- Provision of display space for local arts and design colleges in public buildings or in open sites in town centres/shopping malls.
- Establish neighbourhood or town centre consultative committees representing the local authority, arts organisations, the community and local businesses to promote the whole range of ACE activities at the local level.

Bibliography

Abercrombie, N. (1982) *Cultural Policy in the United Kingdom*, Paris: United Nations Educational Scientific and Cultural Organization (UNESCO).

Abercrombie, P. (1944) *The Greater London Development Plan*, London: HMSO.

Abercrombie, P. and Forshaw, J. H. (1943) *County of London Plan*, London: HMSO.

ACME Housing Association Ltd (1990) *Studios for Artists*, London.

Adams, H. (1970) 'Arts administration in the United States', in Schouvaloff, A. (ed.) (1970) *Place for the Arts*, Manchester: North West Arts Association: 205–6.

Adorno, T. W. (1991) 'Culture and administration', in Adorno, T. W. (ed.) *The Culture Industry: Selected Essays on Mass Culture*, London: Routledge.

Adorno, T. W. (ed.) (1991) *The Culture Industry: Selected Essays on Mass Culture*, London: Routledge.

Adorno, T. W. and Horkheimer, M. (1943) 'The culture industry: enlightenment as mass deception', in *Dialectic of Enlightenment* (trans. Cumming, J.), New York: Seabury, 1972; repr. in During, S. (1993) *The Cultural Studies Reader*, London: Routledge: 29–48.

Adorno, T. W. and Horkheimer, M. (1964) 'L'Industrie culturelle', *Communications* 3: 12–18.

Aitchison, C. (1992) 'Internationalisation and leisure research: the role of comparative studies', paper given at the International VVV/Leisure Studies Association Conference, Tilburg, The Netherlands: LSA.

Akehurst, G., Bland, N. and Nevin, M. (1993) 'Tourism policies in the European Community member states', *International Journal of Hospitality Management* 12: 33–66.

ALA (1991) *Ten Years of Docklands: How the Cake Was Cut*, London: Association of London Authorities.

Albertazzi, D. (1999) 'National vs local cultures: discussing the dream of the "knowable community" in Raymond Williams and the Italian "Lega Nord"', paper given at the Researching Culture Conference, University of North London, September.

Aldous, T. (1992) *Urban Villages: A Concept for Creating Mixed-Use Urban Developments on a Sustainable Scale*, London: Urban Villages Group.

Alexis, W. (1838) 'Berlin in seiner neuen Gestaltung', in Brockhaus, *Conversations — Lexikon der Gegenwart*: 453–63.

Allwood, J. (1977) *The Great Exhibitions*, London: Studio Vista.

AMA (1990) *Tourist Tax 'Green Paper'*, London: Association of Metropolitan Authorities.

Amin, A. and Thrift, N. (eds) (1994) *Globalization, Institutions and Regional Development in Europe*, Oxford: Oxford University Press.

Anderson, B. (1991) *Imagined Communities: Reflections on the Origin and Spread of Nationalism* (2nd revd ed.), London: Verso.

Andra, I. (1987) 'The dialectic of tradition and progress', in *Architecture and Society: In Search of Context*, Sofia: Balkan State Publishing House: 156–8.

Antoniou, J. (1992) 'Europe now', *Building Design*, 10 July: 14–16.

Appadurai, A. (1990) 'Disjuncture and difference in the global economy', *Public Culture* 2: 295–310.

Architect's Journal (1987) 'Use classes order guide', 29 July and 5 August: 57–61.

Architect's Journal (1989) 'Development economics: arts buildings — 1. An economic catalyst', 1 March.

Architect's Journal (1990) 'A Vision for London' [Special Issue], 11(191).

Argyle, M. (1995) *The Sources of Joy*, London: Penguin.

Arnott, J. and Duffield, B. (1989) 'Leisure and community development in rural areas', in Ventris, N. (ed.) *Leisure in Rural Society*, London: Leisure Studies Association.

Artists Space Journal (1992) 'Philadelphia creates its first owner-occupied artists' live/work cooperative', no. 6.

Arts Business Ltd (1991) *The Cultural Economy of Birmingham*, Birmingham: Birmingham City Council.

Arts Council (1955) *Housing the Arts: The Tenth Annual Report of the Arts Council of Great Britain 1954/55*, London: Arts Council of Great Britain.

Arts Council (1959) *Housing the Arts in Great Britain: Part I: London, Scotland, Wales*, London: Arts Council of Great Britain.

Arts Council (1978) *Annual Report and Accounts 1997/8*, London: Arts Council of Great Britain.

Arts Council (1983) *Annual Report and Accounts 1982/3*, London: Arts Council of Great Britain.

Arts Council (1984) *The Glory of the Garden: A Strategy for the Development of the Arts in England*, London: Arts Council of Great Britain.

Arts Council (1985) *A Great British Success Story: An Invitation to the Nation to Invest in the Arts*, London: Arts Council of Great Britain.

Arts Council (1986) *An Urban Renaissance: The Case for Increased Private and Public Sector Co-operation*, London: Arts Council of Great Britain.

Arts Council (1987) *Le Corbusier: Architect of the Century*, London: Arts Council of Great Britain.

Arts Council (1989) *Directory of Arts Centres in the United Kingdom*, London: Arts Council of Great Britain.

Arts Council (1991) *Today's Arts, Tomorrow's Tourists*, Conference Report, Science Museum, London: Arts Council of Great Britain.

Arts Council (1993a) *A Creative Future: National Arts & Media Strategy*, London: HMSO.

Arts Council (1993b) *The Millennium Map: Capital Audit of the Arts in England*, London: Arts Council of Great Britain.

Arts Council (1994) *Tate Gallery St Ives*, National Lottery Broadsheet 1 Arts, London: Arts Council of England.

Arts Council (1995) *Public Attitudes to Local Authority Funding of the Arts*, London: Arts Council of England.

Ashworth, G. (1993) 'Culture and tourism: conflict or symbiosis in Europe?', in Pompl, W. and Lavery, P. (eds) *Tourism in Europe: Structures and Developments*, Wallingford: CAB International.

Ashworth, G. (1994) *Let's Sell Our Heritage to Tourists?*, London: Council for Canadian Studies.

Ashworth, G. and Dietvorst, A. G. J. (1995) *Tourism and Spatial Transformations: Implications for Policy and Planning*, Wallingford: CAB International.

Ashworth, G. and Voogd, H. (1990) 'Can places be sold for tourism?', in Ashworth, G. and Goodall, B. (eds) *Marketing Tourism Places*, London: Routledge.

Atelier-Gesellschaft (1992) 'No arts, no city', in *Artists Need Studios*, Berlin.

Audit Commission (1989a) *The Management of Local Authority Premises*, London: HMSO.
Audit Commission (1989b) *A Review of Urban Programme and Regeneration Schemes in England*, London: HMSO.
Audit Commission (1989c) *Urban Regeneration and Economic Development: The Local Government Dimension*, London: HMSO.
Audit Commission (1989d) *Sport for Whom? Clarifying the Local Authority Role in Sport and Recreation*, London: HMSO.
Audit Commission (1991a) *Local Authorities, Entertainment and the Arts*, Local Government Report No. 2, London: HMSO.
Audit Commission (1991b) *The Road to Wigan Pier: Managing Local Authority Museums and Art Galleries*, Local Government Report No. 3, London: HMSO.
Australia Council (1991) *Local Government's Role in Arts and Cultural Development*, Canberra: Local Government and Arts Taskforce.
BAAA (1988) *Arts and the Changing City: Case Studies*, ed. J. J. Horstman, London: British American Arts Association.
BAAA (1989) *Arts and the Changing City: An Agenda for Urban Regeneration*, London: British American Arts Association.
BAAA (1990) *Investing in the Changing City: Arts Initiatives Beyond Sponsorship*, London: British American Arts Association.
BAAA (1993) *The Artist in the Changing City*, London: British American Arts Association.
Bahrdt, H. P. (1969) *Die Moderne Grofstadt. Soziologische Uberlegungen zum Stadtebau*, Hamburg: Ellert & Richter.
Bailey, P. (ed.) (1986) *Victorian Music Halls*, vol. 1: *The Business of Pleasure*, Milton Keynes: Open University Press.
Bailey, P. (1987) *Leisure and Class in Victorian England: Rational Recreation and the Contest for Control, 1830–1885* (2nd ed.), London: Routledge & Kegan Paul.
Bailey, P. (1989) 'Leisure, culture and the historian: reviewing the first generation of leisure historiography in Britain', *Leisure Studies* 8: 107–27.
Bailleu, A. (2000) 'Purple haze', *RIBA Journal* August: 6–7.
Baird, N. (1976) *The Arts in Vancouver: A Multi-Million Dollar Industry*, Vancouver: Arts Council of Vancouver.
Baird, V. (1999) 'Green cities', *New Internationalist* June: 7–10.
Balchin, P., Sykora, L. and Bull, G. (1999) *Regional Policy and Planning in Europe*, London: Routledge.
Barcelona Future (2000) *BCN Future 2004–2010*, Barcelona: Barcelona Future.
Barker, P. (1999) 'Non-plan revisited: or how cities really grow', *Journal of Design History* 12: 95–110.
Barnett, C. (1999) 'Culture, government and spatiality. Reassessing the "Foucault effect" in cultural policy studies', *International Journal of Cultural Studies* 2: 369–97.
Barrell, S. (1998) 'A short stay in . . . Bilbao', *Independent on Sunday* (London) 7 June: 3.
Barton, A. (1978) 'London comedy and the ethos of the city', *London Journal* 4: 158–80.
Barucci, P. and Becheri, E. (1990) 'Tourism as a resource for developing southern Italy', *Tourism Management*, 11: 227–39.
Bashall, R. and Smith G. (1992) 'Jam today: London's transport in crisis', in Thornley, A. (ed.) *The Crisis of London*, London: Routledge: 37–55.
Bates and Wacker, S. C. (1993) *Community Support for Culture*, Brussels.
Batten, D. F. (1993) 'Venice as a "Theseum" city: the economic management of a complex culture', paper given at the International Arts Management Conference, HEC-Paris, June.
Baud-Bovy, M. and Lawson, F. (1998) *Tourism and Recreation Handbook of Planning and Design*, Oxford: Architectural Press.

Baumol, W. and Bowen, W. (1966) *Performing Arts: The Economic Dilemma*, Cambridge: Twentieth Century Fund.

Building Design (2000) 'All the fun of the fair?' [Comment], 22 September: 13.

Beck, L. W. (ed.) (1988) *Kant: Selections*, New York: Scribner/Macmillan.

Becker, G. S. (1965) A theory in the allocation of time, *Economic Journal* 75: 3.

Becker, G. S. (1976) *The Economic Approach to Human Behavior*, Chicago: University of Chicago Press.

Becker, G. S. (1996) *Accounting for Tastes*, Cambridge, MA: Harvard University Press.

Begg, D., Fischer, S. and Dornbusch, R. (1994) *Economics* (4th ed.), London: McGraw-Hill.

Behr, V. *et al.* (1988) *Kulturwirscafht in bochum Berichte aus dem Institut für Raumplanag*, band 23, Dortmund: Universität Dortmund.

Bell, C. and Bell, R. (1972) *City Fathers: The Early History of Town Planning In Britain*, Harmondsworth: Penguin.

Benedict, B. (ed.) (1983) *The Anthropology of World Fairs*, London: Scolar.

Bennett, T. (1995) 'The multiplication of culture's utility', *Critical Inquiry* 21: 861–89.

Bennett, T. (1998) *Culture: A Reformer's Science*, London: Sage.

Berger-Vachon, C. (1992) *Ateliers: Artists' Studios*, Paris: City of Paris/Mairie de Paris.

Besant, W. (1903) *London in the Eighteenth Century*, London: Black.

Best, G. (1979) *Mid-Victorian Britain 1851–75*, London: Fontana.

BFI (2000) *Film and Television Handbook 2001*, London: British Film Institute.

Bianchini, F. (1987) GLC — RIP: cultural policies in London 1981–1986, *New Formations* 1(1).

Bianchini, F. (1989) 'Cultural policy and urban social movements: the response of the New Left, in Rome (1976–85) and London (1981–86)', in Bramham, P., van der Poel, H. and Mommaas, H. (eds) *Leisure and Urban Processes: Critical Studies of Leisure Policy in Western European cities*, London: Routledge: 18–46.

Bianchini, F. (1991a) *Urban Cultural Policy*, National Arts & Media Strategy, Discussion Document No. 40, London: Arts Council.

Bianchini, F. (1991b) 'Alternative cities', *Marxism Today* June: 36–8.

Bianchini, F. (1993) 'Culture, conflict and cities: issues and prospects for the 1990s', in Bianchini, F. and Parkinson, M. (eds) *Cultural Policy and Regeneration: The West European Experience*, Manchester: Manchester University Press: 199–213.

Bianchini, F. (1994) 'Shaping the cultural landscape', *International Arts Manager* June: 11–16.

Bianchini, F. and Parkinson, M. (eds) (1993) *Cultural Policy and Urban Regeneration: The West European Experience*, Manchester: Manchester University Press.

Bianchini, F. and Schwengel, H. (1991) 'Re-imagining the city', in Corner, J. and Harvey, S. (eds) *Enterprise and Heritage: Crosscurrents of National Culture*, London: Routledge: 212–34.

Bianchini, F., Fisher, M., Montgomery, J. and Worpole, K. (1988) *City Centres, City Cultures: The Role of the Arts in the Revitalisation of Towns and Cities*, Manchester: Manchester Free Press for the Centre for Local Economic Strategies.

Biasni, E. (ed.) *Grands Travaux*, Paris: Connaissance des Arts.

BID (1994) *BID No. 3*, London: Prospect Research, London: Building Intelligence Digest.

Bird, J. *et al.* (eds) (1993) *Mapping the Futures: Local Cultures, Global Change*, London: Routledge.

Bishcof, D. (1985) *Die wirtschaftliche Bedetung der Züricher Kulturinstitute*, Zurich.

Blaser, W. (1990) *Richard Meier, Building for Art*, Berlin: Birkhauser.

Bloomfield, J. (1993) 'Bologna: a laboratory for cultural enterprise', in Bianchini, F. and Parkinson, M. (eds) *Cultural Policy and Urban Regeneration: the West European Experience*, Manchester: Manchester University Press: 73–89.

BMRB (1996) *Survey of Arts and Cultural Activities in Britain*, London: British Market Research Bureau, for the Arts Council.

Boese, M. (2000) 'How "culturally diverse" are Manchester's cultural industries?', paper given at the Cultural Change and Urban Contexts Conference, Manchester, September: 16.

Bohrer, J. and Evans, G. L. (2000) 'Urban parks and green space in the design and planning of cities', in Benson, J. and Rose, M. (eds) *Urban Lifestyles: Spaces, Places, People*, Rotterdam: A. T. Balkema: 147–54.

Bonnemaison, S. (1990) 'City policies and cyclical events', *Celebrations: Urban Spaces Transformed, Design Quarterly* 147, 24–32.

Booth, P. and Boyle, R. (1993) 'See Glasgow, see culture', in Bianchini, F. and Parkinson, M. (eds) *Cultural Policy and Urban Regeneration: the West European Experience*, Manchester: Manchester University Press: 21–47.

Borja, J. and Castells, M. (1997) *Local and Global: Management of Cities in the Information Age*, London: Earthscan.

Borsay, P. (1989) *The English Urban Renaissance: Culture & Society in the Provincial Town, 1660–1770*, Oxford: Clarendon.

Bourdieu, P. (1984) *Distinction: A Social Critique of the Judgement of Taste*, London: Routledge & Kegan Paul.

Bourdieu, P. (1993) *The Field of Cultural Production*, Cambridge: Polity.

Bourdieu, P. and Darbel, A. (1991) *The Love of Art*, London: Polity.

Boyer, C. (1988) 'The return of aesthetics to city planning', *Society* 25: 4–56.

Boylan, P. (1993) 'Museum policy and politics in France, 1959–1991', in Pearce, S. (ed.) *Museums in Europe 1992*, London: Athlone: 87–115.

Boyle, R. and Meyer, P. (1990) 'Local economic development in the USA', *Local Economy* 4(4) [Special issue: 'Lessons from USA, Baltimore, Saint Paul, Chicago']: 272–7.

Braden, S. (1977) *Artists and People*, London: Routledge & Kegan Paul.

Bradley, P. (1993) 'Cultural policy in metropolitan Toronto: creating a culture of access', paper given to the London Arts Conference, 31 March, South Bank Centre.

Brambilla, R., Longo, G. and Dzurinko, V. (1977) *American Urban Malls: A Compendium*, New York: Institute for Environmental Affairs.

Bramham, P., van der Poel, H. and Mommaas, H. (1989) *Leisure and Urban Processes: Critical Studies of Leisure Policy in Western European cities*, London: Routledge.

Braudel, F. (1981) *Capitalism and Material Life 1400–1800*, London: Collins.

Braudel, F. (1985) *The Structures of Everyday Life*, New York: Harper & Row.

Bretton Hall (1999) *Cultural Industry Baseline Study: Yorkshire and Humberside Region*, July, University of Leeds.

Briggs, A. (1990) *Victorian Cities*, London: Penguin.

Brinson, P. (1992) *Arts and Communities: The Report of the National Inquiry into Arts and the Community*, London: Community Development Foundation.

Brownhill, S. (1990) *Developing London's Docklands*, London: Paul Chapman.

Bruton, M. and Nicholson, D. (1987) *Local Planning in Practice*, Cheltenham: Stanley Thornes.

Bruttomesso, R. (ed.) (1993) *Waterfronts: A New Frontier for Cities on Water*, Venice: International Centre Cities on Water.

Bryan, D. *et al.* (1997) *Transmitting the Benefits: The Economic Impact of BBC Wales*, Cardiff: BBC Wales.

Bryan, J. *et al.* (1998) *The Economic Impact of the Arts and Cultural Industries in Wales*, Cardiff: Welsh Economy Research Unit.

BTA (1995) *Overseas Visitor Survey*, London: British Tourist Authority.

Bubha, H. (1994) *The Location of Culture*, London: Routledge.

Buckley, R. (ed.) (1994) 'NAFTA and GATT: the impact of free trade', in *Understanding Global Issues*, Cheltenham: European Schoolbooks.

Buck-Morss, S. (1995) *The Dialectics of Seeing: Walter Benjamin and the Arcades Project*, Cambridge, MA: MIT Press.

Burckhardt, J. C. (1990) *The Civilization of the Renaissance in Italy* (trans. Middlemore, S. G. C.), London: Penguin.

Burgers, J. (1995) 'Public space in the post-industrial city', in Ashworth, G. and Dietvorst, A. G. J. (eds) *Tourism and Spatial Transformations: Implications for Policy and Planning*, Wallingford: CAB International: 147–61.

Burns, J. J. and Mules, T. (eds) (1989) *The Adelaide Grand Prix: The Impact of a Special Event*, Adelaide: Centre for South Australian Economic Studies.

Burtenshaw, D., Bateman, M. and Ashworth, G. J. (1991) *The European City: A Western Perspective*, London: David Fulton.

Burton, T. L. (1971) *Experiments in Recreation Research*, London: George Allen & Unwin.

Byrne, D. (1997) 'Chaotic places or complex places', in Westwood, S. and Williams, J. (eds) *Imagining Cities: Scripts, Signs, Memory*, London: Routledge.

CABE (2001) *The Value of Urban Design*, London: Bartlett School of Architecture & Planning for the Commission for Architecture and the Built Environment.

Cahiers Français (1993) *Culture et societe*, No. 260, Paris: La documentation française.

Calvino, I. (1979) *Invisible Cities*, London: Pan.

Casey, B., Dunlop, R. and Selwood, S. (1996) *Culture as Commodity: The Economics of the Arts and Built Heritage in the UK*, London: Policy Studies Institute.

Castells, M. (1977) *The Urban Question*, London: Edward Arnold.

Castells, M. (1989) *The Informational City: Information Technology, Economic Restructuring and the Urban-Regional Process*, Oxford: Blackwell.

Castells, M. (1996) *The Information Age: Economy Society and Culture*, vol. 1: *The Rise of the Network Society*, Oxford: Blackwell.

CEC (1996) *Report on Economic Cohesion*, Brussels, Committee of the European Commission.

CELTS (Evans, G. L., Shaw, S. and Bertram, J.) (2000) *Visitor Baseline Study of the Jubilee Line Extension*, JLE Impact Study Unit, London: University of Westminster.

CENTEC (1995) *Employment and Training Needs in the Media and Entertainment Industries*, London: Arts Business for Central London Training & Enterprise Council.

Cervero, R. and Landis, J. (1997) 'Twenty years of the Bay Area Rapid Transit System: land use and development impacts', *Transport Research* Part A, 31: 309–33.

Chalklin, C. W. (1980) 'Capital expenditure on building for cultural purposes in provincial England 1730–1830', *Business History* 22: 51–70.

Challans, T. and Sargent, A. (1991) *Local Authorities and the Arts*, Discussion Document No. 16, National Arts & Media Strategy, London: Arts Council.

Chanan, M. (1980) *The Dream that Kicks: The Prehistory and Early Years of Cinema in Britain*, London: Routledge & Kegan Paul.

Chang, T. C. (2000) 'Renaissance revisited: Singapore as a "Global City for the Arts"', *International Journal of Urban and Regional Research* 24(4): 818–31.

Cherry, G. (1972) *The Evolution of British Town Planning (1914–74)*, London: Lawrence Hill.

Cheshire, P. and Hay, A. (1989) *Urban Problems in Western Europe*, London: Allen & Unwin.

Choay, F. (1969) *The Modern City Planning in the Nineteenth Century*, London: Studio Vista.

Christopherson, S. (1994) 'The Fortress City: privatised spaces, consumer citizenship', in Amin, A. (ed.) *Post-Fordism: A Reader*, Oxford: Blackwell: 409–27.

City of Cardiff (1994) *The Economic Importance of the Cultural Industries in Cardiff*, Report to the Economic Development Committee, March.

City of Toronto (1991) *Cityplan '91: Public Art Policy Study*, No. 20, March.

Clammer, J. (1987) *Contemporary Urban Japan*, Oxford: Blackwell.

Clawson, M. and Knetch, J. (1986) *Economics of Outdoor Recreation*, Baltimore: Johns Hopkins University Press.

Clifford, J. (1988) *The Predicament of Culture*, Cambridge, MA: Harvard University Press.

Clifford, J. (1990) 'On collecting arts and culture', in Ferguson, R. (ed.) *Out There: Marginalisation and Contemporary Culture 4*, Cambridge, MA: MIT Press.

Clifford, J. (1997) 'Museums as contact zones', in Clifford, J. (ed.) *Travel and Translation in Late Twentieth Century*, New Haven: Harvard University Press.

Coalter, F. (1990) 'The politics of professionalism: consumers or citizens', *Leisure Studies* 9: 107–19.

Cohen, E. (1999) 'Cultural fusion', in *Values and Heritage Conservation*, Los Angeles: Getty Conservation Institute: 44–50.

Cohen, J.-L. and Fortier, B. (1988) *Paris La Ville et Ses Projets (A City in the Making)*, Paris: Babylone.

Colenutt, B. (1988) 'Local democracy and inner city regeneration', *Local Economy* 3: 119–25.

Comedia (1991a) *The Cultural Industries in Liverpool*, Report to Merseyside Task Force, Liverpool: Comedia.

Comedia (1991b) *London World City: The Position of Culture*, London: London Planning Advisory Committee.

Comedia (1996) *The Social Impact of Arts Programmes*, Working Papers 1 to 6, Stroud: Comedia.

Connolly, D. (1998) 'Paper dreams of the Parisian future', *Architect's Journal* 9 July: 49.

Connolly, S. (1997) 'The measurement of additionality: a case study of the UK National Lottery', in *Business and Economics in the 21st Century*, vol. 1, BES International Conference.

Conway, H. (1989) 'Victorian parks. Part 1', *Landscape Design* September: 21–3.

Cook, A. J. (1981) *The Privileged Playgoers of Shakespeare's London, 1576–1642*, Princeton: Princeton University Press.

Cook, M. (1993) *French Culture Since 1945*, Harlow: Longman.

Cook, R. M. (1972) *Greek Art*, Harmondsworth: Penguin.

Cooke, P. (1990) *Back to the Future*, London: Unwin Hyman.

Cooke, P., Terk, E., Karnite, R. and Blagnys, G. (2000) 'Urban transformations in the capitals of the Baltic States: innovation, culture and finance', in Bridge, G. and Watson, A. (eds) *A Companion to the City*, Oxford: Blackwell.

Coombes, A. (1994) *Reinventing Africa*, New Haven and London: Yale University Press.

Cosh, M. (1990) *The Squares of Islington*, Part 1: *Finsbury and Clerkenwell*, London: Islington Archaeology & History Society.

Council of Europe (1991) *European Parliament VI Resolution on a Community Tourism Policy* 11/6/91, 88/63l, Strasbourg: Council of Europe.

Council of Europe (1992) *European Urban Charter*, Standing Conference of Local and Regional Authorities of Europe (CLRAE), 18 March, Strasbourg.

Coupland, A. (1992) 'Every job an office job', in Thornley, A. (ed.) *The Crisis of London*, London: Routledge.

Coupland, A. (1997) *Reclaiming the City*, London: E & FN Spon.

Cowan, R. and Gallery, L. (1990) 'A vision for London', *Architects' Journal* [Special Issue] 11(191): 29–87.

Cowen, T. (1998) *In Praise of Commercial Culture*, Cambridge, MA: Harvard University Press.

CPRE (1993) *The Lost Land*, London: Council for the Protection of Rural England.

Craig, S. (1991) *Customer Service Audit of Leisure Facilities*, London: Leisure Futures Ltd.

Craig, S. and Evans, G. L. (1995) *The London Millennium Study — A Survey of Events and Projects Planned for the Millennium in London*, London: London First.

Crang, M. (1998) *Cultural Geography*, London: Routledge.

Crickhowell, N. (1997) *Opera House Lottery: Zaha Hadid and the Cardiff Bay Project*, Cardiff: University of Wales Press.

Crimp, D. (1985) 'On the museum's ruins', in Foster, H. (ed.) *Postmodern Culture*, London: Pluto: 43–56.

Crouch, D. (1998) 'The street in the making of popular geographical knowledge', in Fyfe, N. R. (ed.) *Images of the Street: Planning Identity and Control in Public Space*, London: Routledge: 161–75.

Crowhurst, A. J. (1992) 'The music hall, 1885–1922. The emergence of a national entertainment industry in Britain', unpublished PhD thesis, Cambridge: University of Cambridge.

CSP (1999) *Culture and the City — Ten Ways to make a Difference*. Agenda 3.2, Consultation Document from the Cultural Strategy Partnership for London.

Cubeles, X., and Fina, X. (1998) *Culture in Catalonia*, Barcelona: Fundacio Jaume Bofill.

Cuff, P. and Kaiser, B. (1986) *Animation of the City: Washington, DC Downtown Study*, Washington, DC: Partners for Liveable Spaces.

Cullingworth, J. B. (1979) *Town & Country Planning in Britain* (7th ed.), London: George Allen & Unwin.

Cullingworth, J. B. and Nadin, V. (1994) *Town and Country Planning in Britain* (11th ed.) London: Routledge.

Curran, J. and Porter, V. (1983) *British Cinema History*, London: Weidenfeld & Nicholson.

Cwi, D. (1981) *Economic Impact of the Arts & Cultural Institutions*. Case studies in Columbus, Minneapolis/St Paul, St Louis, Salt Lake City, San Antonio, Springfield. Washington, DC: National Endowment for the Arts.

Cwi, D. and Lyall, K. (1977) *Economic Impacts of Arts & Cultural Institutions: A Model for Assessment and a Case Study in Baltimore*, Research Report No. 6, Washington, DC: National Endowment for the Arts: 21–4.

Daly, H. E and Cobb, J. B. (1990) *For the Common Good: Redirecting the Economy towards Community, the Environment and a Sustainable Future*, London: Merlin.

Darton, D. (1985) *Social Change and the Arts*, London: National Association of Arts Centres and Henley Centre for Forecasting.

Davey, P. (ed.) (1993) 'Recreation', *Architectural Review* 194(1157): 4–98.

Davidoff, P. (1965) 'Advocacy and pluralism in planning', *Journal of the American Institute of Planners* 21(4); repr. in LeGates, R. T. and Stout, F. (eds) (1996) *The City Reader*, London: Routledge: 421–32.

Davidoff, P. and Reiner, T. A. (1973) 'A choice theory of planning', in Faludi, A. (ed.) *A Reader in Planning Theory* (1994 repr.), Oxford: Pergamon.

Davies, H. W. E. (1994b) Towards a European planning system?, *Planning Practice and Research* 9: 63–9.

Davies, H. W. E. and Gosling, J. A. (1994a) *The Impact of the European Community on Land Use Planning in the United Kingdom*, London: Royal Town Planning Institute.

Davies, N. (1982) *The Ancient Kingdoms of Mexico*, London: Pelican.

Davies, R. *et al.* (1992) *The Effect of Major Out-of-Town Retail Development*, London: HMSO.

Davies, S. and Selwood, S. (1999) 'English cultural services: government policy and local strategies', *Cultural Trends* 30: 69–110.

Dawson, B. (1999) *Street Graphics India*, London: Thames & Hudson.

DCC (1987) *Urban Development Corporations: Six Years in London's Docklands*, London: Docklands Consultative Committee.

DCC (1990) *The Docklands Experiment: A Critical Review of Eight Years of the London Docklands Development Corporation*, London.

DCMS (1998) *Creative Industries Mapping Document*, London: Department for Culture, Media and Sport.

DCMS (1999) *Draft Guidance for Local Cultural Strategies*, London: Department for Culture, Media and Sport.

DCMS (2000) *Creative Research: A Modernising Government Review of DCMS's Statistical and Social Policy Research Needs*, London: Department for Culture Media and Sport.

De la Durantaye, M. (1999) 'Municipal cultural policies in Quebec and quality of life indicators', in Proceedings of the International Conference on Cultural Policy, Bergen: 275–85.

De Ville, B. and Kinsley, B. (1989) *Cultural Facilities: Oversupply or Undersupply — Guidelines for Increasing Participation*, Ottawa: Department of Communications.

Debord, J. (1983) *The Society of the Spectacle*, Detroit: Black & Red.

Deckker, T. (2000) 'Brasilia: city versus landscape', in Deckker, T. (ed.) *The Modern City Revisited*, London: E & FN Spon: 167–93.

Deegan, J. and Dineen, D. A. (1998) 'Tourism policy and rapid visitor growth: the case of Ireland', paper given at the TOLERN Annual Conference, University of Durham, December.

Deffner, A. (1992a) 'Cultural activities and free-time: social and geographical dimensions', in Maloutas and Economou (eds) *Social Structure and Urban Organisation in Athens*, Athens: Paratitris: 377–442.

Deffner, A. (1992b) 'Cultural spaces in Athens: continuity and change', paper given to the Leisure and New Citizenship, European Leisure and Recreation Association — VIII Congress, Bilbao.

Deffner, A. (1993) 'Cultural activities in Greece: tradition or modernity? (geographical distribution of cultural spaces in Greece)', paper given to the Leisure Studies Association 3rd International Conference, *Leisure in Different Worlds*, 14–18 July, Loughborough University: LSA.

Department de Cultura (1998) *Economy and Culture in Catalonia: Basic Statistics*, Barcelona: Generalitat de Catalunya.

Derrida, J. (1991) *The Other Heading: Reflections on Today's Europe*, Paris: Autre Cap.

Design Museum–Architecture Foundation (2000) *Living in the City*, London: Design Museum–Architecture Foundation.

Dethier, J. and Guiheux, A. (1994) *La ville: arts et architecture en Europe 1870–1993*, Paris: Editions du Centre Pompidou.

DETR (1999) *An Urban Renaissance, Final Report of the Urban Task Force*, for the Department for the Environment, Transport and the Regions, London: Routledge.

Devesa, M. (1999) 'The policy of cultural expenditure in Castilla-Leon (Spain)', paper given to the Incentives and Information in Cultural Economics, FOKUS-ACEI Joint Symposium, Vienna, January.

DMU (1995) *Course Prospectus for MA in European Cultural Planning*, Leicester: De Montfort University.

DNH (1995) *Guidance on Local Authority Arts Provision*, London: Department of National Heritage.

Doak, J. (1993) 'Commercial property boom, gloom, and the way ahead', *Planning Practice and Research*, 8(4).

Dobson, L. C. and West, E. G. (1988) Performing arts subsidies and future generations, *Journal of Cultural Economics* 12: 8–115.

Docklands Forum with Miller, C. (ed.) (1989) *Does the Community Benefit?: What Can the Private Sector Offer? Lessons from the London Docklands*, May, London: Docklands Forum.

DoE (1974) *Structure Plans*, Circular No. 98/74, London: Department of the Environment.

DoE (1977a) *Recreation and Deprivation in Inner Urban Areas*, London: HMSO for the Department of the Environment.

DoE (1977b) *Leisure and Quality of Life Experiments* (2 vols), London: HMSO for the Department of the Environment.

DoE (1984) *The Reallocation of Planning Functions in the Greater London Council (GLC) and Metropolitan County Council (MCC) Areas — Revised Proposals Paper*, 14 June, London: Department of the Environment.

DoE (1985a) *Local Government Act, 1985: Section 5*, London: HMSO.

DoE (1985b) *Streamlining the Cities*, White Paper, Cmnd 9063, London: HMSO.

DoE (1986) *Paying for Local Government*, Cmnd 9714, London: HMSO.

DoE (1987a) *Historic Buildings and Conservation Areas — Policy and Procedures*, Circular No. 8 (87), London: HMSO.

DoE (1987b) *Use Classes Order 13/87*, London: Department of the Environment.

DoE (1988a) *Urban Programme 1986–87: A Report on Operations and Achievements in England*, London: HMSO.

DoE (1988b) *General Development Order*, London: Department of the Environment.

DoE (1989a) *Regional Guidance for the South East*, Planning Policy Guidance No. 9, London: HMSO.

DoE (1989b) *Planning Agreements, Consultation Paper on Section 52*, TCPA, 1971, London: Department of the Environment.

DoE (1990a) *The Town and Country Planning Act*, London: HMSO.

DoE (1990b) *Tourism and The Inner City: An Evaluation of the Impact of Grant Assisted Tourism Projects*, London: HMSO.

DoE (1990c) *This Common Inheritance*, London: HMSO.

DoE (1990d) *Planning Policy Guidance on Archaeology and Planning*, PPG No. 16, London: HMSO.

DoE (1990e) *Planning and Compensation Act*, London: HMSO.

DoE (1991a) *Planning Policy Guidance on Sport & Recreation*, No. 17, London: HMSO.

DoE (1991b) *Strategic Guidance for London*, RPG3, London: HMSO.

DoE (1991c) *Census of Employment*, London: HMSO.

DoE (1992a) *Unitary Development Plans: Public Local Inquiries*, London: HMSO.

Dovey, K. (1989) 'Old scabs/new scars: the hallmark event and the everyday environment', in Syme *et al.* (eds) *The Planning and Evaluation of Hallmark Events*, Aldershot: Avebury: 73–80.

Downey, J. (1999) 'XS 4 All? "Information Society" policy and practice in the European Union', in Downey, J. and McGuigan, J. (eds) *Technocities*, London: Routledge: 121–38.

Downey, J. and McGuigan, J. (eds) (1999) *Technocities*, London: Routledge.

Doxiadis, C. (1968) *Ekistics: An Introduction to the Science of Human Settlements*, London: Hutchinson.

DPA (2000) *Creative Industries Strategy for London*, London: David Powell Associates for the London Development Partnership.

DRV Research (1986) *Economic Impact Study of Tourist and Associated Arts Developments in Merseyside*, The Tourism Study.

Du Gay, P. (ed.) (1997) *Production of Culture/Cultures of Production*, Milton Keynes: Open University.

Dumont, C. (1979) *Cultural Action in the European Community*, CEC, 3/1980, Brussels.

Dunleavy, P. (1980) *Urban Political Analysis*, London: Macmillan.

Dunleavy, P. (1991) *Democracy Bureaucracy & Public Choice: Economic Explanations in Political Science*, Hemel Hempstead: Harvester Wheatsheaf.

During, S. (1993) *The Cultural Studies Reader*, London: Routledge.

Eagleton, T. (2000) *The Idea of Culture*, Oxford: Blackwell.

EC (1997) *The European Observatory for SMEs*, 5th Annual Report. Brussels: European Commission.

EC (1998) *Culture, The Cultural Industries and Employment*, Commission Staff Working Paper, Brussels: European Commission.

EC (1999) *Information Society Technologies for Tourism*, DG XIII, Brussels: European Commission.

EC (2000) *Information Society Technology: Work Programme 2001*, Brussels: European Commission.

Edensor, T. (1998a) 'The culture of the Indian street', in Fyfe, N. R. (ed.) *Images of the Street: Planning, Identity and Control in Public Space*, London: Routledge: 205–24.

Edensor, T. (1998b) *Tourists at the Taj: Performance Meaning at a Symbolic Site*, London: Routledge.

Edgar, D. (1991) 'From Metroland to the Medicis', in Fisher, M. and Owen, U. (eds) *Whose Cities?*, London: Penguin: 19–31.

Ehrenreich, B. and Ehrenreich, J. (1979) 'The professional-managerial class', in Walker, P. (ed.) *Between Labour and Capital*, Boston: South End.

Ellison, M. (1994) 'Orchestras "lack audiences not fans"', *The Guardian* 13 October: 3.

Ellmeier, A. and Rasky, B. (1998) *Cultural Policy in Europe: European Cultural Policy? Nation-State and Transnational Concepts*, Austrian Culture documentation. Vienna: International Archive for Cultural Analysis.

Englefield, D. (1987) *Local Government and Business: A Practical Guide*, London: Municipal Journal Ltd for the Industry and Parliament Trust.

English Heritage (1997) *Maritime Greenwich: Draft Management Plan for Consultation*, October, London: English Heritage.

ÉPAD (1993) *La Défense*, Point Info-Service Communication, Paris: Établissement Public de Aménagement de la Region de La Défense.

EU (1995) *Structural Outline and Current Situation of Cultural Statistics in Member-States of the European Union*, Working Papers (France, Netherlands, Sweden, Italy, Germany), Paris: Ministere Culture Francophonie.

Evans, G. L. (1989) *Survey of Employment in the Arts & Cultural Industries in Islington*, London: London Borough of Islington/Greater London Arts.

Evans, G. L. (1990) *Premises Needs and Problems of Crafts Firms in Clerkenwell*, London: Local Enterprise Research Unit, Polytechnic of North London.

Evans, G. L. (1993a) 'Leisure and tourism investment incentives in the European Community: changing rationales', paper given at the International LSA Conference, Loughborough University.

Evans, G. L. (1993b) *Arts and Cultural Tourism in Europe: Policy and Markets*, Proceedings of the 2nd International Arts Management Conference, HEC-Paris, June.

Evans, G. L. (1993c) *Planning for the Arts and Regeneration in London: An Urban Renaissance?*, PILTS Paper No. 6, London: University of North London Press.

Evans, G. L. (1993d) *An Economic Strategy for the Arts & Cultural Industries in Haringey*, and *Survey of Employment in the Arts & Cultural Industries*, London Borough of Haringey Technical & Environmental Services, London: London Arts Board.

Evans, G. L. (1994) 'Tourism in Greater Mexico and the Indigena — whose culture is it anyway?', in Seaton. A. (ed.) *Tourism: State of the Art*, Chichester: Wiley: 836–47.

Evans, G. L. (1995a) 'The National Lottery: planning for leisure or pay up and play the game?', *Leisure Studies* 14: 225–44.

Evans, G. L. (1995b) 'Tourism & leisure in Eastern Europe: the Westernisation Project', in Leslie, D. (ed.) *Tourism, Culture and Participation*, vol. 1, Publication No. 51, Brighton: LSA: 59–79.

Evans, G. L. (1995c) 'Tourism and education: core functions of museums?', in Leslie, D. (ed.) *Tourism, Culture and Participation*, vol. 1, Publication No. 51, Brighton: LSA: 157–80.

Evans, G. L. (1996a) 'Planning for the British Millennium Festival: establishing the visitor baseline and a framework for forecasting', *Journal of Festival Management and Event Tourism* 3: 183–96.

Evans, G. L. (1996b) *Health, Travel and Tourism*, London: Health Education Authority.

Evans, G. L. (1996c) 'Planning for the arts and culture in London and Toronto: a tale of two cities', paper given at the British Association for Canadian Studies Annual Conference, University of Exeter.

Evans, G. L. (1996d) *Media Services Sector — Report on the Pilot Network in the CILNTEC Area*, London: CELTS for City & Inner London North TEC.

Evans, G. L. (1997) *MultiMedia Sector — Employment and Labour Market Report*, London: CILNTEC.

Evans, G. L. (1998a) 'In search of the cultural tourist and the post-modern Grand Tour', paper given at *Relocating Sociology*, the International Sociological Association — XIV Congress, International Tourism, Montreal, July.

Evans, G. L. (1998b) 'La Demande Européenne en Matiere de Tourisme Culturel', in Rautenberg, M. (ed.) *Le Tourisme Culturel, Phénomène de Société Publics et Marchés*, Lyon: Presses Universitaires de Lyon.

Evans, G. L. (1998c) 'European regional development policy and the arts and urban regeneration', paper given at the UACES European Cultural Policy Conference, City University, London, April.

Evans, G. L. (1998d) 'Urban leisure: edge city and the new pleasure periphery', in Collins, M. and Cooper, I. (eds) *Leisure Management — Issues and Applications*, Wallingford: CAB International: 113–38.

Evans, G. L. (1998e) *Study into the Employment Effects of Arts Lottery Spending in England*, Research Report No. 14, London: Arts Council of England.

Evans, G. L. (1998f) 'Millennium tourism: planning, pluralism and the party', paper given at the 5th World Leisure & Recreation Congress, Sao Paulo, October.

Evans, G. L. (1999a) 'Measuring the arts and cultural industries — does size matter?', in Roodhouse, S. (ed.) *Proceedings of The New Cultural Map: A Research Agenda for the 21st Century* Conference, Bretton Hall, University of Leeds: 26–34.

Evans, G. L. (1999b) 'Leisure and tourism investment incentives in the European Community: changing rationales', in McPherson, G. and Foley, M. (eds) *Sustainability and Environmental Policies* (Vol. 1), Publication No. 50, Brighton: LSA: 1–27.

Evans, G. L. (1999c) 'Heritage tourism: development and diversity', paper given at the 12th World Congress of Conservation and Monumental Heritage, *The Wise Use of Heritage*, ICOMOS Assembly, Mexico.

Evans, G. L. (1999d) 'Last chance lottery and the millennium city', in Whannel, G. and Foley, M. (eds) *Leisure, Culture and Commerce: Consumption and Participation*, Publication No. 64, Brighton: LSA.

Evans, G. L. (1999e) 'The economics of the national performing arts — exploiting consumer surplus and willingness-to-pay: a case of cultural policy failure?', *Leisure Studies* 18: 97–118.

Evans, G. L. (1999f) 'Cultural tourism and cultural policy — identity and the European project', in Proceedings of the International Conference on Cultural Policy Research, University of Bergen.

Evans, G. L. (1999g) 'Cultural change and cultural management in London's Tate Galleries', paper given at the International Arts Management Conference, Helsinki School of Economics, June.

Evans, G. L. (2000a) 'Historic Quebec — capital city, World Heritage City and the Québecois Project', paper given at the BACS 25th Annual Conference, Edinburgh, July.

Evans, G. L. (2000b) 'Measure for measure: evaluating performance and the arts organisation', *Studies in Cultures, Organizations and Societies* 6: 243–66.

Evans, G. L. (2000c) 'Planning for urban tourism: a critique of unitary development plans and tourism policy in London', *International Journal of Tourism Research* 2: 326–47.

Evans, G. L. and Cleverdon, R. (2000) 'Fair trade in tourism — community development or marketing tool?', in Richards, G. and Hall, D. (eds) *Tourism and Sustainable Community Development*, London: Routledge: 137–53.

Evans, G. L. and Foord, J. (1999) 'Cultural policy and urban regeneration in East London: world city, whose city?', in Proceedings of the International Conference on Cultural Policy Research, University of Bergen: 457–94.

Evans, G. L. and Foord, J. (2000a) 'Landscapes of cultural production and regeneration', in Benson, J. and Rose, M. (eds) *Urban Lifestyles: Spaces, Places, People*, Rotterdam: A. T. Balkema: 249–56.

Evans, G. L. and Foord, J. (2000b) 'European funding of culture: promoting common culture or regional growth?', *Cultural Trends* 36: 53–87.

Evans, G. L., Foord, J. and White, J. (1999) *Putting Cultural Activity back into Stepney*, London: London Borough of Tower Hamlets.

Evans, G. L. and Reay, D. (1996) *Arts Culture and Entertainment Park Plan — Topic Study*, Waltham Abbey: Lee Valley Regional Park Authority.

Evans, G. L. and Shaw P. (1992) *Arts Centres Review for Portsmouth*, Portsmouth City Arts/Hampshire County Arts, Winchester: Southern Arts Board.

Evans, G. L. and Shaw, S. (1999) 'Urban tourism and transport planning: case of the Jubilee Line Extension and East London Corridor', paper given at the RGS/IBG Symposium British Tourism: The Geographical Research Frontier, University of Exeter, September.

Evans, G. L., Shaw, P. and White, J. (1997) 'Digest of arts and cultural trends 1987–1996', Pre-Publication Draft, Arts Councils of England, London: ACE Research Publication.

Evans, G. L., Shaw, P., White, J. *et al.* (2000) *Artstat: Digest of Arts Statistics and Trends in the UK 1986/87–1997/98*, London: Arts Council of England.

Evans, G. L. and Smeding, S. (1997) *Survey of Leisure Services Revenue and Capital Budgets 1998/87*, Associations of Metropolitan Authorities, District Councils and County Councils, London: University of North London.

Evans, G. L. and Smith, M. (2000) 'A tale of two heritage cities: Old Quebec and Maritime Greenwich', in Robinson, M. *et al.* (eds) *Tourism and Heritage Relationships: Global, National and Local Perspectives*, Sunderland: Business Education Publishers: 173–96.

Evans, R. (1997a) *Regenerating Town Centres*, Manchester: Manchester University Press.

Evans, R. J. (1997b) *In Defence of History*, London: Granta.

Everitt, A. (1992) 'Homage to the arts', *The Insider* Winter: 6–7.

Fainstein, S. (1984) *The City Builders: Property, Politics and Planning in London and New York*, Oxford: Blackwell.

Fainstein, S., Gordon, I. and Harloe, M. (1992) *Divided Cities*, Oxford: Blackwell.

Fairs, M. (1999) 'Spanish let fly — Barcelona's Royal Gold Medal Winners use RIBA ceremony to deliver broadside at neglect of Britain's cities', *Building Design* 25 June: 1.

Falassi, A. (ed.) (1987) *Time Out of Time: Essays on the Festival*, Albuquerque: University of New Mexico Press.

Faludi, A. (1973) *A Reader in Planning Theory* (1994 repr.), Oxford: Pergamon.

Fanstein, S. (1994) *The City Builders: Property, Politics, and Planning in London and New York*, Cambridge, MA: Blackwell.

Farrell, T. (2000) 'Urban regeneration through cultural masterplanning', in Benson, J. F. and Roe, M. (eds) *Urban Lifestyles: Spaces, Places People*, Rotterdam: A.T. Balkema.

Faubion, J. D. (ed.) (1994) *Michel Foucault: Power, The Essential Works 3*, London: Allen Lane.

Featherstone, M. (1991) *Consumer Culture and Postmodernism*, London: Sage.

Feist, A. (1995) 'A statutory basis for the arts', October, London: Arts Council of England: 24–52.

Feist, A., Fisher, R., Gordon, C. and Morgan, C. (1998) *International Data on Public Spending on the Arts in Eleven Countries*, ACE Research Report No. 13, London: Arts Council of England.

Feist, A. and Hutchison, R. (1989) *Cultural Trends 1*, London: Policy Studies Institute.

Fernandez-Armesto, F. (1996) *Millennium: A History of Our Last Thousand Years*, London: Black Swan.

Field, B. and MacGregor, B. (1987) *Forecasting Techniques for Urban and Regional Planning*, London: Hutchinson.

Fisher, M. (1991) 'Introduction', in Fisher, M. and Owen, U. (eds) *Whose Cities?*, London: Penguin: 1–6.

Fisher, M. and Owen, U. (eds) (1991), *Whose Cities?*, London: Penguin.

Fisher, R. (1993) *The Challenge for the Arts: Reflections on British Culture in Europe in the Context of the Single Market and Maastricht*, London: Arts Council.

Fleming, T. (ed.) (1999) *The Role of Creative Industries in Local and Regional Development*, Sheffield: Government Office for Yorkshire and the Humber.

Focas, C., Genty, P. and Murphy, P. (1995) *Top Towns*, London: Guinness Publ.

Foley, D. L. (1973) 'British town planning: one ideology or three?', in Faludi, A. (ed.) *A Reader in Planning Theory*, Oxford: Pergamon: 69–94.

Fontana, J. (1994) *Europa ante el espejo*, Barcelona: Critica.

Foord, J. (1999) 'Creative Hackney: reflections on hidden art', *Rising East* 3: 38–66.

Forrester, S. (1985) *Arts Activities in Building-based Organisations Throughout Greater London*, London Association of Arts Centres, London: Policy Studies Institute.

Forster, W. (1983) *Arts Centres and Education*, London: Arts Council.

Foster, H. (ed.) (1983) *Postmodern Culture*, London: Pluto.

Foucault, M. (1991) 'Governmentality', in Burchell, G., Gordon, C. and Miller, P. (eds) *The Foucault Effect: Studies in Governmentality*, London: Harvester Wheatsheaf: 87–104.

Fox, C. (1992) *London — World City 1800–1840*, New Haven and London: Yale University Press in association with the Museum of London.

Frampton, K. (1985) 'Towards a critical regionalism: six points for an architecture of resistance', in Foster, H. (ed.) *Postmodern Culture*, London: Pluto: 16–30.

Frangialli, F. (1998) *The Role of Private Financing in Sustainable Cultural Development*, 28 September, Washington, DC: World Bank.

Frey, B. S. and Pommerehne, W. W. (1989) *Muses and Markets: Explorations in the Economics of the Arts*, Oxford: Blackwell.

Frey, R. L. (1976) *Theater und Ökonomie. Eine wirtschafitliche Analyse der Basler Theater von Ökonomiestudenten der Universität Basel.*

Friedmann, J. (1986) 'The World City hypothesis', *Development and Change* 17: 69–83.

Frost, M. and Peterson, G. O. (1978) *The Economic Impact of Non-Profit Arts Organisations in Nebraska 1976–77*, Omaha.

Fuller-Love, N., Jones, A. and Peel, D. (1996) *The Impact of S4C on Small Businesses*, Aberystwyth: ESRC Research Report.

Fyfe, N. R. (ed.) (1998) *Images of the Street. Planning Identity and Control in Public Space*, London: Routledge.

Garcia, S. (ed.) (1993) *European Identity and the Search for Legitimacy*, London: Pintner.

Gardiner, C. (1998) *Box Office Data Report 1997*, London: Society of London Theatres.

Garnham, N. (1983) *The Cultural Industries and Cultural Policy in London*, AR116 and IEC 940, London: GLC.

Garnham, N. (1984) 'Cultural industries: what are they?', *Views*, Independent Film and Video Producers Association, London.

Garreau, J. (1991) *Edge City: Life on the New Frontier*, New York: Anchor.

Getz, D. (1991a) *Festivals, Special Events and Tourism*, New York: van Nostrand.

Getz, D. (1991b) Assessing the economic impact of festivals and events, *Journal of Applied Recreation Research* 19: 61–77.

Getz, D. (1994) 'Event tourism and the authenticity dilemma', in Theobald, W. (ed.) *Global Tourism*, Oxford: Butterworth-Heinemann: 313–29.

Giedon, S. (1963) *Space, Time and Architecture*, Cambridge, MA: Harvard University Press.

Gilhespy, I. (1991) *The Economic Importance of the Arts in the South*, Bournemouth Polytechnic, Winchester: Southern Arts Board.

Girard, A. (1987) 'The Ministry of Culture', in Stewart, R. (ed.) *The Arts Politics, Power and the Purse*, London: Arts Council: 8–12.

GLA (1989) *Study into the Turnover, Salary and Conditions of Key Arts Workers*, Leisure Futures Ltd, London: Greater London Arts.

GLA (1990a) *The Arts Plan for London*, London: Greater London Arts.

GLA (1990b) *A Strategy for the Arts in London*, London: Greater London Arts.

GLA (1990c) *Supporting the Arts in London: Summary of the Arts Plan for London and GLA's Strategy for the Arts 1990–1995*, London: Greater London Arts.

GLA (1990d) *Supporting the Arts in London: Funding Guidelines* (1), London: Greater London Arts.

GLA with Montgomery, J., Evans, G. L. and Gavron, N. (eds) (1991) *Space for the Arts in London — Planning for London's Arts, Culture and Entertainment*, London: Greater London Arts.

Glasgow District Council (1990) *The Economic Importance of the Arts in Glasgow*, Factsheet No. 6, Glasgow: Festivals Unit.

Glasgow, M. and Evans, B. I. (1949) *The Arts in England*, London: Falcon.

Glass, R. (1973) 'The evaluation of planning: some sociological considerations', in Faludi, A. (ed.) *A Reader in Planning Theory*, Oxford: Pergamon: 45–68.

GLC (1967) *Greater London Development Plan*, London: Greater London Council.

GLC (1975) *Greater London Recreation Study*, Research Report 19, London: Greater London Council.

GLC (1985) *State of the Arts or the Art of the State: Strategies for the Cultural Industries*, London: Greater London Council.

Glennie, P. D. and Thrift, N. (1992) 'Modernity, urbanism and modern consumption', *Society and Space* 10: 423–43.

Glennie, P. D. and Thrift, N. (1993) 'Modern consumption: theorizing commodities and consumers', *Society and Space* 11: 603–6.

Gonzalez, J. M. (1993) 'Bilbao: culture, citizenship and quality of life', in Bianchini, F. and Parkinson, M. (eds) *Cultural Policy and Urban Regeneration: The West European Experience*, Manchester: Manchester University Press: 73–89.

Gooch, A. (1998) 'Catalan quotas spark fear of Babel', *The Guardian* 9 July: 20.

Goodey, B. (1983) *Urban Culture at a Turning Point?*, Strasbourg: Council of Europe for the Council for Cultural Co-operation.

Gooding, A. (1995) 'Garden festivals unpacked — ephemeral vistas or prospects for the future?', unpublished MA dissertation, University of North London.

Gormley, A. (1998) *Making an Angel of the North*, London: Booth-Clibborn.

Gorz, A. (1989) *A Critique of Economic Reason*, London: Verso.

Goss, J. (1992) 'The magic of the mall: an analysis of form, function, and meaning in the contemporary retail built environment', *Annals of the Association of American Geographers* 83: 18–47.

Gowland, D. A., O'Neill, B. C. and Reid, A. L. (1995) *The European Mosaic: Contemporary Politics, Economics and Culture*, Harlow: Longman.

Graburn, N. H. (1976) *Ethnic and Tourist Arts: Cultural Expressions from the Fourth World*, Berkeley: University California Press.

Graham-Dixon, A. (1999) *Renaissance*, London: BBC Worldwide.

Grant, A. (1990) 'Out of town leisure developments', paper given at the UK Leisure Property Conference (4:4): London.

Gras, H. K. (1999) 'Myths and statistics, or a clearer view on the nineteenth-century stage', paper given at the Researching Culture Conference, University of North London, September.

Gratz, R. B. and Mintz, N. (1998) *Cities Back from the Edge: New Life for Downtown*, New York: Wiley.

Greed, C. H. (1994) *Women & Planning*, London: Routledge.

Greenhalgh, P. (1988) *Ephemeral Vistas. The 'Expositions Universelles', Great Exhibitions and World's Fairs, 1851–1939*, Manchester: Manchester University Press.

Greenhalgh, P. (1991) 'Lessons from the great international exhibitions', in Vergo, P. (ed.) *The New Museology*, London: Reaktion.

Griffin, E. and Ford, L. (1980) 'A model of Latin American city structure', *Geographical Review* 70: 397–422.

Gujral, R. (1994) 'Opinion', *Architecture Today* 50: 7–8.

Guppy, M. (ed.) (1997) *Better Places Richer Communities*, Sydney: Australia Council.

Habermas, J. (1987) *The Theory of Communicative Action*, vol. 1: *The Critique of Functionalist Reason*, Cambridge: Polity.

Habermas, J. (1992) 'Citizenship and national identity: some reflections on the future of Europe', *Praxis International* 12: 1–19.

Hacon, D., Dwinfour, P. and Jermyn, H. (1998) *A Statistical Survey of Regularly Funded Organisations based on Performance Indicators for 1997/98*, London: Arts Council of England.

Halabi, H. (1987) *The Lowell Cultural Plan: A Study,* Department of Urban Studies and Planning, Cambridge, Mass: MIT.

Hall, C. M. (1988) 'The politics of hallmark events: a review', paper given to the APSA, University of New England, Armidale.

Hall, C. M. (1992) *Hallmark Tourist Events: Impacts, Management, Planning*, London: Belhaven.

Hall, P. (1977) *Europe 2000*, London: Duckworth.

Hall, P. (1988) *Cities of Tomorrow*, Oxford: Blackwell.

Hall, P. (1996) *Cities of Tomorrow: An Intellectual History of Urban Planning and Design in the Twentieth Century*, Oxford: Blackwell.

Hall, P. (1998) *Cities and Civilization: Culture, Innovation, and Urban Order*, London: Weidenfeld & Nicholson.

Hall, S. (1990) *Cultural Identity and Diaspora*, in Rutherford, J. (ed.) *Identity*, London: Lawrence & Wishart.

Hall, T. (1986) *Planung europascher Hauptstadte*, Stockholm: Almqviwst & Wiksell.

Handler, R. (1987) 'Heritage and hegemony: recent works on historic preservation and interpretation', *Anthropological Quarterly* 60: 137–41.

Hannigan, J. (1999) *Fantasy City*, London: Routledge.

Hanru, H. and Obrist, H.-U. (1999) 'Cities on the move', in *Cities on the Move, Urban Chaos and Global Change, East Asian Art, Architecture and Film Now*, London: Hayward Gallery Publ.: 10–15.

Hargreaves, D. H. (1983) 'Dr Brunel and Mr Dunning: reflections on aesthetic knowing', in Ross, M. (ed.) *The Arts: A Way of Knowing*, Oxford: Pergamon.

Harland, J. and Kinder, K. (1999) *Crossing the Line: Extending Young People's Access to Cultural Venues*, London: Calouste Gulbenkian Foundation.

Harman, J., Sharland, R. and Bell, G. (1996) 'Local Agenda 21 in action', *RSA Journal*, April: 41–52.

Harris S. P. (ed.) (1984) *Insights/On Sites — Perspectives on Art in Public Places*, Washington, DC: Partners for Liveable Spaces.

Harris, J. (1994) *Private Lives, Public Spirit: Britain 1870–1914*, London: Penguin.

Harvey, D. (1993) 'Goodbye to all that? Thoughts on the social and intellectual condition of contemporary Britain', *Regenerating Cities* 5: 11–16.

Harvie, C. (1994) *The Rise of Regional Europe*, London: Routledge.

Haverfield (1913) *Ancient Town Planning*, Oxford: Oxford University Press.

Haywood, L. *et al.* (1989) *Understanding Leisure*, Cheltenham: Stanley Thornes.

HCP (1997) *Helsinki Urban Guide*, Helsinki: Helsinki City Planning Department.

Healey, P. (1997) *Collaborative Planning: Shaping Places in Fragmented Societies*, London: Macmillan.

Healey, P. *et al.* (1988) *Land Use Planning and the Mediation of Urban Change: The British Planning System in Practice*, Cambridge: Cambridge University Press.

Heartfield, J. (2000) *Great Expectations: The Creative Industries in the New Economy*, London: Design Agenda.

Heidegger, M. (1971) 'Building, dwelling, thinking', in *Poetry, Language, Thought*, New York: Harper Colophon.

Heilbrun, J. and Gray, C. M. (1993) *The Economics of Art and Culture: An American Perspective*, Cambridge: Cambridge University Press.

Heinich, H. (1988) 'The Pompidou Centre and its public: the limits of an utopian site', in Lumley, R. (ed.) *The Museum Time Machine*, London: Comedia.

Hendry, T. (1985) *Cultural Capital: The Care and Feeding of Toronto's Artistic Assets*, Toronto: Toronto Arts Council.

Henley Centre for Forecasting (1985) *Social Change and the Arts*, National Association of Arts Centres, London: Henley Centre for Forecasting.

Henley Centre for Forecasting (1998) *Leisure and Value for Time*, P. Edwards for the World Tourism Organization, London: Henley Centre.

Henley, J. (2000) 'Artists' luxury squats paint portrait of life in the Seine', *The Guardian* 17 August.

Henriques, B. and Thiel, J. (1998) 'The cultural economy of cities: a comparative study of the audiovisual sector on Hamburg and Lisbon', paper given at the Xth Association of Cultural Economists Conference, Barcelona, June.

Henry, I. (1980) *Approaches to Recreation Planning and Research in the District Councils of England & Wales*, Leisure Studies Association Quarterly, London: LSA.

Henry, I. (1993) *The Politics of Leisure Policy*, London: Macmillan.

Hewison, R. (1990) *Future Tense: A New Art for the Nineties*, London: Methuen.

Hewison, R. (1995) *Culture & Consensus: England, Art and Politics Since 1940*, London: Methuen.

Hibbert, C. (1985) *Rome: The Biography of a City*, London: Penguin.

Hillman, J. (ed.) (1971) *Planning for London*, Harmondsworth: Penguin.

Hillmand-Chartrand, H. and McCaughey, C. (1989) 'The arms-length principle and the arts — an international perspective: past present and future', in Cummings, M. C. and Schuster, J. M. (eds) *Who's to Pay for the Arts: The International Search for Models of Support*, New York: American Council for the Arts.

Hitchcock, M. (1998) 'Cool Britannia and tourism: museums, arts — development for the new millennium', University of North London Professorial Lecture, London.

HMSO (1993) *Treaty on European Union*, Maastricht, 7 February 1992 (entered into force 1 November 1993).

Hobsbawm, E. J. (1971) 'From social history to the history of society' *Daedalus* no. 100: 20–45.

Hobsbawm, E. J. (1977) *The Age of Capital 1848–1875*, London: Abacus.

Hobsbawm, E. J. (1995) *Age of Extremes: The Short Twentieth Century 1914–1991*, London: Abacus.

Hobsbawm, E. J. with Polito, A. (2000) *The New Century* (trans. Cameron, A.), London: Little Brown.

Hogarth, T. and Daniel, W. W. (1988) *Britain's New Industrial Gypsies*, London: Policy Studies Institute.

Holbrook, E. L. (1987) *The Economic Impact of the Arts on the city of Boston*, Boston: ARTS/City of Boston.

Horne, D. (1986) *The Public Culture: The Triumph of Industrialism*, London: Pluto.

Horowitz, H. (1989) 'Cultural change and econmic planning', *Cultural Economics 88: An American Perspective*, Association of Cultural Economists, Ohio: University of Akron: 163–70.

Horstman, J. J. (1994) *Creating Spaces, Unitary Development Plans and the Arts, Culture and Entertainment*, London: London Arts Board.

Hough, M. (1990) *Out of Place: Restoring Identity to the Regional Landscape*, New Haven and London: Yale University Press.

Howard, E. (1898) *Garden Cities of Tomorrow*, London: Faber & Faber.

Howard, E. (1902) *Garden Cities of Tomorrow*, London: Swan Sonnenschein.

Huet, A. *et al.* (1991) *Capitalisme et industries culturelles* (2nd ed.), Grenoble: Presses Universitaires de Grenoble.

Hughes, G. (1989) 'Measuring the economic value of the arts', *Policy Studies* 9: 33–45.

Hummel, M. (1988) *Die volkswirtschaftliche. Gutachen im Auftrag des Bundesminister. Des Inneren*, Munich: Kurzfasung. Ifo Institut für Wirschaftsforschung.

Hummel, M. and Berger, M. (1988) *The National Economic Significance of Culture*, Berlin: Dunker & Humblot.

Hustak, A. (1998) 'Last laugh goes to fest organizers', *Montreal Gazette*, 28 July.

Hutchison, R. and Forrester, S. (1987) *Arts Centres in the United Kingdom*, London: Policy Studies Institute.

Ibrahim, A. (1996) *The Asian Renaissance*, Singapore: Times Books International.

Irvine, A. (1999) *The Battle for the Millennium Dome*, London: Irvine New Agency.

Irving, M. (1998) 'Museum trouble', *Blueprint* no. 156: 26–8.

Jackson, P. (1998) 'Domesticating the street', in Fyfe, N. R. (ed.) *Images of the Street: Planning, Identity and Control in Public Space*, London: Routledge.

Jacob, M. A. (1995) *Culture in Action*, Seattle: Bay Press.

Jacobs, J. (1961) *The Death and Life of Great American Cities*, Harmondsworth: Penguin.

Jacobsen, S. (2000) 'Indonesia on the threshold: towards an ethnification of the nation?', *International Institute for Asian Studies* Newsletter, 22 June: 22.

Janne, P. (1970) *Facilities for Cultural Democracy*, Council for Cultural Co-operation, Strasbourg: Council of Europe.

Jardine, L. (1996) *Worldly Goods: A New History of the Renaissance*, London: Macmillan.

Jay, M. (1973) *The Dialectical Imagination: A History of the Frankfurt School and the Institute of Social Research 1923–50*, London: Heinemann Educational.

Jeannotte, S. (1999) 'Cultural policies and social cohesion: perspectives from Canadian research', in *Proceedings of the International Cultural Policy Research Conference*, Bergen, November: vol. II, 623–41.

Jencks, C. (1996) 'The city that never sleeps', *New Statesman*, 28 June: 26–8.

Jenkins, S. (1995) *Accountable to None: The Tory Nationalization of Britain*, London: Hamish Hamilton.

Jevons, W. S. (1883) *Methods of Social Reform*, New York: Augustus M. Kelley.

Johnson, P. (2000) *The Renaissance*, London: Weidenfeld & Nicholson.

Jones, B. and Keating, M. (1995) *The European Union and the Regions*, Oxford: Clarendon.

Jones, J. (2000) 'Come friendly bombs . . .: the Eurocrats want to boost the arts by creating a few cities of culture', *The Guardian*, 8 January.

Jordan, G. and Weedon, C. (1995) *Cultural Politics: Class, Gender, Race and the Postmodern World*, Oxford: Blackwell.

Jordan-Bychov, T. G. and Domosh, M. (1999) *The Human Mosaic: A Thematic Introduction to Cultural Geography*, New York: Addison-Wesley.

Judge, D., Stoker, G. and Wolman, H. (eds) (1995) *Theories of Urban Politics*, London: Sage.

Juneau, A. (1998) 'Impact Economique des Activites du Secteur de la Culture, des Cinq Regions du Montreal Metropolitan et de la Region de L'ile de Montreal', Montreal.

Kahn, A. (1998) 'From the ground up: programming the urban site', *Harvard Architecture Review* 10 [*Civitas/What City?*]: 54–71.

Kant, I. (1790) *Critique of the Faculty of Judgement* (excerpts trans. Bernard, J. H.) (London 1892); with revisions in *Kant: Selections* (1988), ed. Beck, L. W., New York: Scribner/Macmillan.

Kate ten K. (1994) 'Claws for thought: people power is fashionable with developers and planners once more. But does it work?', *The Guardian*, 10 June: 16.

Kauffman, T. D. (1995) *Court, Cloister and City: The Art and Culture of Central Europe 1450–1800*, London.

Kelly, A. (1998) *The Brief History of Western Philosophy*, Oxford: Blackwell.

Kelly, A. and Kelly, M. (2000) *Impact and Values — Assessing the Arts and Creative Industries in the South West*, Bristol: Bristol Cultural Development Partnership.

Kelly, O. (1984) *Community, Art and the State: Storming the Citadels*, London: Comedia.

Keynes, M. (1930) *A Treatise on Money*, London: Macmillan.

King, A. D. (ed.) (1991) *Culture, Globalization and the World-System*, Basingstoke: Macmillan.

King, A. D. (1990) *Global Cities: Post-Imperialism and the Internationalization of London*, London: Routledge.

King, A. D. (1991) 'The global, the urban and the world', in King, A. D. (ed.) *Culture Globalization and the World System*, Basingstoke: Macmillan: 149–54.

Kitto, H. D. F. (1951) *The Greeks*, Harmondsworth: Penguin.

Kloosterman, R. C. and Elfring, T. (1991) *Werken in Nederland*, Schoonhoven: Academic Service.

Knight, L. C. (1937) *Drama and Society in the Age of Jonson*, London: Chatto & Windus.

Knott, C. (1994) *Crafts in the 1990s*, London: Crafts Council.

Knuttson, K. E. (ed.) (1998) *Culture and Human Development*. Report on a conference on Culture, Cultural Research and Cultural Policy, August 1997, Stockholm: Royal Academy of Letters, History and Antiquities.

Kong, L. and Yeoh, B. S. A. (forthcoming) *Landscapes and Construction of a 'Nation'*, Syracuse: Syracuse University Press.

Konstepidemin (1993) *Working Studios*, Gothenburg: Galleri Konstepidemin, Haraldsgatan.

Kostof, S. (1991/99) *The City Shaped: Urban Patterns and Meanings Through History*, London: Thames & Hudson.

Kotowski, B. and Frohling, M. (1993) *No Art, No City*, Atelier-Gesellscahft, Kulturwerkl des Berufsverbandes Bildender Kunstler, Berlin: BBK.

KPMG (1994) *The Arts; A Competitive Advantage for California*, Sacramento: California Arts Council.

Kreisbergs, L. (ed.) (1979) *Local Government and the Arts*, New York: American Council for the Arts.

Kreitzman, L. (1999) *The 24 Hour Society*, London: Profile.

Krieger, K. (1989) 'Community cultural planning in Massachusetts', *Cultural Economics 88: An American Perspective*, Association of Cultural Economists, Ohio: University of Akron: 171–82.

Kroller, E.-M. (1996) 'EXPO '67: Canada's Camelot?', in *Proceedings of the British Association for Canadian Studies Annual Conference*, April, University of Exeter: 6.

Kruger, H. P. (1969) 'The German theatre today: some reflections on the promotion of the arts in the Federal Republic of Germany', in Schouvaloff, A. (ed.) (1970) *Place for the Arts*, Liverpool: Seel House: 201–3.

LAAC (1984) *Forum on the Arts in London*, Greater London Council, ICA, London: London Association of Arts Centres.

LAB (1992a) *The Arts and Urban Policy*, National Arts & Media Strategy Seminar, London: London Arts Board.

LAB (1992b) *London and the Arts: The City's Role and Contribution*, Report on Consultative Seminar for the National Arts and Media Strategy, Arts Council, 13th January, London: London Arts Board.

LAB (1993) *Annual Report 1992/3*, London: London Arts Board.

LAB (1999) *Arts and the City*, Quarterly News from the London Arts Board, no. 1.

Lacroix, J.-G. and Tremblay, G. (1997) 'The information society and cultural industries theory', *Current Sociology* (Trend Report, trans. Ashby, R.), 45.

Laffin, M. and Young, K. (1985) The changing roles and responsibilities of Local Authority Chief Officers, *Public Administration* 63.

Landry, C. (1998) 'Culture and cities', *Urban Age* September: 8–10.

Landry, C. (2000) *The Creative City. A Toolkit for Urban Innovators*, London: Earthscan.

Landry C. *et al.* (1997a) *The Economic Importance of Cultural Industries to the London Borough of Tower Hamlets*, Stroud: Comedia.

Landry C. *et al.* (1997b) *Cultural Industries Strategy for Tower Hamlets*, Stroud: Comedia.

Lane, J. (1978) *Arts Centres — Every Town Should Have One*, London: Paul Elek.

Lane, R. (1998) 'The Place of Industry', *Harvard Architecture Review* 10 [*Civitas/What City?*]: 151–61.

Laperièrre, H. and Latouche, D. (1996) 'So far from culture and so close to politics: the new art facilities in Montreal', *Culture et Ville* no. 96–8, Montreal: INRS.

Laperrière, H. and Latouche, D. (1999) *Nous Sommes Tous Des Quebecois: La Representation Des Regions Du Quebec Dans La Capitale*, Montreal: INRS.

Lash, S. and Urry, J. (1994) *Economies of Signs and Spaces*, London: Sage.

Latouche, D. (1994) *Les arts et les industries culturelles dans la region de Montreal: bilan et enjeux*, Montreal: INRS-Urbanisation.

Law, C. M. (1992) 'Urban tourism and its contribution to economic regeneration', *Urban Studies* 29: 599–618.

Law, C. M. (1993) *Urban Tourism: Attracting Visitors to Large Cities*, London: Mansell.

Lawless, P. and Gore. T. (1999) 'Urban regeneration and transport investment: a case study of Sheffield 1992–96', *Urban Studies* 36: 527–45.

LCC (1987) *An Arts and Cultural Industries Strategy for Liverpool: A Framework*, Planning Department, Liverpool: Liverpool City Council.

Le Corbusier (Jeanneret, J. C.) (1929) 'A contemporary city', in *The City of To-morrow and its Planning*, London: John Rodher.

Le Gales, P. and Lequesne, C. (eds) (1998) *Regions in Europe*, London: Routledge.

Le Grand, J. (1998) 'Social exclusion in Britain today', ESRC Seminar discussion paper, London: London School of Economics.

Leadbeater, C. (2000) *Living on Thin Air: The New Economy*, London: Penguin.

Lee, A. (1991) *Consultation with Aboriginal & Ethno-Racial Communities*, Metro's Role in Arts and Culture, Municipality of Metro Toronto.

Lee, J. (1965) *A Policy for the Arts: The First Steps*, Cmnd 2601, London: HMSO.

Lee, L. (1969) *As I Walked Out One Midsummer Morning*, London: Penguin.

Lee, M. (1997) 'Relocating location: cultural geography, the specificity of place and the City of Habitus', in McGuigan, J. (ed.) *Cultural Methodologies*, London: Sage.

Lefebvre, H. (1974) *The Production of Space* (trans. Nicholson-Smith, D.), Oxford: Blackwell.

LeGates, R. T. and Stout, F. (eds) (1996) *The City Reader*, London: Routledge.

Leisure Consultants (1996) *Leisure Forecasts 1996–2000*, Sudbury: Leisure Consultants.

Leisure Opportunities (2000) *Heron City*, 24 January: 18–20.

Lejeune, J.-F. (1996) 'The city as landscape', *Journal of Decorative and Propaganda Arts* [Cuba Theme Issue 1875–1945].

Leonnard, M. (1998) 'Cool Britannia', *Sunday Times*, 26 April: 9.

Leslie, D. and Muir, F. (1996) *Local Agenda 21, Local Authorities and Tourism: A United Kingdom Perspective*, Glasgow: Glasgow Caledonian University.

Leventhal, L. M. (1990) 'The best for the most: CEMA and state sponsorship of the arts in wartime, 1939–1945', *Twentieth Century British History*, 1: 293–303.

Levine, J., Lockwood, C. and Worpole, K. (1997) 'Rethinking regeneration', *World Architecture* 58(4) [Special Issue: Urban Regeneration].

Levine, M. and Megida, A. (1989) 'Is the party over for Baltimore?', *Baltimore Jewish Times*, 14 July: 54–60; in Giloth, R. (1990) 'Beyond common sense: the Baltimore renaissance', *Local Economy* 4: 291.

Lewis, J. (1990) *Art, Culture and Enterprise*, London: Routledge.

Lewis, J., Morley, D. and Southwood, R. (1987) *Art — Who Needs It?: An Audience for Community Arts*, Leisure Report No. 1, London: Comedia.

Ley, D. and Olds, K. (1988) 'Landscape as festival: world's fairs and the culture of heroic consumption', *Environment and Planning D: Society and Space* 6: 191–212.

Lichfield, D. (1992) *Urban Regeneration for the 1990s*, DLA, London: London Planning Advisory Committee.

Lim, H. (1993) 'Cultural strategies for revitalizing the city: a review and evaluation', *Regional Studies* 27: 589–95.

Lingayah, S., MacGillivray, A. and Raynard, P. (1997) *The Social Impact of Arts Programmes — Creative Accounting: Beyond the Bottom Line*, Working Paper 2, Stroud: New Economics Foundation and Comedia.

Lintner, V. and Mazey, S. (1991) *The European Community: Economic and Political Aspects*, London: McGraw-Hill.

Lipjhart, A. (1977) *Democracy in Plural Societies: A Comparative Exploration*, New Haven and London: Yale University Press.

LIRC (2000) *Leisure Forecasts 2000–2005*, Sheffield: Leisure Industry Research Centre.

Lissitzky, E. (1970) *Russia: An Architecture for World Revolution* [Vienna, 1930] (trans. Dluhosch, E.), London: Lund Humphries.

Loftman, P. and Nevin, B. (1993) *Urban Regeneration and Social Equity: A Case Study of 1986–1992 Birmingham*, Birmingham: University of Central England in Birmingham.

Loman, P. *et al* (1989) *The European Communities and Cultural Policy: A Legal Analysis*, Zeist.

London Borough of Enfield (1993) *Unitary Development Plan*, 13.3.3., London: Enfield Environmental Services.

London Borough of Greenwich (1998) *The Greenwich Cultural Plan — A Framework for Development*, April, London.

London Borough of Haringey (1991) *Urban Design Action Team — Alexandra Palace and Wood Green Report*, London: Urban Design Group.

Longman, P. (1999) *Director's Report*, The Theatres Trust 22nd Annual Report Year ended 31 July 1999, London: Theatres Trust.

Looseley, D. L. (1997) *The Politics of Fun. Cultural Policy and Debate in Contemporary France*, Oxford: Berg.

Lopez, R. S. (1971) *The Commercial Revolution of the Middle Ages, 930–1350*, Englewood Cliffs: Prentice-Hall.

Lopez, R. S. (1952) 'The trade of medieval Europe: the south', in Postan, M. and Rich, E. E. (eds) *The Cambridge Economic History of Europe*, Cambridge: Cambridge University Press: 257–354.

Lopez, R. S. (1959) 'Hard times and investment in culture', in Dannenfeldt, K. H. (ed.) *The Renaissance: Medieval or Modern*, Boston: DC Heath: 50–61.

Lowyck, E. and Wanhill, S. (1992) 'Regional Development and tourism within the European Community', in Cooper, C. and Lockwood, A. (eds) *Progress in Tourism, Recreation and Hospitality Management*, London: Belhaven: 227–44.

LPAC (1988) *Strategic Planning Advice for London*, London: London Planning Advisory Committee.

LPAC (1990a) *Strategic Planning Policies for the Arts, Culture and Entertainment*, Report No. 18/90, London: London Planning Advisory Committee.

LPAC (1990b) *Model UDP Policies for the Arts, Culture and Entertainment Activities*, London: London Planning Advisory Committee.

LPAC (1991) *London: World City Moving into the 21st Century*, London: HMSO.

LPAC (1992a) *Strategic Planning Issues for London: A Discussion Document*, London: London Planning Advisory Committee.

LPAC (1992b) *Review of the relationship between UDPs and Strategic Advice and Guidance*, Report No. 22/93, London: London Planning Advisory Committee.

LPAC (1993) *Draft 1993 Advice on Strategic Planning Guidance for London*, June, London: London Planning Advisory Committee.

LSE (1996) *The Arts and Cultural Industries in the London Economy*, London: Group for the London Arts Board, London School of Economics.

Lumley, R. (ed.) (1988) *The Museum Time Machine*, London: Comedia.

Lynch, K. (1960) *The Image of the City*, Cambridge, MA: MIT Press.

Lynch, K. (1972) *What Time is This Place?*, Cambridge, MA: MIT Press.

MacCannell, D. (1996) *Tourist or Traveller?*, London: BBC Education.

MacClancy, J. (1997) 'The museum as a site of contest. The Bilbao Guggenheim', *Focaal Journal of Anthropology* 1: 271–8.

Macdonald, I. (1986) *Arts, Education and Community*, London: London Association of Arts Centres.

MacKeith, J. (1996) *The Art of Flexibility: Art Centres in the 1990s*, The Arts Council of England Research Report No. 8, London: ACE.

Mackin, M., Johnson, D. and Edmund, J. (1998) *The Cultural Sector: A Development Opportunity for Tourism in Northern Ireland*, Northern Ireland Tourist Board.

Mackrell (1995) *Working for Dance*, London: Arts Council of England.

Mairet, E. (1933) *Rural Industries Magazine*, Rural Industries Bureau.

Mairie de Paris (1993) *Studio-Flat Combinations (Ateliers-Logements)*, Paris: Mairie de Paris.

Malraux, A. (1966) 'For a Maison de la Culture', speech made at the opening of the Maison de la Culture at Amiens on 19 March; in Schouvaloff, A. (ed.) (1970) *Place for the Arts*, Manchester: North West Arts Association: 134–6.

Malraux, A. (1978) *The Voices of Silence*, Princeton: Princeton University Press.

Manchester Polytechnic (1989) *The Culture Industry, The Economic Importance of the Arts & Cultural Industries in Greater Manchester*, Manchester: Centre for Employment Research.

Mango, C. (1998) *Byzantium: The Empire of the New Rome*, London: Phoenix.

Manley, L. (1995) *Literature and Culture in Early Modern London*, Cambridge: Cambridge University Press.

Mariani, M. A. (1998) 'Arts and tourism: enterprise development in the cultural and environmental sector', paper given at the Xth International Conference on Cultural Economics, Barcelona, June.

Marquand, D. (1994) 'Prospects for a Federal Europe. Reinventing federalism: Europe and the left', *New Left Review* 203: 17–26.

Marshall, A. (1925) *Principles of Economics* (8th ed.), London: Macmillan.

Marshall, A. H. (1974) *Local Government and the Arts*, Institute of Local Government Studies, Birmingham: University of Birmingham.

Marwick, A. (1991) *Culture in Britain Since 1945*, Institute of Contemporary British History, Oxford: Blackwell.

Marx, K. (1973) *Grundrisse*, London: Penguin.

Maslow, A. H. (1954) *Motivation and Personality*, London: Harper.

Mason, P. (1998) *Bacchanal!: The Carnival Culture of Trinidad*. Philadelphia: Temple University Press.

Massey, D. (1984/95) *Spatial Division of Labour* (2nd ed.), Basingstoke: Macmillan.

Massey, D. (1994) *Space, Place and Gender*, Cambridge: Polity.

Massey, D., Allen, J. and Pile, S. (1999) *City Worlds*, London: Routledge.

May, E. (1931) 'City building in the USSR', *Das Neue RuBland* 8–9: 703–4.

Mayfield, T. L. and Compton, J. L. (1995) 'Development of an instrument for identifying community reasons for staging a festival', *Journal of Travel Research*, Winter: 37–44.

McGuigan, J. (1996) *Culture and the Public Sphere*, London: Routledge.

McNulty, R., Leo Penne, R. and Jacobson, D. (1986) *The Return of the Liveable City, Learning from America's Best*, Washington, DC: Acropolis.

Meller, H. E. (1976) *Leisure and the Changing City*, London: Routledge & Kegan Paul.

Mennell, S. (1976) *Cultural Policy in Towns: A Report on the Council of Europe's 'Experimental Study of Cultural Development in European Towns'*, Council for Cultural Co-operation, Strasbourg: Council of Europe.

Middleton, P. (1994) *Urban Tourism 90's Style — Or a New Search for Pixie Dust!*, British Urban Regeneration Association News 5/88–9.

Midwest Research Institute (1980) *Economic Impact of the Performing Arts on Kansas City*, Kansas City: Midwest Research Institute.

Miles, M. (1997) *Art Space and the City. Public Art and Urban Futures*, London: Routledge.

Mills, C. W. (1959) 'The cultural apparatus', *The Listener* 61: 552–6.

Ministry of Cultural Affairs (1995) *New Zealand Cultural Statistics*, Wellington, New Zealand.

Ministry of Education (1959) *Standards of Public Library Services*, London: HMSO.

Mitterrand, F. (1989) 'Preface', in Biasni, E. (ed.) *Grands Travaux*, Paris: Connaissance des Arts.

Modi, A. (1998) *Theatrical Traditions in India*, WLRA Congress, Sao Paulo, October.

Mokre, M. (1998) *EU Cultural Intervention in Area Regeneration processes*, UACES European Cultural Policy Conference, City University, London, April.

Molotoch, H. (1996) 'LA as design product: how art works in a regional economy', in Scott, A. J. and Soja, E. (eds) *The City: Los Angeles and Urban Theory at the End of the Twentieth Century*, Berkeley: University of California Press: 225–75.

Montgomery, J. (1989) *Socio-Economic Profile of the Southern Arts Region*, Winchester: Southern Arts Board.

Montgomery, J. and Gavron, N. (1991) *Paper on Arts Infrastructure*, London: London Arts and Urban Regeneration Group: 1–4.

MORI (1998) *The West End Theatre Audience*, Research Study conducted for the Society of London Theatre, November 1996–November 1997, London: MORI.

Morin, E. (1987) *Penser Europe*, Paris: Gallimard.

Morin, E. (1991) *Europa Denkem*, Frankfurt.

Morris, E. (1994) 'Heritage and culture. A capital for the new Europe', in Ashworth, G. J. and Larkham, P. J. (eds) *Building a New Heritage. Tourism, Culture and Identity in the New Europe*, London: Routledge: 229–59.

Morrison, W. and West, E. (1986) Child exposure to the performing arts: the implications for adult demand, *Journal of Cultural Economics*, 10: 17–24.

Mostafavi, M. (1999) 'Cities of distraction', in *Cities on the Move, Urban Chaos and Global Change, East Asian Art, Architecture and Film Now*, London: Hayward Gallery Publ.: 7–9.

Mulder, P. (1991) *European Integration and the Cultural Sector*, Discussion Document No. 15, National Arts & Media Strategy, London: Arts Council.

Mulgan, G. and Worpole, K. (1986) *Saturday Night or Sunday Morning? From Arts to Industry — New Forms of Cultural Policy*, London: Comedia.

Mulhern, F. (1993) 'A European home?', in Bird, J. *et al.* (eds) *Mapping the Futures: Local Cultures, Global Change*, London: Routledge.

Mulryne, R. and Shewring, M. (1995) *Making Space for Theatre. British Architecture and the Theatre since 1958*, Stratford-upon-Avon: Mulryne & Shewring Ltd.

Mumford, L. (1940) *The Culture of Cities*, New York: Secker & Warburg.

Mumford, L. (1945) *City Development*, London: Harcourt Brace Jovanovich/Harvest.

Mumford, L. (1961) *The City in History: Its Origins, Its Transformation, Its Prospects*, Harmondsworth: Penguin.

Munro, T. (1967) *The Arts and their Interrelations* (2nd ed.), Cleveland: Western Reserve University Press.

Museum of Finnish Architecture (1978) *Alvar Aalto: 1898–1976*, Helsinki.

Myerscough, J. (1988) *The Economic Importance of the Arts in Britain*, London: Policy Studies Institute.

Myerscough, J. (1989) *Economic Strategy for the Arts in Hampshire*, Winchester: Hampshire County Council.

Myerscough, J. (1990) 'The economic contribution of the arts', paper given at *Tourism and the Arts* Conference, Science Museum, June, London: English Tourist Board.

NACCCE (1999) *All Our Futures: Creativity, Culture and Education*, Report of the National Advisory Committee on Creative and Cultural Education, London, May.

Nagata, C. (1991) 'Consultation with area municipalities', in *Metro's Role in Arts and Culture*, Toronto: Municipality of Metro Toronto.

Nasution, K. S. (1998) 'The challenge of living heritage', *Urban Age*: 28.

National Building Museum (1998) *Building Culture Downtown: New Ways of Revitalizing the American City*, Washington, DC.

National Playing Fields Association (1971) *Outdoor Play Space Requirements*, London: NPFA (under review).

Negrier, E. (1993) 'Montpellier: international competition and community access', in Bianchini, F. and Parkinson, M. (eds) *Cultural Policy and Urban Regeneration: The West European Experience*, Manchester: Manchester University Press.

Newman, A., and McLean, F. (1998) 'Heritage builds communities: the application of heritage resources to the problems of social exclusion', *International Journal of Heritage Studies* 4: 143–53.

Newman, P. and Thornley, A. (1994) *A Comparison of London, Paris and Berlin*, Department of Land Management and Development, Reading: University of Reading.

Nicholson Lord, N. (1994) *Ecology, Parks and Human Need*, Working Paper No. 4, Stroud: Comedia.

Nicholson, G. (1990) 'The campaign for messy government; or perfect structures don't work', paper given at the Vision for London Conference 'Preparing Unitary Development Plans', 19 March, London: Association of London Authorities.

Nicholson, G. (1992) 'The rebirth of community planning', in Thornley, A. (ed.) *The Crisis of London*, London: Routledge: 119–34.

Norquist, J. O. (1998) *The Wealth of Cities: Revitalising the Centers of American Life*, Reading, MA: Addison-Wesley.

O'Brien, J. (1997) *Arts Centres in England: A Statistical Appendix*, London: Arts Council of England.

O'Brien, J. and Feist, A. (1995) *Employment in the Arts and Cultural Industries: An Analysis of the 1991 Census*, ACE Research Report No. 2, London: Arts Council of England.

O'Connor, J. (2000) 'Markets and customers', in Roodhouse, S. (ed.) *Proceedings for The New Cultural Map: A Research Agenda for the 21st Century*, Bretton Hall: University of Leeds: 16–25.

O'Connor, J. and Wynne, D. (1996) *From the Margins to the Centre: Cultural Production and Consumption in the Post-Industrial City*, Aldershot: Arena.

O'Hagan, J. (1998) *The State and the Arts: An Analysis of Key Economic Policy Issues in Europe and the United States*, Gloucester: Edward Elgar.

Observer, The (2001) 'Century city', *The Observer* 1 February–29 April.

Olds, K. (1995) 'Globalization and the production and new urban spaces: Pacific Rim megaprojects in the late 20th century', *Environment and Planning A*: 1713–43.

Olsen, D. J. (1982) *Town Planning in London: The Eighteenth and Nineteenth Centuries*, New Haven and London: Yale University Press.

Office for National Statistics (ONS) (1999) *Social Trends 29*, London: HMSO.

Owusu, K. and Ross, J. (1988) *Behind the Masquerade: The Story of the Notting Hill Carnival*, London: Arts Media Group.

PACEC (1990) *An Evaluation of Garden Festivals*, Inner Cities Research Programme, Department of the Environment, London: PA Cambridge Economic Consultants.

Parkinson, M. and Bianchini, F. (eds) (1993) 'Liverpool: a tale of missed opportunities?', in *Cultural Policy and Urban Regeneration: The West European Experience*, Manchester: Manchester University Press.

Parry, N. and Parry, J. (1989) 'Meritocrats' last stand', *Times Higher Education Supplement* 15 December: 17.

Patten, D. (2000) Artist's residencies and social exclusion, *Public Art Journal* 1: 41–8.

Peacock, A., Shoesmith, E. and Milner, G. (1984) *Cost Inflation in the Performed Arts*, London: Arts Council of Great Britain.

Pearce, D. (1998) 'Tourism development in Paris: public intervention', *Annals of Tourism Research* 5: 457–76.

Pennybacker S. (1989) '"The millennium by return of post": reconsidering London progressivism, 1889–1907', in Feldman, D. and Stedman Jones, G. (eds) *Metropolis London*, London: Routledge: 129–62.

Percival, S. (1991) 'Visions of artists and mechanics of funding', in *A Creative City*, London: Greater London Arts/Public Art Development Trust.

Perloff, H. S. (1979) *The Arts in the Economic Life of the City of Los Angeles*, New York: American Council for the Arts.

Peters, J. (1982) 'After the fair: what Expos have done for their cities', *Planning* 18: 13–19.

PHPC (1992) *Artists' Space Journal*, International Edition, no. 6, June, Philadelphia Historic Preservation Corporation.

Pick, J. (1980) *The State of the Arts*, Eastbourne: City Arts/John Offord.

Pick, J. (1985) *The Theatre Industry*, London: Comedia.

Pick, J. (1988) *The Arts in a State: A Study of the Government Arts Policies from Ancient Greece to the Present*, Bristol: Bristol Classical.

Pick, J. (1991) *Vile Jelly: The Birth, Life and Lingering Death of the Arts Council of Great Britain*, Doncaster: Brymill.

Pick, J. (1999) 'A critique of the cultural industries', in Roodhouse, S. (ed.) Proceedings of *The New Cultural Map: A Research Agenda for the 21st Century*, Breton Hall: University of Leeds: 5–7.

Pick, J. and Anderton, M. (1996) *Arts Administration* (2nd ed.), London: E & FN Spon.

Pickvance, C. G. (1976) *Urban Sociology: Critical Essays*, London: Tavistock.

Pirenne, H. (1925) 'City origins and cities and european civilization', in *Medieval Cities* (trans. Halsey, F.), Princeton: Princeton University Press.

Pomeroy, S. B., Burstein, S. M., Donlan, W. and Tolbert Roberts, J. (1999) *Ancient Greece: A Political, Social, and Cultural History*, New York: Oxford University Press.

Population Reference Bureau (1995) *World Population Data Sheet*, Washington, DC.

Port Authority of New York (1983) *The Arts as an Industry: Their Economic Importance to the New York–New Jersey Metropolitan Region*, New York: PANY/NJ.

Port Authority of New York (1993) *The Arts as an Industry: Their Economic Importance to the New York–New Jersey Metropolitan Region*, New York: PANY/NJ.

Porter, R. (1982) *English Society in the Eighteenth Century*, London: Pelican.

Portsmouth City Council (1991) *The Arts in Portsmouth: A Current Situation Review*, Portsmouth: Portsmouth Arts Museums and Archives Committee.

Potter, R. B. and Lloyd-Evans, S. (1998), *The City in the Developing World*, Harlow: Longman.

Pratt, A. (1997) *The Cultural Industries Sector. Its Definition and Character from Secondary Sources on Employment and Trade, Britain 1984–91*, London: LSE.

Pratt, A. (1998) 'A "Third Way" for creative industries? Hybrid cultures: the role of bytes and atoms in locating the new cultural economy and society', *International Journal of Communications, Policy and Law*, Issue 1: Web-Doc 4-1-1998.

Punter, J. (1992) 'Classic carbuncles and mean streets: contemporary urban design and architecture in central London', in Thornley, A. (ed.) *The Crisis of London*, London: Routledge: 69–89.

Raeburn, M. and Wilson, V. (eds) (1987) *Le Corbusier: Architect of the Century*, London: Arts Council.

Rasmussen, S. E. (1937) *London: The Unique City* [1948] (revd 1982 ed.), Cambridge, MA: MIT Press.

Read, H. (1964) *The Philosophy of Modern Art*, London: Faber & Faber.

Rearick, C. (1985) *Pleasures of the Belle Epoque: Entertainment and Festivity in Turn of the Century France*, New Haven and London: Yale University Press.

Redhead, S. (ed.) (1999) *Rave Off. Politics and Deviance in Contemporary Youth Culture*, Aldershot: Ashgate.

Reich, R. (1991) *The Work of Nations*, New York: Knopf.

Richards, G. (ed.) (1996) *Cultural Tourism in Europe*, Wallingford: CAB International.

Richards, G. and Hall, D. (eds) (2000) *Tourism and Sustainable Community Development*, London: Routledge.

Richie, A. (1998) *Faust's Metropolis: A History of Berlin*, London: Harper Collins.

Rietveld, H. (1999) Living the dream, in Redhead, S. (ed.) *Rave Off. Politics and Deviance in Contemporary Youth Culture*, Aldershot: Ashgate: 41–4.

Ritchie, J. R. (1984) Assessing the impact of hallmark events: conceptual and research issues, *Journal of Travel Research* 23: 2–11.

Ritchie, J. R. and Smith, B. S. (1991) 'The impact of a mega-event on host region awareness: a longitudinal study', *Journal of Travel Research*, 27: 3–10.

Roberts, R. (1974) 'Planning for leisure', *Building* 15: 98–102.

Robertson, R. (1990) 'After nostalgia: wilful nostalgia and the phase of globalization', in Turner, B. (ed.) *Theory Culture and Society*, London: Sage: 45–61.

Robins, K. (1993) 'Prisoners of the city', in Carter, E. (ed.) *Space and Place, Theories of Identity and Location*, London: Lawrence Wishart.

Robins, K. (1996) 'Collective emotion and urban culture', in Brandner, B., Mattl, S. and Ratzenbock, V. (eds) *Kulturpolitik und Restrukturierung der Stadt*, Vienna: 73–96.

Rogers, R. and Fisher, M. (1992) *A New London*, London: Penguin.

Rojas, E. (1998) 'Financing urban heritage conservation in Latin America', in *Proceedings of the City, Space and Globalization Conference*, University of Michigan, Ann Arbor, 26 February–1 March.

Rojas, E. (1999) *Old Cities New Assets. Preserving Latin America's Urban Heritage*, Washington, DC: Inter-American Development Bank.

Rolfe, H. (1991) *Arts Festivals in the UK*, London: Policy Studies Institute.

Rosenzweig, R. and Blackmar, E. (1992) *The Park and the People: A History of Central Park*, New York: Cornell University Press.

Rosler, M. (1994) 'Place, position, power, politics', in Becker, C. (ed.) *The Subversive Imagination*, New York: Routledge.

Roth, L. (1998) 'The benefits of the European Union Structural Funds for the development of the South of Italy', unpublished MA dissertation, University of North London.

Rowntree, B. S. and Lavers, G. R. (1951) *English Life and Leisure: A Social Study*, London: Longman, Green & Co.

RSA (1993) *Ideas Across Frontiers*, London: Royal Society for the Encouragement of Arts, Manufactures & Commerce.

Rustin, M. (1994) 'Unfinished business: from Thatcherite modernisation to complete modernity', in Perryman, M. (ed.) *Altered States: Postmodernism, Politics, Culture*, London: Lawrence & Wishart: 73–93.

Ryan, R. (2000) 'New frontiers', in *Tate*, Tate Modern Special Issue no. 21, London: 90–6.

Ryan, R. (2001) 'Urban generations', in *Tate* no. 24 (spring), London: 23–31.

Rydell, R. (1993) *World of Fairs: The Century-of-Progress Expositions*, Chicago: Chicago University Press.

Rydell, R. W. (1984) *All the World's a Fair*, Chicago: University of Chicago Press.

Rydin, Y. (1993) *The British Planning System: An Introduction*, Basingstoke: Macmillan.

Rykwert, J. (2000) *The Seduction of Place: The City in the Twenty-First Century*, London: Weidenfeld & Nicholson.

SAC (1992) *The Social Impact of the Arts in Scotland* (ed. Shaw, P.), Edinburgh: Scottish Arts Council.

SAC (1995a) *The Social Impact of the Arts*, Edinburgh: Scottish Arts Council.

SAC (1995b) *The Arts in Scotland's Urban Areas*, Edinburgh: Scottish Arts Council.

Sacco, G. (1976) 'Morphology and culture of European cities', in van Hulton, M. (ed.) *Europe 2000*, Project 3, The Hague: Nijhoff: vol. 1, 162–87.

Said, E. W. (1978) *Orientalism: Western Conceptions of the Orient*, London: Routledge.

Said, E. W. (1994) *Culture and Imperialism*, London: Vintage.

San Francisco Art Commission (1990) *San Francisco Arts Economy*, Joint Study by SF Planning Department and SF State University Public Research Institute, San Francisco: SFAC.

Sassatelli, M. (1999) 'Imagined Europe. The European cities of culture and the shaping of a European cultural identity: the case of Bologna', in *Proceedings of the International Conference on Cultural Policy*, Bergen: 593–607.

Sassen, S. (1991) *Global City: New York, London, Tokyo*, Princeton: Princeton University Press.

Sassen, S. (1994) *Cities in a World Economy*, Thousand Oaks: Pine Forge.

Sassen, S. (1996) 'Rebuilding the global city: economy, ethnicity and space', in King, A. (ed.) *Re-presenting the City: Ethnicity Capital and Culture in the 21st-Century Metropolis*, London: Macmillan: 23–42.

Sassen, S. and Roost, F. (1999) 'The city: strategic site for the global entertainment industry', in Judd, D. R. and Fainstein, S. S. (eds) *The Tourist City*, New Haven and London: Yale University Press: 143–54.

Scalbert, R. (1994) 'Have the *Grands Projets* really benefited Paris?', *Architect's Journal* 3(200): 20.

SCC (1988) *Southampton Cultural Industries Audit Brief*, Southampton: Economic Development Unit, Southampton City Council.

Schmidjell, R. and Gaubinger, R. B. (1980) 'Quantifizierung der externen Effekte des Kuntsektors am Beispiel der Salzburger Festspiele', *Wirtschaftspolitische Blätter* 27: S89–97.

Schouvaloff, A. (ed.) (1970) *Place for the Arts*, North West Arts Association, Liverpool: Seel House.

Schuster, J. M. (1994) 'Funding for the arts & culture through dedicated state lotteries — Part 1: The twin issues of additionality and substitution', *Journal of European Cultural Policy* 1: 21–41.

Schuster, J. M. (1995) *Supporting the Arts; An International Comparative Study*, Washington, DC: US Government Publishing Office.

Schuster, J. M. (1996) 'Thoughts on the art and practice of comparative cultural research', in *Cultural Research in Europe 1996*, Amsterdam: Boeckman Foundation/CIRCLE.

Scott, A. (2000) *The Cultural Economy of Cities*, London: Sage.

SCP (1996) *Social and Cultural Report 1996 The Netherlands*, The Hague: Social and Cultural Planning Office.

Screen Digest (1994a) 'UK multiplex cinemas: phase 1 nears maturity', February.

Screen Digest (1994b) 'Cinema gross box office', September.

Seabrook, J. (1996) *In the Cities of the South*, London: Verso.

Searle, M. S. and Brayley, R. E. (1993) *Leisure Services in Canada*, Pennsylvania State College: Venture.

Selwyn, T. (1993) 'It's not even Londoners who love London', *In Focus*, Roehampton: Tourism Concern: 10–11.

Selwyn, T. (1995) 'Landscapes of liberation and imprisonment: towards an anthropology of the Israeli landscape', in Hirsch, E. and O'Hanlon, M. (eds) *The Anthropology of Landscape: Perspectives on Place and Space*, Oxford: Clarendon.

Sennett, R. (1970) *Families Against the City: Middle Class Homes of Industrial Chicago, 1872–1890*, Cambridge, MA: Harvard University Press.

Sennett, R. (1986) *The Fall of Public Man*, London: Faber & Faber.

Sennett, R. (1994) *Flesh and Stone: The Body and City in Western Civilization*, London: Faber & Faber.

Senter, A. (1998) 'Taking the waters: the Sadler's Wells story', in *Sadler's Wells: A Celebration 1683–1998*, London: Sadler's Wells Appeal Fund.

Seregeldin, M. (1999) 'Preserving the historic urban fabric in a context of fast-paced change', in *Values and Heritage Conservation*, Los Angeles: Getty Conservation Institute: 51–8.

Serota, N. (2000) *Experience or Interpretation: The Dilemma of Museums of Modern Art*, London: Thames & Hudson.

Shackley, M. (1998) *Visitor Management: Case Studies from World Heritage Sites*, Oxford: Butterworth-Heinemann.

Sharp, D. *et al.* (1992) 'Europe now: planning in the European Community', *Building Design* 3 July: 14–18.

Shaw, P. (1989) *The Public Art Report: Local Authority Commissions of Art for Public Places*, London: Public Art Forum.

Shaw, P. (1990a) *The Public Art Report: Commissions by Local Authorities*, London: Public Arts Development Trust.

Shaw, P. (1990b) *Percent for Art: A Review*, London: Arts Council.

Shaw, P. (1996) *Artist's Fees and Payments in the UK*, November, National Artists Association.

Shaw, P. (1999) *The Arts and Neighbourhood Renewal: A Research Report*, Policy Action Team 10, London: Department for Culture, Media and Sport.

Shelton Trust (1986) *Culture and Democracy Manifesto*, London: Comedia; cited in Lewis, J. (1990) *Art, Culture and Enterprise*, London: Routledge: 111.

Sherlock, H. (1991) *Cities Are Good For Us*, London: Paladin.

Shore, C. (1993) 'Inventing the "People's Europe": critical perspectives on European Community cultural policy', *Man. Journal of the Royal Anthropological Institute* 28(4): 779–800.

Shurmer-Smith, L. and Burtenshaw, D. (1990) 'Urban decay and rejuvenation', in Pinder, D. A. (ed.) *Western Europe: Challenge and Change*, London: Belhaven.

Sillitoe, K. K. (1969) *Planning for Leisure*, London: HMSO.

Sinfield, A. (1989) 'Changing concepts of the arts: from the leisure élite to Clause 28', *Leisure Studies* 8: 129–39.

Sitte, C. (1965) *City Planning According to Artistic Principles* (trans. Collins, G. R. and Collins, C. C.), London: Phaidon.

Sjoeberg, G. (1960) *The Pre-Industrial City*, London: Free Press.

Sklair, L. (1991) *Sociology of the Global System*, Hemel Hempstead: Harvester Wheatsheaf.

Smith, A. D. F. (1992) 'National identity and the idea of a European unity', *International Affairs* 68: 55–76.

Smith, B. (2000) 'Modernism in its place', in *Tate*, Tate Modern Special Issue no. 21, London: 79–83.

Smith, M. P. (1991) *City, State and Market. The Political Economy of Urban Society*, Oxford: Blackwell.

So, F. S. and Getzels, J. (1988) *The Practice of Local Government Planning*, Washington, DC: International City Management Association.

Social Data Research Ltd (1990) *Housing and Workspace Needs of Toronto's Artists and Artisans*, City of Toronto Housing Department.

Social Exclusion Unit (2000) *National Strategy for Neighbourhood Renewal: A Framework for Consultation*, London: Cabinet Office.

Soja, E. W. (2000) *Postmetropolis*, Blackwell: Oxford.

Solesbury, W. (1998) *Good Connections: Helping People to Communicate in Cities*, Working Paper No. 9, Stroud: Comedia/Demos.

Southern Arts Board (1991) *The Arts in All Our Lives: A Strategy for the Arts in the South 1990–95*, Winchester: SAB.

Southern, R. (1962) *The Seven Ages of the Theatre*, London: Faber & Faber.

SPACE (2000) 'Temporary contemporary', *The Guardian*, 22 June.

Sports Council (1968) *Planning for Sport — Report of a Working Party on Scales of Provision*, London: CCPR.

Sports Council (1972) *Provision for Sport, Indoor Swimming Pools, Indoor Sports Centres, Golf Courses*, London: HMSO.

Sports Council (1977) *Provision for Sport, Indoor Swimming Pools, Indoor Sports Centres, Golf courses* (update from 1972), London: HMSO.

Sports Council (1978) *Provision for Swimming Pools, A Guide to Planning*, London: HMSO.

Stadt Köln (1985) *Kulturelle Grossveranstaltungen in Köln 1981–82*: Cologne: Kölner Statistische Nachrichten.

Stanworth, J., Purdy, D. and Kirby, D. (1992), *The Management of Success in 'Growth Corridors'*, Small Firms, Small Business Research Trust, Milton Keynes: Open University.

Stark, P. (1984) *The Unplanned Arts Center as a Base for Planned Growth in Arts Provision*, London: City University.

Stark, P. (1994) *Strengthening Foundations: A Report and Proposal from the Voluntary Arts Network* (Officers Draft, December 1993), Newcastle: VAN.

STB (1996) *Tourism 21: Vision of a Tourism Capital*, Singapore: Singapore Tourist Board.

Steele, J. (1983a) *Planning for Leisure in London: Overview and Annotated Bibliography*, Papers in Leisure Studies No. 10, ed. Veal, A. J., London: Polytechnic of North London.

Steele, J. (1983b) *Leisure Planning and Information Needs in the London Local Authorities*, Papers in Leisure Studies No. 9, London: Polytechnic of North London.

Stephen-Wells, J. (1991) *A Roof Over the Arts: A Special Study of Issues Pertaining to Facilities, Workspaces and Live/Work Spaces for the Arts in Metro*, Toronto: MMT.

Stewart, F. (1990) *The Economics of Leisure*, The UK Leisure Property Conference, London 2: 4–24.

Stewart, R. (1987) 'The arts, politics, power and the purse', in *Report of an International Conference on the Structure of Arts Funding*, March, London: Arts Council.

Stoker, G. (1995) 'Regime theory and urban politics', in Judge, D., Stoker, G. and Wolman, H. (eds) *Theories of Urban Politics*, London: Sage: 54–71.

Stoker, G. and Mossberger, K. (1994) 'Urban theory in comparative perspective', *Government and Policy* 12: 195–212.

Stone, C. (1993) 'Urban regimes and the capacity to govern: a political economy approach', *Journal of Urban Affairs* 15: 1–28.

Stone, N. (1972) *The Causes of the English Revolution 1529–1642*, London: Routledge.

STTEC (1993) *Research into the Live Entertainment and Mass Communication Sectors in London*, Research Brief, London: South Thames Training & Enterprise Council.

Stungo, N. (1994) 'An American in Paris: Frank Gehry's American Center at Bercy', *The Independent on Sunday*, 29 May: 18–19.

Stungo, N. (2000) 'A return to Victorian values', *The Observer*, 9 January.

Style, S. (2000) 'Community regeneration in Chiapas: the Zapatista struggle for autonomy', *City* 4: 263–70.

Sudjic, D. (1993) *The 100 Mile City*, London: Flamingo.

Sudjic, D. (2001) 'The city that never sleeps', *The Observer Review*, 4 February: 10.

Summerfield, B. (1968) *Business in the Middle Ages*, New York: Cooper Square.

Sunderland City Council (2000) *Local Cultural Strategy: Project Brief*, Sunderland.

Sutcliffe, A. (1970) *The Autumn of Central Paris: The Defeat of Town Planning 1850–1970*, London: Edward Arnold.

Syme, G. T., Shaw, B. J. and Fenton, D. M. (1989) *The Planning and Evaluation of Hallmark Events*, Aldershot: Avebury.

Symon, P. and Verhoeff, R. (1999) *The New Arts in Birmingham: A Local Analysis of Cultural Diversity*, International Conference on Cultural Policy Research, Bergen, November.

TAC (1988) *No Vacancy: A Cultural Facilities Policy for the City of Toronto*, Toronto: Toronto Arts Council.

TAC (1992a) *Metro's Role in Arts and Culture: A Discussion Paper for the Municipality of Metropolitan Toronto*, Toronto: Toronto Arts Council.

TAC (1992b) *The Arts and Economic Development*, February, Toronto: Toronto Arts Council.

Tate (2001) *The Urban Myth*. Century City Special Issue no. 24, Spring, London: Tate Gallery.

Tauhmann, W. and Behrens, F. (1986) *Economic Impacts of the cultural facilities in Bremen*, Bremen: University of Bremen.

Taylor, R. (1998) *Berlin and its Culture*, New Haven and London: Yale University Press.

Teitz, M. (1968) 'Toward a theory of urban public facility location', *Regional Science Association* 21: 35–51.

The Economist (1991) 'Let the town halls decide about Mozart', *The Economist*, 20 April: 18.

Theatres Trust (1993) *The Care and Maintenance of Theatres*, Pilot study of the condition of theatres in England 1989/90, March, London: Department of National Heritage.

Thomas, C. J. and Bromley, D. F. (2000) 'City-centre revitalisation: problems of fragmentation and fear in the evening and night-time city', *Urban Studies* 37: 1403–29.

Thomas, M. and Roberts, G. (1997) *The Multimedia Industry in Wales*, Cardiff: WDA.

Thompson, R. (1994) 'Opening the door to Europe', *Planning Week*, 26 May: 18.

Thorold, P. (1999) *The London Rich: The Creation of a Great City from 1666 to the Present*, London: Viking.

Thorpe Committee, Ministry of Housing and Local Government (1969) *Thorpe Report of the Environmental Committee of Enquiry into Allotments*, London: HMSO.

Tibbalds, F. (1992) *Making People Friendly Towns*, London: Tibbalds Partnership.

Tietz, M. (1968) 'Toward a theory of urban public facility location', *Regional Science Association* 21: 35–51.

Timbart, O. (1984) 'The financing of culture in France', in *Funding the Arts in Europe*, Strasbourg: Council of Europe.

Titmuss, R. M. (1974) *Social Policy: An Introduction*, London: Allen & Unwin.

Tomkins, A. (1993) 'The city cultures of London: renewal or decline?', paper given to the London Arts Conference, South Bank Centre, 31 March, London: LAC.

Tomlinson, J. (1999) *Globalization and Culture*, Oxford: Blackwell.

Towse, R. (1995) *The Economics of Artists' Labour Markets*, London: Arts Council of England.

TRaC (2000) *Social Exclusion and the Provision and Availability of Public Transport*, London: Department for the Environment, Transport and the Regions.

Trevelyan, G. M. (1967) *English Social History: A Survey of Six Centuries, Chaucer to Queen Victoria* [1942], Penguin: Harmondsworth.

Trienekens, S. J. (2000) Cultural diversity in cultural consumption: exploring the separate and spatially divided cultural circuits', paper given at the *Cultural Change and Urban Contexts* Conference, Manchester, September: 62.

TRRU (1979) 'Leisure & community development in rural areas', in Arnott, J. and Duffield, B. (eds) *Leisure and Rural Society*, Edinburgh: Tourism and Recreation Research Unit.

Truman, H. (1934) *The Official History of the Royal Society of Arts*, London: RSA.

Tuan, Yi-Fu (1976) 'Humanistic geography', *Annals of the Association of American Geographers*, 66: 276.

Tuan, Yi-Fu (1977) *Space and Place: The Perspective of Experience*, Minneapolis: Minnesota University Press.

Tzonis, A. and Lefaivre, L. (1981) 'The grid and the pathway. An introduction to the work of Dimitris and Susana Antonakakis', *Architecture in Greece* 15.

Ulldemolins, J. R. (2000) 'From "Chino" to Raval. Art merchants and the creation of a cultural quarter in Barcelona', paper given at the *Cultural Change and Urban Contexts* Conference, Manchester, September: 19.

UNDP (1995) *Human Development Report*, Oxford: Oxford University Press.

UNESCO (1969) *Cultural Policy: A Preliminary Study*, Paris: UNESCO.

UNESCO (1970) *Some Aspects of French Cultural Policy*, Studies and Research Department of the French Ministry of Culture, Paris: UNESCO.

UNESCO (1972) *Convention Concerning the Protection of the World Cultural and Natural Heritage*, Paris: UNESCO.

Unwin, R. (1909) *Town Planning in Practice: An Introduction to the Art of Designing Cities and Suburbs*, London: T. Fisher Unwin.

Urban Cultures Ltd (1994) *Prospects and Planning Requirements of the Creative Industries in London*, London: London Planning Advisory Committee.

URBED (1988) *Developing the Cultural Industries Quarter in Sheffield*, Sheffield: Sheffield City Council.

Urry, J. (1995) *Consuming Places*, The International Library of Sociology, London: Routledge.

Uysal, M., Gahan, L. and Martin, B. (1993) 'An examination of event motivations: a case study', *Festival Management and Event Tourism* 1: 5–10.

VAN (1994) *Survey of Local Authorities Arts Audits and Plans*, Arts Business Ltd, Newcastle: Voluntary Arts Network.

Van Eyck, A. (1962) 'A step towards a configurative discipline', *Forum* 16: 81–9.

Van Puffelen, F. *et al.* (1986) *More Than One Billion Gilders. The Economic Significance of the Professional Arts in Amsterdam*. Hrsg: Amsterdams Uit-Buro, Stichting voor Economisch Onderzoek der Universiteit van Amsterdam, Amsterdam.

Vasari, G. (1550) *The Lives of Artists* (first published in Italian, enlarged version published in 1568, trans. 1970), London: A. B. Hinds.

Vaughan, D. R. (1990) *The Economic Impact of the Arts & Residents in Portsmouth*, Bournemouth Polytechnic, Winchester: Southern Arts Board.

Vaughan, R. (1992) *The Arts and the Residents of Portsmouth*, Bournemouth Polytechnic, Winchester: Southern Arts Board.

Veal, A. J. (1975) *Recreation Planning in New Communities: A Review of the British Experience*, Research Memo 46. Birmingham: University of Birmingham, Centre for Urban and Regional Studies.

Veal, A. J. (1982) *Planning for Leisure: Alternative Approaches*, Papers in Leisure Studies No. 5, May, Department of Extension Studies, London: Polytechnic of North London.

Veal, A. J. (ed.) (1983) *Planning for Leisure in London: Overview and Annotated Bibliography*, Papers in Leisure Studies No. 10 (Steele, J. ed.), London: Polytechnic of North London.

Veal, A. J. (1993) 'Planning for leisure: past, present and future', in Glyptis, S. (ed.) *Leisure and the Environment: Essays in Honour of Professor J. A. Patmore*, London: Belhaven: 85–95.

Venturi, R. (1966) *Complexity and Contradiction in Architecture*, New York: Museum of Modern Art.

Verwijnen, J. and Lehtovuori, P. (1999) *Creative Cities: Cultural Industries, Urban Development and the Information Society*, Helsinki: University of Art and Design Press.

Vickers, G. (1999) *Key Moments in Architecture: The Evolution of the City*, London: Hamlyn.

Vigar, M. (1991) *Cultural Diversity, Cypriot Cultural Interest and Aspirations*, Discussion Document No. 7A, National Arts & Media Strategy, London: Arts Council.

Von Eckardt, W. (1982) *The Good Life: Creating Human Community Through the Arts*, New York: American Council for the Arts.

Wainwright, M. (1993) 'London? Just a tiny piece in the . . .', *The Guardian*, 17 August: 1–2.

Wall, C. (1998) *The Literary and Cultural Spaces of Restoration London*, Cambridge: Cambridge University Press.

Wall, G. and Purdon, M. (1987) *Economic Impact of the Arts in Ontario*, University of Waterloo, Ontario: Ontario Arts Council.

Wall Street Journal (1985) 'Old New England city heals itself . . .', *Wall Street Journal* 1 February.

Wallace, N. (1993) 'Introductory paper', given to the Symposium on the Future of London Arts Centres, Drill Hall, 13 September, London: London Arts Board.

Walsh, A. (1986) *Recreation Economic Decisions*, Pennsylvania State College: Venture.

Walvin, J. (1978) *Leisure and Society 1830–1950*, London: Longman.

Walvin, J. (1984) *English Urban Life (1776–1851)*, London: Hutchinson.

Wangermée, R. (1991) *Cultural Policy in France, European Programme for the Appraisal of Cultural Policies*, Council for Cultural Co-operation, Strasbourg: Council of Europe.

Wanhill, S. (1997) 'Peripheral area tourism: a European perspective', *Progress in Tourism and Hospitality Research* 3: 47–70.

Ward, B. and Dubos, R. (1972) *Only One Earth*, New York: Norton.

Ward, S. (1998) *Selling Places: The Marketing and Promotion of Towns and Cities 1850–2000*, London: E & FN Spon.

Wasserman, B., Sullivan, P. and Palermo, G. (2000) *Ethics and the Practice of Architecture*, New York: Wiley.

Waters, B. (1987) 'Planning: use Class Order 2 Application', *Architect's Journal*, 5 August: 57–9.

Weber, M. (1964) *The Theory of Social and Economic Organisation (Wirtschaft und Gesellschaft)*, New York: Free Press.

Weightman, G. (1992) *Bright Lights, Big City: London Entertained 1830–1950*, London: Collins & Brown.

Weiner, D. (1989) 'The people's palace: an image for East London in the 1880s', in Feldman D. and Stedman Jones G. (eds) *Metropolis London: Histories and Representations Since 1800*, London: Routledge: 40–55.

Werthner, H., Nachira, F., Orests, S. and Pollock, A. (1997) *Information Society Technology for Tourism. Report of the Strategic Advisory Group on the 5th Framework Program on Information Society*, 8 December, Brussels.

White, E. (1969) *Arts Centres in Great Britain*, London: Arts Council of Great Britain.

Wilding, R. (1989) *Supporting the Arts — Review of the Structure of Arts Funding*, London: Office of Arts and Libraries.

Wilkinson, P. F. (1973) 'The use of models in predicting the consumption of outdoor recreation', *Journal of Leisure Research* 5: 34–47.

Williams, R. (1958) *Culture and Society 1780–1950*, London: Chatto & Windus.

Williams, R. (1961) *The Long Revolution*, London: Pelican.

Williams, R. (1975) *The Country and the City*, St Albans: Paladin.

Williams, R. (1981) *Culture*, London: Fontana.

Williams, R. (1983) *Towards 2000*, London: Pelican.

Williams, R. H. (1982) *Dream Worlds: Mass Consumption in Late Nineteenth Century France*, Berkeley: University of California Press.

Willis, F. (1948) *101 Jubilee Road. A Book of London Yesterday*, London: Phoenix House.

Willis, P. (1991) *Towards a New Cultural Map*, Discussion Document No. 18, National Arts & Media Strategy, London: Arts Council.

Wilson, D. M. (1989) *The British Museum: Purpose and Politics*, London: BMP.

Wilson, E. (1988) *Politics and Leisure*, London: Unwin Hyman.

Wilson, E. (1991) *The Sphinx in the City: Urban Life, the Control of Disorder, and Women*, London: Virago.

Wislocki, P. (2000) 'House of harmonies', *Building Design*, 21 January: 18–19.

Wolff, J. (1981) *The Social Production of Art*, Basingstoke: Macmillan.

Wolmar, C. (1989) 'Follow the red brick road', *The Weekend Guardian*, 8 April: 5.

World Bank (1998) *Culture and Development at the Millennium: The Challenge and the Response*, Washington, DC.

Worpole, K. (1988) 'The urban desert', *Good Housekeeping*, April: 114–19.

Worpole, K. (1991) 'Trading places: the city workshop', in Fisher, M. and Owen, U. (eds) *Whose Cities?*, London: Penguin: 142–52.

Worpole, K. (1992) 'Cities: the buzz and the burn', *The Guardian*, 25 May: 21.

Worpole, K. (1994) 'The new "City States"?', in Perryman, M. (ed.) *Altered States*, London: Lawrence & Wishart: 157–73.

Worpole, K. (2000) *Here Comes the Sun: Architecture and Public Space in Twentieth-Century European Culture*, London: Reaktion.

Worpole, K., Curson, T., Evans, G. L. and Shaw, S. (1999) *Interim Report on the Applicability of Standards for Assessing Demand for Open Space in London*, London: CELTS for the London Planning Advisory Committee.

Worpole, K., Curson, T., Evans, G. L. and Shaw, S. (2000) *Report on the Applicability of Standards for Assessing Demand for Open Space in London*, London: W. S. Atkins for the London Planning Advisory Committee.

Worpole, K. and Greenhalgh, L. (1999) *The Richness of Cities: Urban Policy in a New Landscape – Final Report*, Stroud: Comedia/Demos.

Wright, P. (1993) 'A train of thought', *The Guardian*, 14 August: 13–14.

WTO (1998) *Tourism 2020 Vision: Executive Summary*, Madrid: World Tourism Organization.

WTO (1999) *Changes in Leisure Time: The Impact on Tourism*, Madrid: World Tourism Organization.

WTO (2000) *Tourist Taxation*, Madrid: World Tourism Organization.

Wu, F. (2000) 'The global and local dimensions of place-making: remaking Shanghai as a world city', *Urban Studies* 37: 1359–77.

Wulf-Mathies, M. (1999) 'European Commission support for culture', *Official Journal of the European Communities* (C182/55) 28 June.

Wynne, D. (1992) *The Culture Industry: The Arts in Urban Regeneration*, Aldershot: Avebury.

Yamada, H. and Yasuda, H. (1998) 'The economic impacts of cultural industries mainly in the Tokyo Metropolitan Area: an interregional and interindustrial analysis', paper given at the *Xth International Conference on Cultural Economics*, Barcelona, June.

Yeo, E. and Yeo, S. (1981) *Popular Culture and Class Conflict 1590–1914*, Brighton: Harvester.

Young, K. (1984) 'Metropolitan government and the development of the concept of reality', in Leach, S. (ed.) *The Future of Metropolitan Government*, Institute of Local Government Studies, Birmingham: University of Birmingham.

Younge, G. (2000) 'Harlem — the new theme park', *The Guardian Saturday Review*: 1–2.

Zallo, R. (1988) *Economica de la communicacion y la cultura*, Madrid: Akal.

Zeidler, E. H. (1983) *Multi-use Architecture in the Urban Context*, New York: van Nostrand Reinhold.

Zelinsky, W. (1992) *The Cultural Geography of the United States* (2nd ed.), Englewood Cliffs: Prentice-Hall.

Zimmer, A. and Toepler, S. (1996) 'Cultural policies and the welfare state: the cases of Sweden, Germany and the United States', *Journal of Arts Management, Law and Society* 26: 167–93.

Zimmern, A. (1961) *The Greek Commonwealth: Politics and Economics in Fifth-Century Athens*, London: Oxford University Press.

Zukin, S. (1988) *Loft Living: Culture and Capital in Urban Change*, London: Radius.

Zukin, S. (1995) *The Cultures of Cities*, Cambridge, MA: Blackwell.

Zukin, S. (1996) 'Space and symbols in an age of decline', in King, A. D. (ed.) *Re-Presenting the City: Ethnicity, Capital and Culture in the 21st Century Metropolis*, London: Macmillan: 43–59.

Index